Behavioural Ecology

AN EVOLUTIONARY APPROACH

Behavioural Ecology

AN EVOLUTIONARY APPROACH

EDITED BY

J. R. KREBS and N. B. DAVIES
Edward Grey Institute of Field Ornithology
Department of Zoology
University of Oxford

BLACKWELL SCIENTIFIC PUBLICATIONS
OXFORD LONDON EDINBURGH MELBOURNE

Osney Mead, Oxford, OX2 0EL
8 John Street, London, WC1N 2ES
9 Forrest Road, Edinburgh, EH1 2QH
P.O. Box 9, North Balwyn, Victoria, Australia

British Library Cataloguing in Publication Data

Behavioural ecology.
1. Animals, Habits and behavior of
2. Ecology 3. Evolution
I. Krebs, J R II. Davies, N B
591.5 QL751

ISBN 0-632-00285-9

ISBN 0 632 00113 5 (cloth)
ISBN 0 632 00285 9 (paper)
First published 1978

Published in the U.S.A. and Canada by
Sinauer Associates, Inc.
Sunderland, Massachusetts

Set by Enset Ltd.
Midsomer Norton, Bath and
Printed in Great Britain by
Billing & Sons Ltd, Guildford.
Bound by
Kemp Hall Bindery, Oxford

Contents

Part 2. Sex, Mating and Signals

Part 3. Strategies in Space and Time

List of Authors

BRIAN C. R. BERTRAM King's College Research Centre, King's College, Cambridge University, Cambridge, England

NICHOLAS B. DAVIES Edward Grey Institute, Department of Zoology, Oxford University, Oxford, England, OX1 3PS

RICHARD DAWKINS Animal Behaviour Research Group, Department of Zoology, Oxford University, Oxford, England, OX1 3PS

STEPHEN T. EMLEN Section of Neurobiology and Behavior, Division of Biological Sciences, Cornell University, Ithaca, N.Y. 14853, U.S.A.

PAUL J. GREENWOOD School of Biological Sciences, University of Sussex, Falmer, Brighton, Sussex, England, BN1 9QG

TIMOTHY R. HALLIDAY Department of Biology, The Open University, Milton Keynes, Buckinghamshire, England

PAUL H. HARVEY School of Biological Sciences, University of Sussex, Falmer, Brighton, Sussex, England, BN1 9QG

BERND HEINRICH Department of Entomological Sciences, University of California, Berkeley, California 94720, U.S.A.

HENRY S. HORN Department of Biology, Princeton University, Princeton, New Jersey 08540, U.S.A.

JOHN R. KREBS Edward Grey Institute, Department of Zoology, Oxford University, Oxford, England, OX1 3PS

JOHN MAYNARD SMITH School of Biological Sciences, University of Sussex, Falmer, Brighton, Sussex, England, BN1 9QG

ROBIN H. McCLEERY Edward Grey Institute, Department of Zoology, Oxford University, Oxford, England, OX1 3PS

GEOFFREY A. PARKER Department of Zoology, University of Liverpool, Liverpool, England, L69, 3BX

LINDA PARTRIDGE Department of Zoology, University of Edinburgh, West Mains Road, Edinburgh, Scotland, EH9 3JT

Preface

This book came about because we wanted to be able to tell our students where they could read good summaries of the latest ideas in behavioural ecology. We felt that the sort of book we had in mind would have to be written by a number of authors currently working in the field rather than as a single author text. In retrospect we are confident that the slight lack of continuity of style which is inevitably found in a multi-author volume is more than offset by the excitement and insight conveyed by people writing in their own field. It was our intention that the book should be used primarily by higher level undergraduates but most of the chapters contain enough by way of new ideas and reviews to be of interest to research workers and teachers in behaviour and ecology.

Our choice of subject matter is unashamedly biased; some important areas are not covered, for example migration, the evolution of learning, and co-evolution. We have picked on the areas in which ethology, ecology and the theory of natural selection have come together in the last few years to create new and stimulating ideas. There are a number of basic theoretical concepts, for example kin selection and optimality theory, which appear so often throughout the book that they could not be allocated a chapter of their own without a great deal of repetition. We have tried to fill in the gaps to some extent in Chapter 1, where we also point to the places later in the book where more detailed discussions of the general theoretical ideas can be found. There is inevitably still a small amount of overlap between chapters, and some ideas are discussed twice (e.g. marginal value theorem in chapters 2 and 8, ideal free theory in chapters 11 and 12). This repetition is probably a good thing as it shows that the same theory can be viewed from different angles and applied to different problems. We asked authors to emphasise theoretical ideas as well as data, but as far as possible we have kept complex mathematical theory out of the book, using graphical solutions instead.

A student with virtually no mathematical background should be able to cope with the whole book, even though some of the ideas are quite difficult.

We have been most fortunate in choosing cooperative authors who not only made an excellent job of the writing, but also delivered their chapters on time!

Edward Grey Institute
Oxford, February 1978

John Krebs
Nicholas Davies

Acknowledgements

We are especially grateful to Robert Campbell of Blackwell Scientific Publications for all his expert advice and enthusiasm. Christopher Perrins and the students at the Edward Grey Institute provided the stimulating and critical environment in which this book was born, Maggie Norris has converted mountains of illegible scrawl into organised typescript and Mary Stubbs prepared the index. The following co-operated generously over requests for information and gave permission to quote unpublished results: Barbara Cockrell, Robin Cook, Graham Pyke and Jeff Waage (Chapter 2); John R. G. Turner and Paul W. Sherman (Chapter 5); Richard Howard (Chapter 7); Glen Woolfenden (Chapter 9) and Tim Clutton-Brock and Sievert Rohwer (Chapter 10). The following are gratefully acknowledged for help with specific chapters: Richard Wrangham (Chapter 3); Bert Hölldobler (Chapter 4); John Gittleman, Georgina Mace, Peter Greig-Smith, David Bousfield, John Maynard Smith (Chapter 5), together with the Science Research Council for financial support; Alasdair Houston and Richard Sibly (Chapters 7 and 13); Miss A. Callaghan (Chapter 8); the John Simon Guggenheim Foundation, the National Geographic Society, the Chapman Fund of the American Museum of Natural History, and the National Science Foundation for their financial support, and Natalie J. Demong for stimulating comments and criticism throughout (Chapter 9); Timothy Birkhead and Michael Brooke (Chapter 11); P. Ashmole, M. Dow, A. Ewing, V. French and Miss T. MacKay (Chapter 12); Pat Searle (Chapter 13); Bob May and Betty Horn (Chapter 14).

Chapter 1
Introduction: Ecology, Natural Selection and Social Behaviour

NICHOLAS B. DAVIES AND JOHN R. KREBS

1.1 Introduction

David Lack once recalled two events which made a lasting impression on him. One was the publication, when he was an undergraduate at Cambridge, of R.A. Fisher's 'The genetical theory of natural selection'. Lectures on 'evolution' in the Cambridge zoology course of the time consisted of comparative morphology and phylogeny, and Fisher's explicit statement of the link between genetics and evolutionary theory was a revelation to Lack which marked the start (by him and many others) of a whole new approach to the study of natural selection. The second event was a meeting of the British Ecological Society in 1944, at which Lack presented, with reference to his work on Galapagos finches, the idea that competition was a major force leading to the ecological separation of related species in a community during evolution. This idea met with widespread scepticism, yet now it is an accepted dogma of community ecology. The story of these two events shows how the climate of ideas has changed dramatically in one generation. The links between population genetics, evolution and ecology are now so self-evident that no one would even begin to express doubts about them. More often than not the three subjects are taught in one undergraduate course entitled 'population biology'. The subject matter of this book takes modern population biology as its foundation and explores areas in which ethology and population biology overlap. Evolutionary thinking permeates the whole book and although pure population genetics and ecology only rarely appear in explicit form they obviously lurk just beneath the surface throughout.

Behavioural ecology as we understand it has emerged from four schools of thought which developed primarily in the early 1960s. In England J.H. Crook, followed by D. Lack, pioneered the 'comparative approach' linking social organisation in birds and primates with ecological factors; W.D. Hamilton and J. Maynard Smith made explicit the ideas of kin selection and inclusive fitness; and N. Tinbergen's group developed the procedure of testing questions about the survival value of behaviour with simple field experiments or observations (e.g. Tinbergen *et al.* 1967, Patterson 1965). At the same time the whole face of ecology was changed by R.H. MacArthur and his followers in North America. From the viewpoint of behavioural ecology, MacArthur's contribution was to firmly establish the notion that hypotheses about evolutionary questions in ecology could be couched in precise mathematical terms. A similar type of approach although less quantitative was developed in parallel by others in North America, notably G.H. Orians (e.g. Orians 1969) and J.L. Brown (e.g. Brown 1934). MacArthur made many important contributions to mathematical ecology, but two ideas which he used are particularly pertinent to behavioural ecology: optimality theory (MacArthur & Pianka 1966) and the concept of an evolutionary stable strategy (ESS). The second of these ideas has since been very much developed by J. Maynard Smith (see Chapters 6 and 10).

1.2 Optimality models and ESSs

MacArthur first used the idea of optimal choice in the context of foraging behaviour. He proposed that one could work out from first principles a set of rules for the behaviour of an efficient predator, an idea which has since produced a substantial body of work (Chapter 2). Subsequently the same sort of approach has been used to ask questions about territory size (Chapter 11), group size (Chapter 3), life history strategies (Chapter 14), and mating strategies (Chapter 8). One of the most recent developments is the use of optimality theory to explore how internal motivational factors and external ecological circumstances combine to influence the way in which animals decide what to do next (Chapter 13).

The argument for using optimality theory in behavioural ecology is that natural selection should tend to produce animals which are maximally efficient at propagating their genes and therefore at doing all other activities, which of course subserve this function in the end. One of the many possible criticisms of the use of optimality arguments is that there is often no single best solution to a problem. This is especially true when the value of one solution depends on the behaviour of other individuals in the population. For example, the best place for a male dungfly to search for females depends on where the other males are (Chapter 8); the best thing to do in a fight depends on what the opponent does (Chapter 10); and the best sex ratio for a mother to have in her offspring depends on the overall ratio of males to females in the population (Chapter 6). In such cases there is often no single optimum strategy which all individuals should adopt. Thus the solution to the male dungfly's problem is to sometimes search in the grass around a dung pat and sometimes on the grass itself; in a fight it is sometimes best to give in easily and sometimes better to fight viciously, and so on. To work out the theoretical solution to these problems we have to find a strategy or mixture of strategies which is not susceptible to invasion by an hypothetical alternative. For example when the population sex ratio (at least in species with equal sized males and females) is 50:50 no alternative strategy (e.g. 60:40 sex ratio, 100:0 sex ratio) can do better (Chapter 6). In short, the technique for this sort of analysis is to search for an evolutionary stable strategy (ESS). As with optimality theory, this technique appears or underlies the arguments in many of the chapters including discussions of sex ratio (Chapter 6), ritualised combat (Chapter 10), leks (Chapter 11), and habitat selection (Chapter 12).

1.3 The comparative approach

1.3.1. *Weaver birds*

Although the idea of comparing related species in order to gain insights into the selection pressures influencing social organisation was not originated by J.H. Crook (see for example Winn's (1958) work on darter fish), Crook's study of weaver birds (Ploceinae) has become established as the model for this approach. In essence, Crook's idea was to relate differences between closely related species in their social organisation to differences in ecology. The term social organisation refers primarily to the following characteristics of a species: breeding and feeding group size (e.g. solitary, flocking), spatial organisation (e.g. territorial, overlapping ranges), and mating system (e.g. monogamy, polygyny). The main ecological factors considered by Crook were distribution and availability of food and nest sites, and the impact of predators.

Crook (1964) compared the social organisation and ecology of the 90 species of ploceine weaver birds from Africa and Asia. He showed that there is a striking dichotomy into species inhabiting evergreen forests, which are primarily insectivorous, live in monogamous pairs and defend large territories, and savannah dwelling species which are graminivorous (seed eating), nest in large colonies, feed in flocks and are usually polygynous. The correlation between social organisation and diet or habitat is clear, but the causal network which results in these relationships is still to some extent in dispute.

The difference in nesting and feeding dispersion is probably related both to food dispersion and to the influence of predators. The forest insectivores often build cryptic nests and rely on spacing out as a defence against predators of the eggs and young. They also exploit a food supply which is relatively scattered, predictable and not superabundant: these features in combination mean that the food is economically defendable (Chapter 11). Hence both pressure from predators and the dispersion of food favour spaced out nests and territorial defence. The savannah species, in contrast, build conspicuous nests (they are bulky, probably to provide thermal insulation) in inaccessible or protected places such as the tips of branches of spiny acacia trees, and feed on ephemeral but locally superabundant food supplies. They therefore to some extent escape predation on their nests and young, and further they cannot easily secure food by territorial defence. Birds, including the graminivorous ploceines, which eat clumped and unpredictable food tend to feed in flocks, because each individual can take advantage of the searching ability of others in finding clumps of food (Chapter 3). If the clumps are sufficiently large, competition within them is not severe enough to offset the advantage of being in a group. Colonial nesting in the savannah species occurs because there are rather few well-protected nesting sites. Colonies also tend to be near locally abundant food, and further, birds in colonies may benefit by following others to good feeding places, as they do when feeding in flocks.

The difference in mating system between forest and savannah species can also be explained in terms of diet and habitat. The reproductive success of forest insectivores is probably limited (as in most passerines) by the ability of the parents to bring food to the nest. In these conditions neither parent would benefit by desertion since the remaining individual would be unable to rear the young, and monogamy is the rule. Most tropical graminivorous birds are monogamous, so the tendency towards polygyny in seed eating ploceines is probably not directly related to diet (Lack 1968). Polygyny probably arises because some individual males succeed in defending a number of safe nest sites in the colony while other males get none. Females would clearly do better to mate polygynously with the former than monogamously with the latter (see also Chapter 11). Associated with the differences in mating system, the savannah species are more sexually dimorphic than forest species, and they often have elaborate courtship displays centred on the nest.

These interpretations of the effect of predation and food supply on social organisation are supported by the fact that species with intermediate diets consisting of a mixture of seeds and insects are also intermediate in social organisation, some resembling graminivores and others forest insectivores. The grassland graminivores also lend support to the argument. They nest on the ground in tall grass or reed beds, breed in loose colonies and feed in flocks. The nests are accessible to predators, so crypsis and spacing out is advantageous, but as outlined above seed eating is usually associated with colonial nesting and flock feeding.

Crook's study illustrates at the same time the strengths and weaknesses of the comparative approach. The great step forward resulting from this approach was firstly to focus attention on the role of predators and food supply in moulding social organisation, and secondly to show how a whole constellation of characters such as nesting dispersion, feeding dispersion, mating system, nest architecture and plumage dimorphism may all be inter-related and come under the influence of the same ecological pressures. Crook's analysis of weaver bird social organisation has been extended to include all bird species, by Crook (1965) and Lack (1968), showing that similar selection pressures act over a wide taxonomic range. The limitations of comparative studies are partly methodological: it is hard to do statistical analyses of the trends outlined above because one cannot be sure of the appropriate independent unit of observation. For example, in Crook's study, thirteen of the grassland species breeding in loose colonies belong to the single genus *Euplectes*: there is no clear rule for deciding whether they should be treated as thirteen independent observations or as one observation. More serious is the difficulty in disentangling cause and effect when it comes to interpreting results. To give a simple example, a close correlation between feeding group size and diet could be interpreted in two quite different ways. Either group size has evolved in direct response to diet, for example seeds may be more efficiently exploited by flocks, or group feeding is advantageous in avoiding predators, but only some species eat the sort of food which is sufficiently common to allow them to forage in groups. Despite these problems, comparative studies have provided considerable insight into the selective forces acting on social organisation not only in birds, but also in primates (Crook & Gartlan

1966, Clutton Brock & Harvey 1977), and ungulates (Jarman 1974), carnivores (Kruuk 1975) and some coral reef fishes (Fricke 1975).

1.3.2 *Primates*

Primates, which have a bewildering variety of social organisations, have proved especially fertile ground for comparative studies. Two general points about the comparative approach which have emerged from primate work are that different aspects of social organisation (e.g. home range size and sex ratio) may be influenced by different ecological pressures; and differences in social structure between species living in the same habitat and apparently with similar ecology may be explained by rather subtle differences in diet.

Once again the pioneering work was due mainly to J.H. Crook (Crook & Gartlan 1966). Crook and Gartlan's orginal attempt to classify all primates into five ecological 'grades' of social organisation proved too simple, and led some people to suggest that phylogenetic heritage rather than present day ecological conditions might determine social structure (e.g. Struhsaker 1969). However the most recent and detailed comparative study of primate social systems (Clutton Brock & Harvey 1977) has largely vindicated Crook's orginal work, albeit in a greatly modified form. The main methodological steps forward made by Clutton Brock and Harvey were that they measured the various aspects of social organisation such as group size on continuous scales rather than as discrete categories, and they used multivariate statistics to sort out the effect of different ecological factors. Their analysis showed, for example, that nocturnal primates which are small, arboreal, and eat fruit or insects, live in small groups and have small home ranges. At the other extreme, diurnal terrestrial primates are large, live in large groups and have large home ranges. Arboreal fruit eaters and leaf eaters tend to be intermediate between these extremes, with frugivores living in larger groups than folivores, and occupying larger home ranges.

How do these patterns relate to the ecological pressures of feeding and predation? Clutton Brock and Harvey suggest that because nocturnal species feed by crawling onto small branches they are constrained to be small in body size, and therefore rely on crypsis to avoid predators: hence they are solitary and inconspicuous. At the other extreme diurnal monkeys, especially those living on the ground, are inevitably conspicuous, so they rely on large body size and group defence as protection against predators. Thus pressure from predators may have had contrasting effects on nocturnal and diurnal primates. However this cannot be the whole story, as there are great differences in group and range size within the diurnal arboreal and terrestrial categories. An example of how detailed studies of feeding ecology might explain such differences is Clutton Brock's (1974) work on the red colobus (*Colobus badius*) and black and white colobus (*Colobus geuereza*). Both these species are forest dwelling arboreal leaf-eaters, but the red colobus lives in groups of about 40 in an undefended range of 67 ha, while the black and white colobus defends a small territory of 15 ha and lives in groups of about 11 individuals. The key to these differences seems to be in the diet: the red colobus specialises on shoots, fruit, and flowers of a variety of trees, while the black and white feeds on leaves of all ages on one or two tree species. The food supply of the red colobus occurs, therefore, in large ephemeral clumps, the exact location depending on when and where a tree comes into leaf, flowers or fruits. In contrast the black and white colobus exploits a more predictable and evenly scattered food supply. This difference in diet seems to account for the difference in size of home range. The red colobus has to wander over a wide area in order to find trees at just the right stage of growth to provide food. Further, the food occurs in large clumps, so that upper limit to group size, at which competition would become intense, is high. Clutton Brock argues that the advantage of living in large groups is increased awareness of and defence against predators. This advantage also applies to the black and white colobus, but the fact that its food is rather thinly scattered sets a low upper limit to group size. The food supply is also rather more predictable, and hence defendable, than that of the red colobus. These interpretations are plausible, but it will be apparent that there are other ways of linking the various facets of social organisation and ecological factors. The comparative approach, as emphasised earlier, cannot sort out cause and effect. Is large home range size a consequence of large group size or vice versa? Is large group size an adaptation for finding clumped food sources or is it a consequence of feeding on clumps of food? Is territorial defence a consequence of diet or is diet an effect of territorial defence and a limited home range? Although the multiple regression analysis used by Clutton Brock and Harvey allows us to see how much of the variance in, say, home range size is explained by group size and diet, it cannot sort out which factors are cause and which are effect.

To summarise, the comparative approach has shown that social organisation is influenced by ecological pressures and it has pinpointed diet and predation as two especially important factors. However, our present understanding of how the behaviour of individuals within social groups is adaptive stems from a rather different approach.

1.4 Altruism and group selection

Living in groups may be favoured by particular ecological conditions, but the benefits of group living are often not equally shared between individuals: some may gain more protection from predators or more food than others. It is not enough to argue that group living is 'good for the species' or 'advantageous to the group'. We now know that selection acts primarily at the level of the individual, or to be more precise at the level of the gene, and not at the level of the group (Williams 1966). It is ironical that the final death of 'group selection' explanations of social behaviour was heralded by a book which strongly advocated group selection, V.C. Wynne-Edwards' 'Animal dispersion in relation to social behaviour' (1962). Wynne-Edwards' thesis was that animals have evolved physiological and behavioural adaptations such as low birth rates to prevent overpopulation. He realised that such a system could not evolve by individual selection. To take the simplest example, if all individuals in a population 'voluntarily' restrict themselves to two young each, a 'cheater' which has three young puts more genes into the next generations than a conforming individual, and therefore the 'cheater' genotype will soon sweep through the population. A great deal of mathematical expertise and sophistication has gone into trying to show how group selection could work in opposition to individual, or gene selection, and the general conclusion is that it simply cannot do so (Maynard Smith 1976a) except under very special (and hence unlikely) circumstances. Individuals die more often than groups, so group selection is a slow lumbering process compared with the 'rapid cut and thrust of individual selection' (Dawkins 1976). Wynne Edwards (1977) himself has come round to this inevitable conclusion and while he still maintains (in our view quite correctly) that social behaviour may influence population size (see Chapter 11), he has abandoned the idea that social behaviour evolved for this reason. Even if Wynne-Edwards was wrong to place so much emphasis on group selection, he did the great service of forcing people to think carefully about the evolution of social behaviour, especially altruism. Altruism in everyday language means 'unselfish regard for others', but in biology we have to distinguish between phenotypic and genotypic altruism. An individual can be said to be phenotypically altruistic when it appears to help others at its own expense, but in fact makes a genetic profit out of the apparent altruism. Only genotypic altruism, in which an individual decreases its relative gene contribution to future generations, poses a problem for the theory of natural selection.

1.5 Inclusive fitness and gene selection

The most familiar example of phenotypic altruism is, of course, parental care. We are not at all surprised when we see a parent hard at work collecting food for its young or protecting them from predators. This behaviour is genotypically selfish because the offspring are related to their parents. Parental behaviour poses no problem for evolutionary theory because, by definition, natural selection favours individuals who maximise their gene contribution to future generations. In terms of gene survival, a gene that happens to cause the parent to behave altruistically towards its young, which are likely to contain a copy of the same gene, becomes more numerous in the gene pool.

It is possible to quantify the degree of relatedness between parent and offspring. During gamete formation there is a reduction division (meiosis) in which any given gene has a 50 per cent chance of going into any one sperm or egg. (There is likewise a 50 per cent chance that its allele on the homologous chromosome will go into the gamete.) Thus half the eggs or sperms will have a copy of a particular gene and half will not. Therefore we can say that, in diploid species, the probability that a parent and its offspring will share a given gene is exactly 0·5. This quantity is called r, the coefficient of relatedness.

In fact the exact definition of coefficient of relatedness is rather tricky. Although one usually reads that it is 'the probability of sharing a given gene', or equivalently 'the proportion of genes shared' this is not strictly true. It is estimated, for example, that the single copy DNA (genes which are only represented once) of man and chimpanzees differs by only one per cent of the base pairs (Britten 1977). While this difference could reflect tiny differences in every gene, it is much more likely that many if not most genes are identical. Nor is it enough to refer only to genes which are shared from a common ancestor (identical by descent), since if one goes back far enough we presumably have a common ancestor with a particular chimpanzee. There seem to be two ways to define the definition of coefficient of relatedness which will get us out of this problem. One is to refer only to rare genes (which are not likely to be found outside the circle of close relatives); the other is to limit the definition to genes shared directly from a recent ancestor such as a great-grandfather.

Now offspring are not the only individuals that share genes in common with an individual—so do other relatives. Once again we can calculate the probability that a gene present in a given individual will be present in its brother ($r=0\cdot5$), sister ($r=0\cdot5$), grandchildren ($r=0\cdot25$) or cousins ($r=0\cdot125$). In diploid species, without inbreeding, $r=\sum(\frac{1}{2})^L$ where L is the number of generation links in a lineage between the two individuals. It was W.D. Hamilton (1964) who was the first to realise that this means that just as we show no surprise when parents are altruistic towards their offspring, so we should also expect individuals to be altruistic towards other relatives. What is maximised by natural selection is not individual fitness but a quantity that Hamilton termed 'inclusive fitness'. That is to say, the fitness of an individual depends both on its own survival and reproductive success and that of its kin.

Another way of expressing this idea is to abandon the term fitness altogether and instead of talking about the inclusive fitness of an individual, frame our argument in terms of gene survival (Dawkins 1976). Genes can assist the survival of replicas of themselves that are present in other individuals. Thus what appears as individual altruism is brought about by gene selfishness (in the language of Dawkins) or by selection for individuals to maximise their inclusive fitness (in the language of Hamilton). Is there any advantage to 'selfish gene' terminology, as opposed to inclusive fitness? One obvious advantage is that genes (even rare ones) may be shared between non-relatives. If two individuals share a gene for altruistic behaviour and can recognise this by a phenotypic marker, they are, as far as that gene is concerned, 'relatives'. By using selfish gene language one is not led into thinking that 'kin selection' can take place only between relatives. Apart from such special cases, the differences between the two terminologies are unimportant and they can be used interchangeably.

As a simple example of how apparent altruism at the individual level can be brought about by gene selfishness, consider a gene which programs an individual to die in order to save the life of relatives. One copy of the gene will be lost in the death of the altruist, but the gene will increase in frequency if the altruistic act saves the life of more than two brothers (or sisters), or more than four grandchildren or more than eight cousins, and so on. In more general terms, any altruistic act will benefit (in genetic terms) the actor as long as the benefit to the recipient is greater than the cost to the actor by a factor equal to $\frac{1}{r}$.

We will now discuss an example of apparent altruism in more detail.

1.5.1. *The social Hymenoptera*

In these social insects (bees, wasps and ants), a colony consists of reproductive females (queens), reproductive males (drones) and sterile females (workers). There is typically only one queen per colony and most of the individuals are workers which do not breed at all. Instead, they spend all their lives working to help the queen produce babies. At first sight this looks like the most dramatic case of altruistic behaviour in the whole of the animal kingdom. How could such an astonishing system have evolved? Four main hypotheses have been proposed.

1. *The supraorganism theory.* Emerson (1949) suggested that the colony is equivalent to a multicellular organism and the workers are no more than appendages of the queen. We can reject this hypothesis because not all of the individuals in a colony are genetically identical (in contrast to the cells of a multicellular organism) and therefore the explanation invokes a form of group selection.

2. *Workers are maximising their inclusive fitness.* Hamilton (1964) realised brilliantly that the key to an understanding of this problem is the rather odd means of sex determination in the Hymenoptera. Some of the eggs laid by the queens are unfertilized—these all develop into males, which are therefore haploid individuals. Fertilized eggs all develop into females, either sterile workers or queens, depending on the type of food provisioning given to the larvae.

This has rather interesting consequences for the coefficients of relatedness between individuals in the colony. Consider the queen first of all. She is equally related to her sons and to her daughters; $r=0 \cdot 5$ in each case. She is a diploid individual so the probability that a particular gene in her will be present in one of her children is exactly 0·5, irrespective of whether the young one is a male or female.

Now consider the worker females. If the queen only mates once and stores the sperm, all the workers will have an identical set of paternal genes because males are haploid and thus all sperms carry identical genes. (In effect, the male has only one set of genes to put into any given sperm so the probability that a sister shares any particular paternal gene is 100 per cent.) As usual, there is a 50 per cent chance that any given maternal gene will be present in a sister. Thus, if a worker inherited a given gene from her father, it would certainly be present in her sister, whereas if she inherited it from her mother there would be a

50 per cent chance that it would also be present in her sister. This means that for full sisters, $r=0{\cdot}75$.

Therefore the remarkable consequence is that sisters are more closely related to each other ($r=0{\cdot}75$) than they would be to their own children ($r=0{\cdot}5$). So the daughters of a queen who have, so to speak, the evolutionary choice of rearing their own daughters or helping their mother to rear their younger sisters, would do better (in terms of gene replication) by doing the latter. This is possible because in the social Hymenoptera generations overlap. Thus worker females are in fact maximising their inclusive fitness by helping their mother to rear offspring rather than having offspring themselves. It is the unusual means of sex determination in the Hymenoptera that has predisposed the workers to show apparent altruism. This may explain why sterile worker castes have evolved at least eleven times in this group but only once (termites) in the whole of the rest of the class Insecta, where both sexes are diploid and more usual coefficients of relatedness apply.

But there is more to the story than this. It is not simply a case of the workers showing altruism towards the queen so as to maximise their inclusive fitness. As with all relationships between parents and offspring, the interests of the two parties are not identical (Trivers 1974, see (3) below). The optimum sex ratio (or more accurately investment ratio) of reproductive offspring for the queen in the colony is 1:1 (see Chapter 6). However a sterile worker female has an r of only $0{\cdot}25$ with her younger brothers. If she devoted equal effort to rearing her younger brothers as to her younger sisters ($r=0{\cdot}75$), then her average r to her siblings would be $0{\cdot}5$, and the worker would do just as well, in terms of gene replication, by having her own offspring. To make a genetic profit by helping the queen, the workers have to manipulate the investment ratio away from the 1:1 value which is optimal for the queen. The preferred investment ratio from the worker's point of view should be biased by 3:1 in favour of reproductive sisters (Trivers & Hare 1976). This follows from the fact that workers are $\frac{1}{3}$ as closely related to drones as reproductive sisters (future queens).

Suppose for a moment that brothers and sisters are equally costly to rear. Then the optimum sex ratio for workers is 3:1 in favour of reproductive sisters, because with this ratio their younger brothers are $\frac{1}{3}$ as common as younger reproductive sisters. This gives the brothers three times the average expected reproductive success of reproductive sisters (since on average there will be three females to every male in the population), and hence the genetic profit to workers from brothers and sisters is equal. Brothers provide three times as many nephews and nieces, but they are each a third as valuable in genetic terms as those from reproductive sisters. The queen, being equally related to sons and daughters, favours an equal sex ratio so that the two sexes have equal average expected reproductive success.

There is thus a conflict in the colony between the queen who is selected to invest equally in the two sexes of her offspring and the workers who are selected to manipulate the ratio to 3 female reproductives to 1 male reproductive. (Although there is no genetic difference between workers and queens, the same gene could be selected to do different things in different bodies.) Who wins the conflict? Trivers and Hare looked at the nests of monogynous (one queen per colony) ant species and measured the ratio of investment (in terms of dry weight) in male and female offspring. They found that in 21 ant species the ratio was close to 3:1 in favour of females, just as predicted by the simple genetic model if the workers were in control. Remarkably, in two species of slave-making ants (where the queen's brood is reared by workers who are stolen from other species) the investment ratio was close to 1:1. In these two cases, Trivers and Hare suggest the queen gets her way because the workers are unrelated to her and are not adapted to respond to whatever chemical or other signals would enable them to detect the male larvae and manipulate the sex ratio away from the queen's optimum.

Alexander and Sherman (1977) propose another explanation for the female bias in investment in offspring. Once again the idea stems from the work of W.D. Hamilton (1967) who pointed out that Fisher's theory of the evolutionary stable sex ratio of 1:1 does not hold when there is inbreeding and siblings compete with each other for matings. For example, in the extreme instance where all the queen's daughters are fertilised by her own sons, then the queen should produce just enough sons to ensure that all her daughters are fully fertilised. Because a male can produce enough sperm to fertilise several females, the queen should produce many more daughters than sons if she is to maximise her reproductive output. The ESS argument is no longer applicable because the expected reproductive success of the sons does not depend on what happens in the rest of the population. This effect of competition between related individuals for mates, say Alexander and Sherman, may account for the female bias in offspring investment. So they suggest that the queen is getting her optimum after all!

One of the problems with the data analysed by Trivers and Hare is that some of the assumptions used in calculating the coefficients of relatedness in their model are often violated. As Alexander and

Sherman emphasise, sometimes the queen mates more than once, and sometimes the workers themselves lay (unfertilised) eggs that develop into males. Although this may help explain why the sex ratios in different colonies are so incredibly variable (even within the same species), it is still a remarkable fact that the ratios approximate, on average, to the prediction made by Trivers and Hare, assuming the workers were in control. It is unfair to expect their simple model to explain every case in detail. What Hamilton's revolutionary theory does provide is a genetical framework for interpreting all social behaviour. His idea underlies the thinking on almost every page of this book.

3. *Maternal manipulation.* A parent has only a limited amount of resources to devote to offspring. How should it spread these resources between its various children? Trivers (1972) defined parental investment as 'any investment by the parent in an individual offspring that increases the offspring's chances of surviving (and hence reproductive success) at the cost of the parent's ability to invest in other offspring'. The investment in any one baby, therefore, is measured in terms of decreased ability to invest in future children.

When parental investment is defined in this way, it can be seen that there is a conflict between the parent and any one offspring over the amount of investment that should be devoted to it (Trivers 1974). When future offspring are full siblings, the parent and the present offspring have equal genetic stakes in them—$r=0{\cdot}5$ in both cases. Thus the cost of investment, measured in decrement to future offspring of the parent, is the same for the parent and the present offspring. However the benefit to the child will be twice that to the parent, because the child is twice as related to itself ($r=1$) as it is to its parent ($r=0{\cdot}5$). In fact the child will be selected to carry on demanding care until the cost to the mother is equal to twice the benefit to the mother. Thus there is a period of conflict between the time that the benefit and the cost to the mother are equal (when mother wants her child to be independent) and the time that the benefit to the mother equals twice the cost (when the child wants to become independent). Just as we would predict from Trivers' model, in both mammals (Hinde 1974) and birds (Davies 1976b, 1978a), parents play an important role in promoting the independence of their offspring.

Trivers' (1974) model just says that there will be a conflict between parents and offspring; it does not predict who will win. According to Trivers and Hare (1976) it is the offspring (workers) who win the conflict in the social Hymenoptera (in this case over the sex ratio of reproductives in the brood), perhaps because they are in a better practical position to act against the parent's (the queen's) wishes. By contrast, Alexander (1974) has argued that the parents will always be able to dominate the offspring. As an extreme example, we may imagine that, in the social Hymenoptera, the queen has imposed sterility on the workers and there is just nothing that they can do to alter their miserable role in life.

Alexander argues the case on two levels, an ultimate genetic level and in terms of a proximate mechanism. His genetic argument is in fact fallacious (Dawkins 1976). He suggested that a gene which made offspring manipulate parents could never be selected for because it would lose any advantage gained through manipulation, when the offspring itself became a parent. The argument could equally well be made in reverse: genes for parental manipulation could never be selected for because they suffered a disadvantage when the parent was an offspring! This not to say that parental manipulation cannot occur, but there is no fundamental genetic asymmetry which dictates that it has to occur. Whether parents manipulate their offspring or vice versa depends on proximate factors. Alexander argues that parents are bigger and stronger and have the chance to physically control their offspring, but as we saw in the last section, this does not seem to be true for the social Hymenoptera studied by Trivers and Hare.

4. *The 'hopeful reproductive' hypothesis.* Hamilton's $\frac{3}{4}$ relatedness argument cannot explain the evolution of sociality in the termites (which do not have the haploid-diploid difference between the sexes) or, for that matter, sociality in many vertebrate species. The 'hopeful reproductive' hypothesis has been developed to explain the evolution of cooperative breeding in birds (see Chapter 9) and M.J. West-Eberhard (in prep.) favours a similar idea for the social Hymenoptera.

Whenever an individual's whole chance of survival and reproduction depends on the integrity of the group, then it is in the individual's own interests to cooperate with others to maintain the group structure, irrespective of whether they are related or not. Each individual is a hopeful reproductive and pays for permission to live in the group. In many social Hymenoptera it is only one individual per colony who eventually fulfills its hopes and actually breeds.

It is important to note that Hamilton's $\frac{3}{4}$ relatedness argument, and Alexander's maternal manipulation idea offer possible genetic bases for eusociality, but they do not say anything about the ecological circumstances which have produced eusociality in some hymenoptera but not others. The 'hopeful reproductive' hypothesis, in contrast, is an ecological argument: sometimes ecological circumstances (e.g. a shortage of nests) may dictate that the best strategy on average for gene propagation is to stay in a group, even though sometimes someone else ends up doing all the breeding.

1.6 Reciprocal Altruism

Hamilton's (1964) model has shown us how altruism to one's kin is explicable as selfish behaviour on the basis of natural selection of genes. Trivers (1971) has presented a model which attempts to explain how altruism towards very distant relatives or even towards other species can still be selected for during evolution.

Whenever the benefit of the altruistic act to the recipient is greater than the cost to the actor, then as long as the help is reciprocated to the actor at some later date, both participants will gain. Trivers imagined the hypothetical example of a rescuer and a drowing man. If the drowning man had a 50 per cent chance of dying and the rescuer only risked, say, a 10 per cent chance of death in successfully rescuing him, then the rescuer would benefit if the recipient of the good deed likewise rescued him at some later date. Each participant has, through reciprocal altruism, traded a 50 per cent chance of dying for a 10 per cent chance.

To distinguish reciprocal altruism from kin selection, we have to know the genetic relationship between the individuals concerned. Packer (1977a) has reported a convincing example of reciprocal altruism in olive baboons (*Papio anubis*). When a female baboon comes into oestrus, a male forms a consort relationship with her prior to mating. A male who does not have a female sometimes enlists the help of another (unrelated) male. This solicited male engages the consort male in a fight and while they are busy doing battle, the male who enlisted help goes off with the female. Packer was able to show that those males that frequently gave aid were those that most frequently received aid and that there was reciprocation involved.

Reciprocal altruism will only evolve if there is discrimination against cheating individuals, who accept help but refuse to repay it. Hence we may expect this form of altruism especially in those species, like ourselves, where individual recognition is possible and where there is scope for the altruistic act to be repaid. In fact Trivers goes so far as to suggest that human characteristics such as friendship, gratitude, sympathy, trust and guilt are all part of the psychological system that regulates reciprocal altruism, controlling cheating and the detection of cheating.

1.7 Summary

The two themes which we have discussed in this introductory chapter form the basis for the evolutionary approach to behavioural ecology in this book. Firstly, natural selection maximises gene survival, and individuals, who are no more than temporary vehicles for genes, will be expected to behave in such a way as to maximise their inclusive fitness. Secondly, the optimal behaviour for an individual to achieve this goal will depend on both the behaviour of other individuals and also on the ecological circumstances which mould the animal's way of life. As far as possible, the authors have aimed to show how theoretical models based on these principles have been tested by observation and experiment. Tinbergen's experimental approach to the study of evolutionary questions is evident in many of the chapters.

The book is divided into three sections. We have imagined that to achieve the goal of maximising the survival of its genes, the animal will be selected to optimise the solution to three sorts of problems it faces during its life. Its first problem is how to organise its feeding behaviour; how and where to search for food and what type of food to eat (Chapter 2), whether to forage alone or in a group (Chapters 3 and 4), and how to avoid becoming food itself (Chapters 3 and 5). The second section tackles the problem of how the animal uses this food to propagate its genes into future generations; should it engage in sexual reproduction (Chapter 6); if so, what sort of mate should it choose (Chapter 7) and how should it search for a mate (Chapter 8)? What ecological factors promote cooperative breeding (Chapter 9)? Chapter 10 considers the more general question of why and how animals communicate not only in courtship but in other contexts as well. The final section examines how an animal deploys its behavioural options in space (Chapters 11 and 12) and time. Earlier chapters considered the question of, given that an animal had chosen to do a certain behaviour (e.g. forage, find a mate, defend a territory), how should it behave? Chapter 13 considers how an animal decides to engage in a particular activity at any one time and how it allocates its time between these conflicting demands. Finally the long term view of an animal's strategy in time is discussed in Chapter 14, which examines life history strategies.

This book is not comprehensive. We have not attempted a global synthesis on the scale of Wilson's (1975) monumental tome. Rather, we have focused on those subjects that we find interesting and where new, exciting ideas are emerging. Perhaps our most notable omission is the evolution of learning and culture. When is learning better than inheritance as a means of acquiring information? How can learning be treated as a component of the optimality models discussed in Chapters 2 and 13? What determines why some cultural traits but not others become established? Although these questions are not discussed in the book, they are an essential part of any attempt to understand how behaviour is influenced by natural selection.

Part 1
Predators and Prey

Introduction

Look out of a window onto a lawn or playing field, and the chances are that you will see a flock of starlings busily probing in the grass for leatherjackets. It is a commonplace sight, but pause to think: Why are they on that particular bit of lawn? How are they searching; randomly, in a straight line, or what? Are they eating every single food item as they go? Why are they in a flock? Why not a bigger flock?

Part 1 of this book considers these questions; it is all about how animals harvest food, and how they avoid being eaten. Chapter 2 deals with the individual animal, and asks whether there are any general decision rules for foraging animals. What criteria do starlings and other animals use in deciding where to forage, which prey to eat, how to search? The argument is that animals are designed by natural selection to harvest food with maximal efficiency. One can devise simple models based on this assumption to predict how a predator should behave, and already the models have met with a fair amount of success, particularly in laboratory experiments. There are, however, many unexplored problems, two of which are mentioned in Chapter 2. We know very little about how predators sample a fluctuating environment, and how they assess prey which differ in nutrient quality rather than simply in energy value.

Chapter 3 takes the story from individuals to groups. Individuals may benefit from group living because the group is better at locating and catching food, or because groups are better at detecting, deterring, and confusing predators. The selfish individual benefits both as a predator and as a potential prey, but group living also has its costs. Chapter 3 discusses some of these with particular reference to birds and mammals: individuals in a group compete for food and mates, and they may run the risk of inbreeding.

Permanent groups of long lived vertebrates may contain closely related individuals, which provides a genetic mechanism, kin selection, to account for the cooperation seen for example, in, lions, mongooses, and hunting dogs. Cooperation and the effects of kin selection are, of course, much more highly developed in the social insects. Although both a heron colony and a honey bee nest can act as 'information centres' in aiding individuals to find good patches of food, in the former, the information transfer is probably best considered as a mild form of parasitism, while in the latter, information transfer is achieved in a more cooperative way by 'bee dances', one of the most elaborate communication systems in the animal kingdom.

Nevertheless social insect and vertebrate groups respond to similar ecological pressures. Chapters 3 and 4 describe how in both tropical fish and stingless bees group aggression may be a means of acquiring food. In both social insects and vertebrates group foraging is often advantageous when a species exploits patchy and unpredictable food, while individual foraging occurs when food is more evenly scattered and predictable. Chapter 4 discusses how seed harvesting ants can switch from solitary to group foraging when the food supply becomes more clumped. Seed gathering and leaf cutting ants may also use sophisticated harvesting strategies which involve avoiding revisits to a place before the food supply has renewed itself.

Chapter 5 focuses on animals as prey. If you were lucky enough to see a hawk trying to attack the starling flock outside your window, the first starling to see the predator would give a call and the birds would all dash for cover. Why should one bird give information to the others while possibly putting itself at risk? Chapter 5 discusses some of the possible answers, including the only detailed study of alarm calls to date, in which kin selection is implicated. Warning colouration also poses a problem for natural selection: how does the individual which is eaten and teaches the predator a lesson perpetuate its genes? Equally problematic are mobbing behaviour of small birds and rump patch signalling in deer, both of which are discussed in Chapter 5.

Is there a general conclusion? Predators seem to be optimally efficient (or quite close to it) at exploiting their food, while prey are highly skilled at escaping, avoiding and defending. Perhaps the problem we are left to consider is: What are the rules governing the evolutionary arms race between predators and prey?

Chapter 2
Optimal Foraging: Decision Rules for Predators

JOHN R. KREBS

2.1 Introduction: Efficient Predators

When an animal harvests food it has to make decisions: choices about where to hunt for food, which kinds of prey to eat, when to try a new place, and so on. This chapter discusses how animals make these choices, not from the point of view of detailed behavioural mechanisms, but by considering general strategic rules which might apply to a wide range of animals. In brief, the rationale is this: animals will, as a result of evolutionary selection pressures, tend to harvest their food efficiently, so if we can work out in theory the decision rules which would maximise the animal's efficiency, these rules ought to predict how the predator makes its choices. [Note that the words 'decision' and 'choice' are not intended to imply anything about conscious thought, they are a short-hand way of saying that an animal is designed to follow certain rules. The term 'predator', is used to include more than just carnivorous animals: to a seed, a finch is as much of a predator as is a shark to us.]

Why should evolution favour efficient predators? In the long term, natural selection favours individuals with the highest fitness (see 1.5) but short-term objectives such as maximising efficiency in harvesting food are certainly important in contributing to eventual reproductive success. This is most conspicuously true in animals such as the titmice studied by Gibb (1960) which have to catch an insect every three seconds throughout a winter day just in order to stay alive; but even a predator such as a lion which spends much of its time resting (Schaller 1972) may be under strong pressure to hunt efficiently, because hunting competes for time with activities such as defending a territory, mating, resting and so on (see Chapter 13). As long as a predator could improve its survival or reproductive success by hunting more efficiently, natural selection will favour efficient predators. The term efficient could mean one of several things. For example, achieving an instantaneous maximum net rate of food (or energy) intake; a maximum rate of intake of some specific nutrient such as a vitamin or essential amino-acid; minimising the fluctuations in rate of intake of total food or specific nutrients; or achieving a maximum net intake over a long time period such as a whole season. In other words, even if we were to accept that optimal foragers maximise foraging efficiency in one sense or another, we still have to test which of several plausible alternative goals is being maximised. Most of the examples in this chapter will be testing models based on the hypothesis that efficient predators make decisions which maximise their net rate of food intake while foraging. The models say nothing about when a predator should or should not forage (see Chapter 13): they assume that the animal has decided to search for food, and ask how it could maximise its net rate of intake during a foraging period. At the end I will briefly come back to the question of alternative goals for optimal foragers.

In order to work out hypotheses about optimal decision rules, it is necessary to think of the problems which face a wide variety of predators, regardless of their specialised tactics for gathering prey. The three types of choice which are likely to be made by many actively foraging animals (as opposed to filter feeders or sit-and-wait predators) are which types of food to eat, where to hunt for food, and what type of searching path to use in hunting for prey or foraging places.

2.2 Optimal choice of food types

2.2.1 *Predators choose profitable prey*

Any prey item eaten by a predator has a cost in terms of the time taken to subdue and eat the item, and a benefit in terms of its net food value (net value means the gross value minus the energy costs of handling and digesting the food). The net food value divided by the handling time is a measure of the profitability of a prey type, and it is a prerequisite for a model of optimal prey choice that predators should be able to distinguish between items of differing profitability and select the more profitable types. Figure 2.1 shows three examples of studies in which the gross food yield per unit handling time for different prey types has been measured and the predator has been shown to prefer the most profitable types. In none of these studies did the prey differ appreciably in conspicuousness, palatability, or nutrient quality. These points are discussed in sections 2.2.4 and 2.2.5.

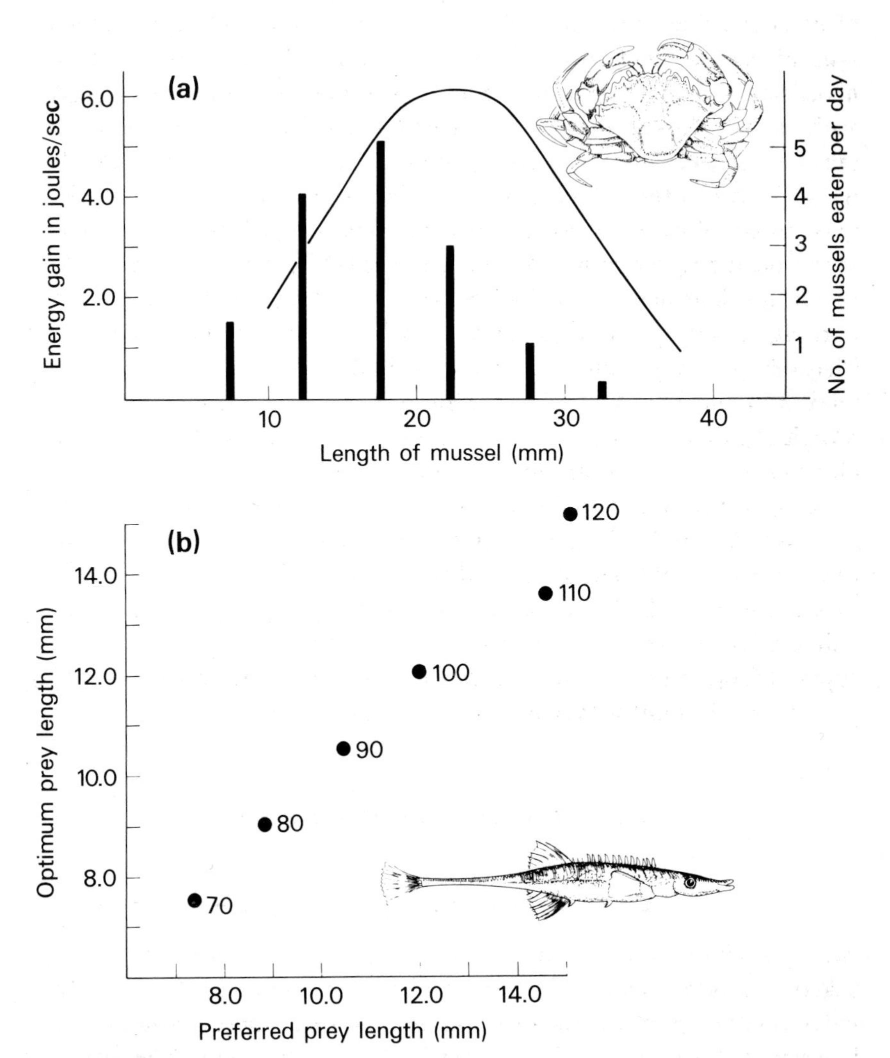

Fig. 2.1. Predators choose profitable prey. (a) Histogram of numbers of mussels of different size classes eaten by 6·0–6·5 cm shore crabs with unlimited prey. The curve shows the energy gain per unit prey breaking time (Elner & Hughes 1978). (b) Preferred length of *Neomysis integer* eaten by different sizes of 15-spined sticklebacks in the wild, plotted against optimum prey length (determined by dry weight of prey/handling time). The figures by the dots refer to fish size (length in mm) (Kislaliogu & Gibson 1976). (c) Selection of flies by pied wagtails. The upper histograms show the available and preferred distributions of flies, the lower curve, (d), shows the profitability of different sizes of prey in calories per second of handling time (Davies 1977a).

(d)

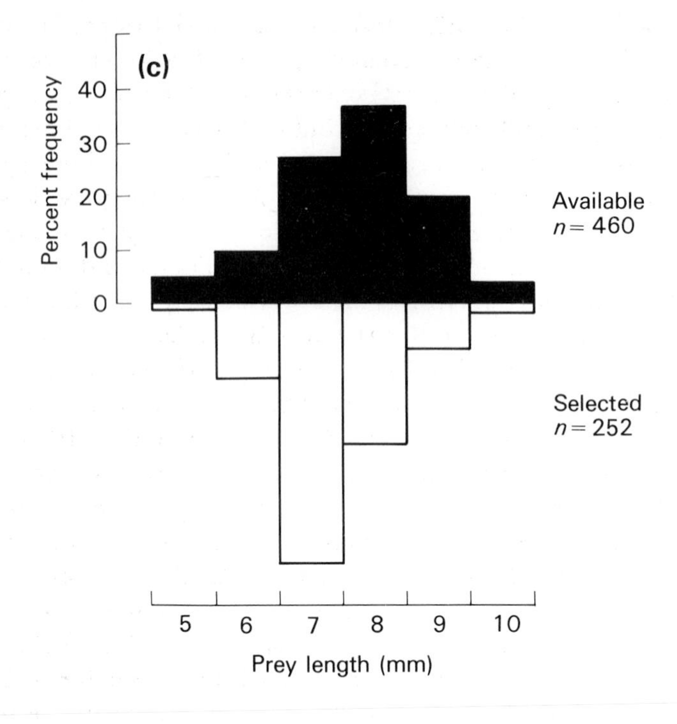

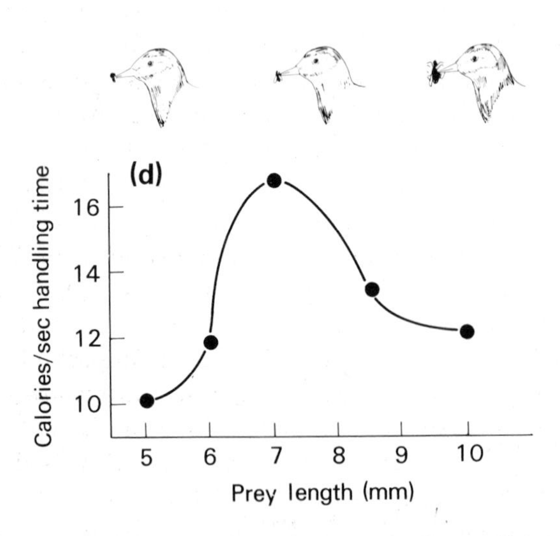

2.2.2 *Optimal breadth of diet depends on prey availability*

While it is easy to see that an optimal predator should prefer the most profitable prey, it is less obvious to what extent it should also include less profitable items in its diet. Assume that a predator spends its foraging time either searching for or handling prey. Intuitively one can see a trade-off: if the predator selects only the best items it has a high rate of food intake per unit handling time, but it also has to spend a relatively long time searching for each item. A totally non-selective predator would spend little time searching since the effective prey density is high, but have a low rate of intake while actually handling prey, because its diet is made up of a mixture of profitable and unprofitable prey. This trade-off is shown graphically in Fig. 2.2a, which is modified from the original model of MacArthur and Pianka (1966). The graph shows how travel time and the average profitability decrease with increasing breadth of diet. The travel time, assumed to be inversely proportional to prey density, falls off more rapidly than the curve of average profitability, the shape which is based on the assumption of a normal distribution of profitability (as for example in Fig. 2.1a). The graphs in Fig. 2.2a and b make two points: first, combining the profitability and travel time curves to calculate the total food intake per unit time gives the optimal number of prey types, and second, that if the availability of high ranking prey increases, the optimal diet includes fewer prey because the travel time for a selective predator is shorter. Another way of making this second point is to say that in a habitat where profitable prey are very common it will not pay the predator to stop and eat a prey which has a low profitability. This is illustrated in Fig. 2.2c which describes the problem facing a predator when it encounters a new type of prey. Suppose the predator has been eating only prey within the optimal set (5 prey types in 2.2a, 3 in 2.2b) and it encounters a new prey. Figure 2.2c shows the predator's current rate of food intake (including travel time) as the slope of a solid line, and the profitability of 2 new prey types as the slope of two broken lines. The food yield per unit handling time from prey type *X* is higher than the predator's current rate of food intake, so that predator would clearly benefit by eating *X*. The reverse is true for *Y*. The important point to note is that prey type *Y* should not be eaten no matter how common it is, because food yield per unit handling time from *Y* is lower than could be achieved from a diet which does not include *Y*. It is also worth remembering that the decision to eat or reject *Y* should be all or nothing: if the broken line for the profitability of a new prey falls above the solid line of net intake for prey already in the diet, the predator should always eat it; if below, it should always ignore it. This simple model of optimal diets was first presented by MacArthur and Pianka (1966) and has subsequently been presented in different algebraic

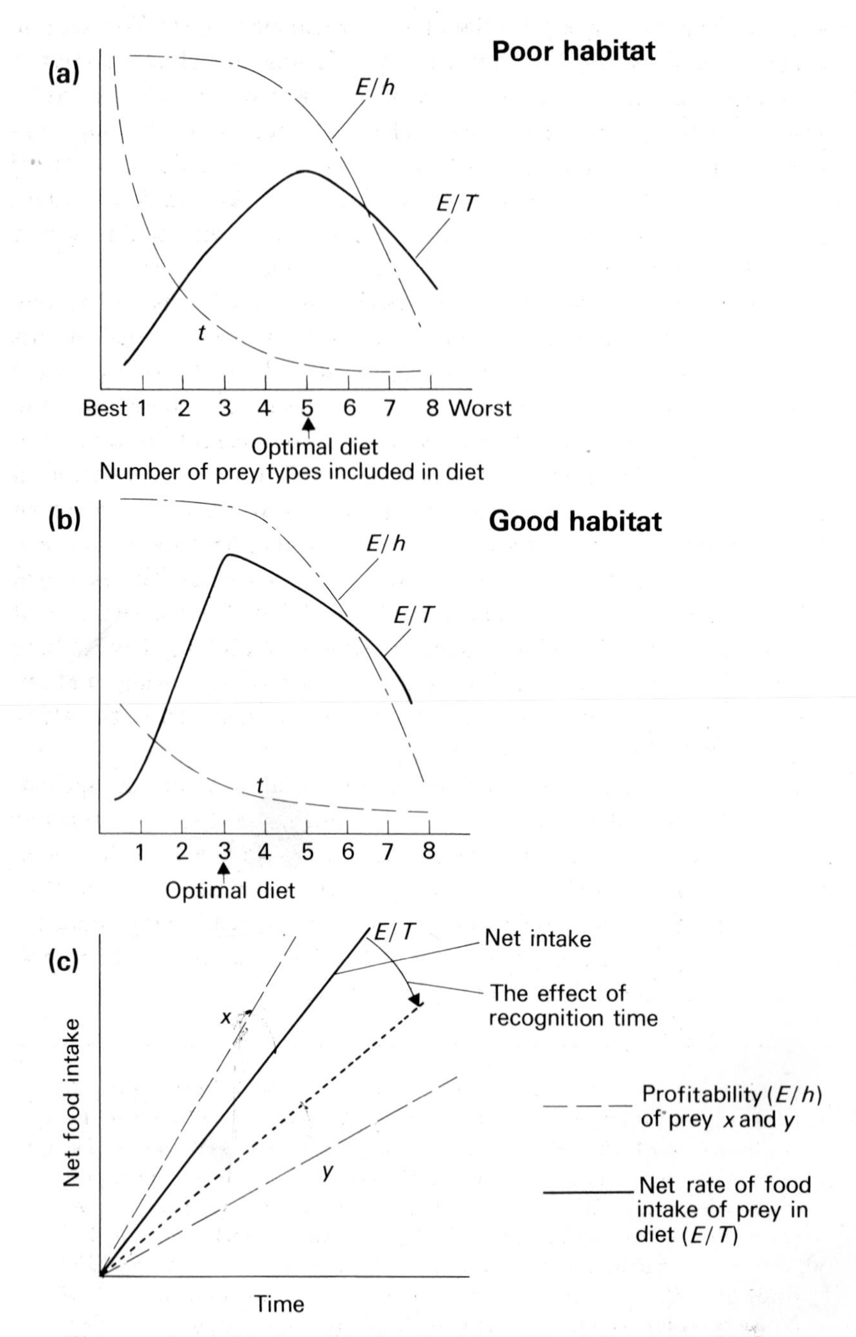

Fig. 2.2. A graphical model of optimal breadth of diet. (a) & (b) show, for a poor and good habitat respectively, the consequence of including different numbers of prey types (ranked in order of profitability) in the

forms by Charnov (1976a), Pulliam (1974), Schoener (1971), Werner and Hall (1974) and others. The important predictions, which can be tested by observations or experiment, are that predators should (i) prefer more profitable prey, (ii) be more selective when profitable prey are common, (iii) ignore unprofitable prey which are outside the optimal set regardless of how common they are. By measuring the food values, handling time and encounter rates for prey type, one could predict exactly how many prey types should be included in the diet.

The model as presented so far ignores the time taken to recognise prey and the possibility that the predator will make incorrect identifications and pursue low ranking prey before recognising them. These important points are discussed by Hughes (in press). If the total handling time for a prey type is made up of handling and recognition time, the recognition time being spent even on prey which are then rejected, a high encounter rate with unprofitable prey will have the effect of lowering the predator's overall rate of food intake, because an appreciable proportion of foraging time will be spent in rejection. This is shown in Fig. 2.2c, as a clockwise rotation of the E/T line. The consequence of this is easy to see: when the recognition time is at all long, low ranking prey could move into the optimal set as a result of increasing in abundance even when more profitable prey are common. Thus the third prediction listed earlier would not apply.

The next section describes four experimental tests of the optimal diet model, three of them with visually hunting vertebrates, and one with arthropod which hunts by chemical and tactile cues. In the visual predators recognition time is probably short (essentially zero) so that we would expect these predators to ignore unprofitable prey independently of their abundance, but this is not to be expected for the fourth predator, which has a longer recognition time.

---

diet. As more prey are added, travel time (t) decreases, but so does the average profitability of prey eaten (E/h). The curve of net food intake per unit time ($E/T = E/t+h$) rises to a peak and then declines. The peak is the optimal diet breadth: in the poor habitat, where good prey are scarce so that travel time is long, the optimal diet includes more prey than in the good habitat. (c) If the predators net intake is E/T (solid line), it should not eat any prey such as Y with a value of E/h below the solid line. Even if prey such as Y are very common, the slope of the solid line cannot be increased by adding these prey to the diet. On the other hand a prey type such as X should be eaten as it increases the value of E/T. If prey such as Y have a recognition time, they might, when common, lower E/T (solid line slope reduced to dotted line).

2.2.3 *Laboratory and field tests of optimal diets*

As with any model, the most important predictions of the optimal diet model are those which are precisely quantitative and hence unlikely to be explicable by another hypothesis. In this section I will describe four studies which have attempted to test the optimal diet model quantitatively.

Bluegill sunfish

Werner and Hall (1974) allowed a group of ten bluegill sunfish (*Lepomis macrochirus*) to hunt for three size classes of *Daphnia* in a large aquarium (approximately 350 litres). They mixed together different size classes of *Daphnia* in the tank and then allowed the fish to hunt for a short time. They counted the number of prey of each class eaten by the fish during a foraging session by sacrificing the fish and analysing their stomach contents. The handling time for all size classes was similar, so that their relative profitability simply depended on the relative sizes of prey classes. In addition to the profitabilities of each prey type, Werner and Hall had to estimate the fishes' encounter rate (a precise estimate of the search time) with each type, which involved calculations of the relative visibility of each size class, based on simple assumptions about the visual field of the fish. Werner and Hall's most important result is shown in Fig. 2.3a. When the prey mixture was presented at a low density (20 of each class), the fish ate the three sizes according to how often they were encountered (i.e. there was no selection). In contrast, when the number of each type of prey was increased to 350, the fish ate almost exclusively the largest and most profitable size class. At an intermediate density of 200 per size class, the fish ate the two largest size classes, so that as the overall prey density increased, low ranking prey were dropped successively from the fishes' diet. In terms of Fig. 2.2c, as the slope of the solid line increases more prey types fall out of the optimal set. Figure 2.3a shows the expected composition of the diet based on the optimal foraging model, as well as the expected diet if the fish were unselective, showing that the model predicted fairly accurately which prey were eaten in each treatment.

Werner and Hall did not test the prediction that the fish should ignore the smallest *Daphnia* regardless of their abundance when the density of large *Daphnia* is high enough for E/T to be larger than (E/h) small, as it was in the intermediate and high density treatments. This

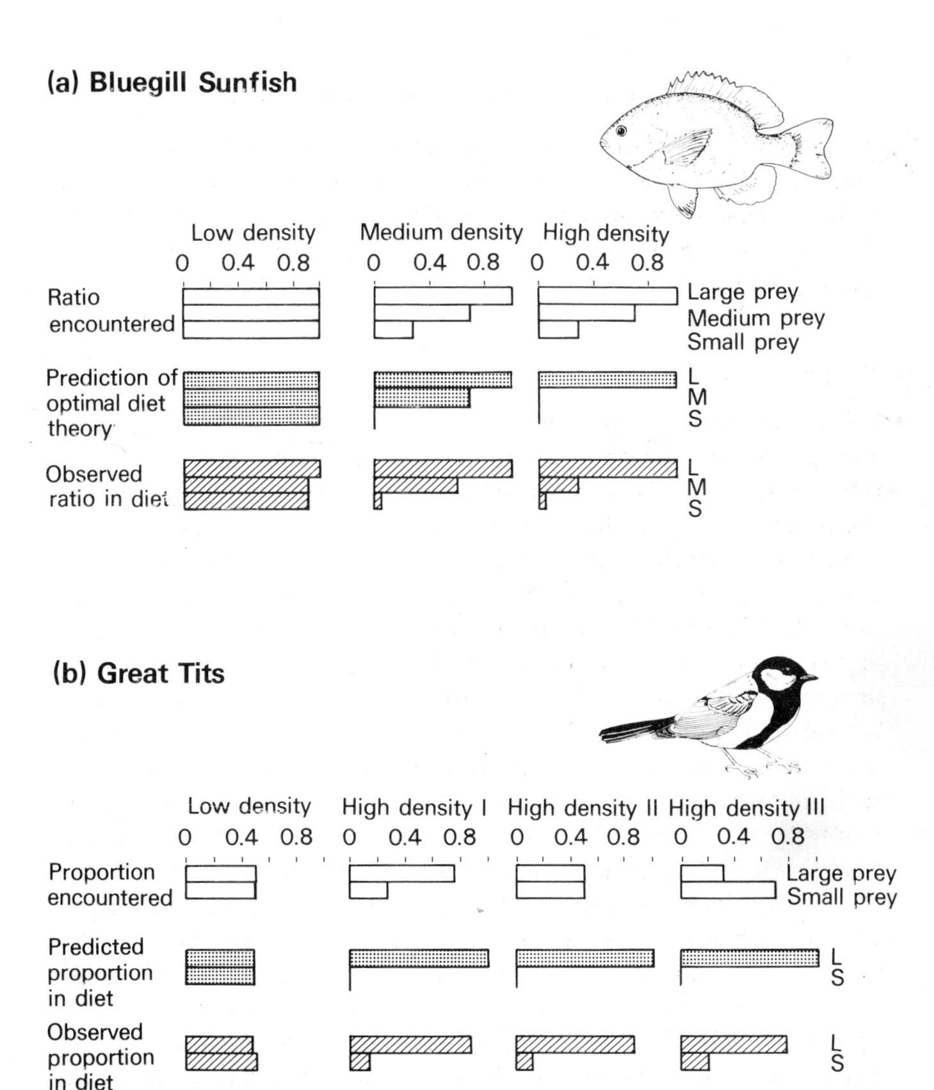

Fig. 2.3. Four studies of optimal choice of diet. (a) Bluegill sunfish (Werner & Hall 1974) preying on different size classes of *Daphnia*: the histograms show ratio of encounter rates with each size class at three different densities together with the predicted and observed ratios in the diet. (b) Great tits preying on large and small mealworm pieces (Krebs *et al.* 1977). The histograms in this case refer to proportion of the two types chosen. (c) Redshank eating worms (Goss-Custard 1977a). The top graph shows that large worms are eaten in proportion to their abundance, and the bottom graph shows that small worms are less likely to be eaten when large ones are common. (d) Shore crabs eating mussels (Elner & Hughes in press): explanation same as for (a).

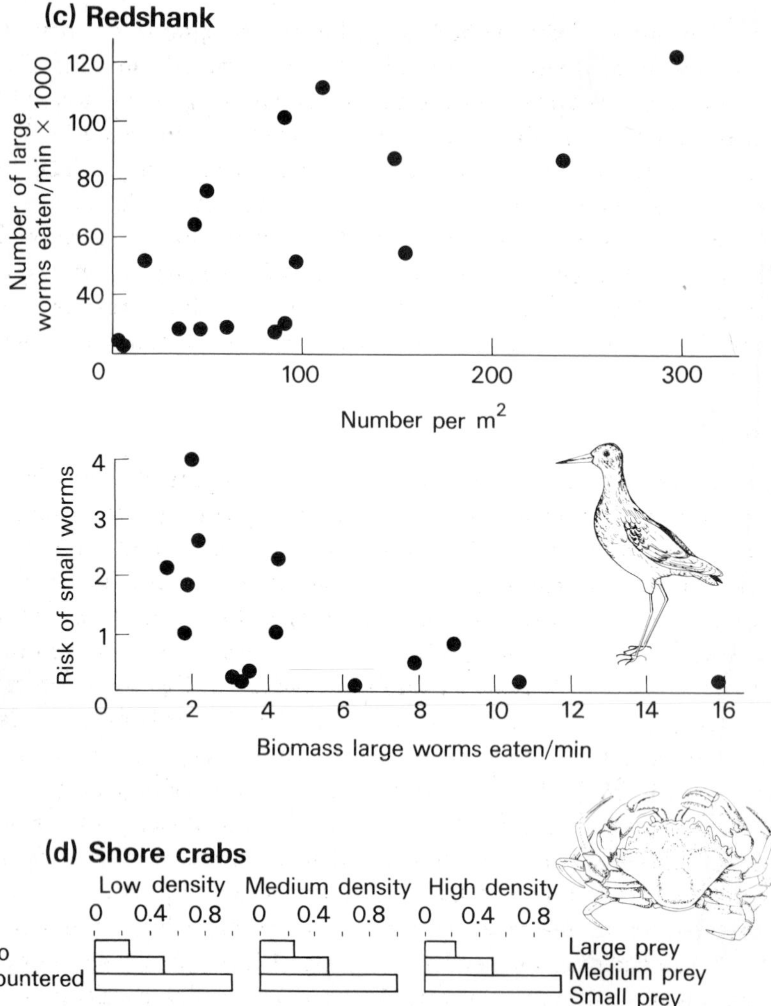

prediction is important because, as O'Brien *et al.* (1976) have shown, Werner and Hall's results could be explained in terms of a different model: the fish simply snaps up whichever *Daphnia* appears largest at that moment in time. A 'large' *Daphnia* could be a small one close by

non-recog
non-selectivity

or a big one slightly further away, but as the overall density of prey increases, it will become more likely that there will be a big *Daphnia* close enough to appear to the fish to be the largest one available. In other words the 'go for the largest' rule would predict exactly the same trend as that observed by Werner and Hall (Fig. 2.3a). The critical observation which could distinguish the two models is the prediction of the optimal diet model that small but very abundant prey should be ignored if large prey are above a certain density (assuming a zero recognition time). In summary, Werner and Hall's results are consistent with, but not exclusive evidence for, the optimal diet model.

Great tits

One study which set out specifically to examine the prediction of ignoring common small prey is that of Krebs *et al.* (1977), using caged great tits as predators, with large (profitable) and small (unprofitable) pieces of meal-worm as prey. They estimated the profitability of the two prey types as (weight/handling time) and controlled the encounter rate with each type very precisely by presenting them to the predator on a moving belt. The results are summarised in Fig. 2.3b: when large and small prey were present at a low density the birds were unselective as predicted, but when the density of large prey was increased to a level at which the birds could do better by ignoring small prey, they became highly selective. Finally, keeping the density of large prey constant, Krebs *et al.* increased the density of small prey so that they were twice as common as large ones. The birds, as predicted, remained highly selective and essentially ignored the small but abundant prey. As Fig. 2.3b shows, the results disagreed with the predictions of optimal foraging in two ways. As with Werner and Hall's sunfish, the great tits did not completely ignore unprofitable prey, and further, the proportion of unprofitable prey taken was not completely independent of their encounter rate. The second divergence could be explained if the great tits required time to recognise small prey (as was discussed earlier), and the first effect might be attributed either to detection errors, or to 'sampling'; the predator has to sacrifice a certain amount of efficiency to acquire information about the relative profitability of each type.

Redshank

There are many qualitative observations from field studies which weakly support the optimal diet model, for example tawny owls

(Herrera 1975) and starfish (Menge 1972) are more selective when food is abundant than when it is scarce (see, however, Smith *et al.* in press), but only one published field study so far has tested the optimal diet model quantitatively. Goss-Custard (1977a) studied the selection by redshank (*Tringa totanus*) (a shorebird) of different sized polychaete worms (*Nereis*, and *Nephthys*) on mudflats. As with the previously discussed laboratory studies, Goss-Custard measured the profitability and availability of different size classes of prey. By comparing the rate of feeding by the redshank on large and small worms at various different study sites, Goss-Custard was able to show that the largest, most profitable, prey were eaten in direct proportion to their own density, while the smallest worms were not taken in relation to their own density, but at a rate inversely proportional to the density of large worms (Fig. 2.3c). In other words, as the density of large worms increased, the redshank became more selective, and further, they tended to ignore small prey regardless of their own density as long as large worms were common. One possible interpretation of these results is that large and small worms interact somehow so that when large ones are common, small ones are hard to catch. Goss-Custard eliminated this interpretation by showing that redshank show the same pattern of selection in controlled laboratory conditions. Goss-Custard used his field results to test the optimal diet model by building a simulation model into which he incorporated his observed relationships between prey choice, prey density, walking speed, and peck time. The model showed that the redshank could not have done better (achieved a higher rate of food intake) by any alternative strategy of choice.

Shore crabs

Elner and Hughes (in press) studied optimal choice of diet by a non-visual predator, the shore crab (*Carcinus maenas*), which hunts for mussels (*Mytelus edulis*) using chemical and tactile cues. They measured the profitability of various size mussels by calculating the energetic yield per unit breaking and eating time. When presented with a low density mixture of mussels of three size classes the crabs were unselective (the experiments lasted three days but the prey were replaced to maintain the initial proportions). At medium and high densities the crabs almost totally ignored the smallest, least profitable size class. Although they showed a preference for the largest size class in high density treatment, they did not totally ignore the intermediate class as predicted by the optimal foraging model (Fig. 2.3d). In a further set of tests using just the two most profitable size classes at high density but in varying proportions, Elner and Hughes showed that crabs included the less profitable type in the diet as its abundance increased. They suggest, but do not show quantitatively, that this is an effect of rejection time: the crabs take 1–2 seconds to reject an unprofitable mussel and as discussed earlier, this could lead to the inclusion of common, but low ranking prey in the diet.

2.2.4 *Prey quality as a complicating factor*

The results discussed so far fit quite well the predictions of optimal diet models, but they have ignored several important factors. In particular, they have all involved selection between different sizes of the same prey type, so that differences in digestibility, nutrient quality and energetic costs of handling have been ignored (Pulliam 1975a). The benefit has been measured purely in terms of gross food intake. It is well known that animals may select prey according to nutrient quality. Red grouse prefer 3-4 year old heather which is especially rich in nitrogen and phosphorus (Moss *et al.* 1972) and rats can select a balanced diet if presented with an array of foods containing different nutrients (Rozin 1976). Requirements of specific nutrients may mean that predators do not choose prey which yield the highest total food reward per unit handling time. For example, Goss-Custard (1977b) found that the same redshank which make an optimal choice of different sized worms, prefer the smaller and less profitable *Corophium* (an amphipod) to all sizes of worms. *Corophium* yields less food per unit handling time than worms, but may contain more of some essential nutrient. Figure 2.4 illustrates how redshank may choose *Corophium* on the basis of nutrient value. Assume that the redshank needs two components of food: energy (a) and nutrients (b), and that worms contain more a, but *Corophium* are richer in b. If the bird eats only worms, its intake is a mixture of a and b shown by line W, while if it takes *Corophium* it gets a mixture shown by C. The curved dashed lines are isoclines of benefit to the redshank of different combinations of a and b. The bird ought to choose whichever combination of worms and *Corophium* gives it the highest total benefit (crosses the highest isocline), but it is constrained in its choice by the total time available to gather prey. The total time budget is represented by the dotted line in Fig. 2.4, the slope of which represents the relative price of eating worms and *Corophium*. Worms are much more efficient in terms of energy intake (shown by the ratio of W_E:C_E), but *Corophium* yields much more of another essential

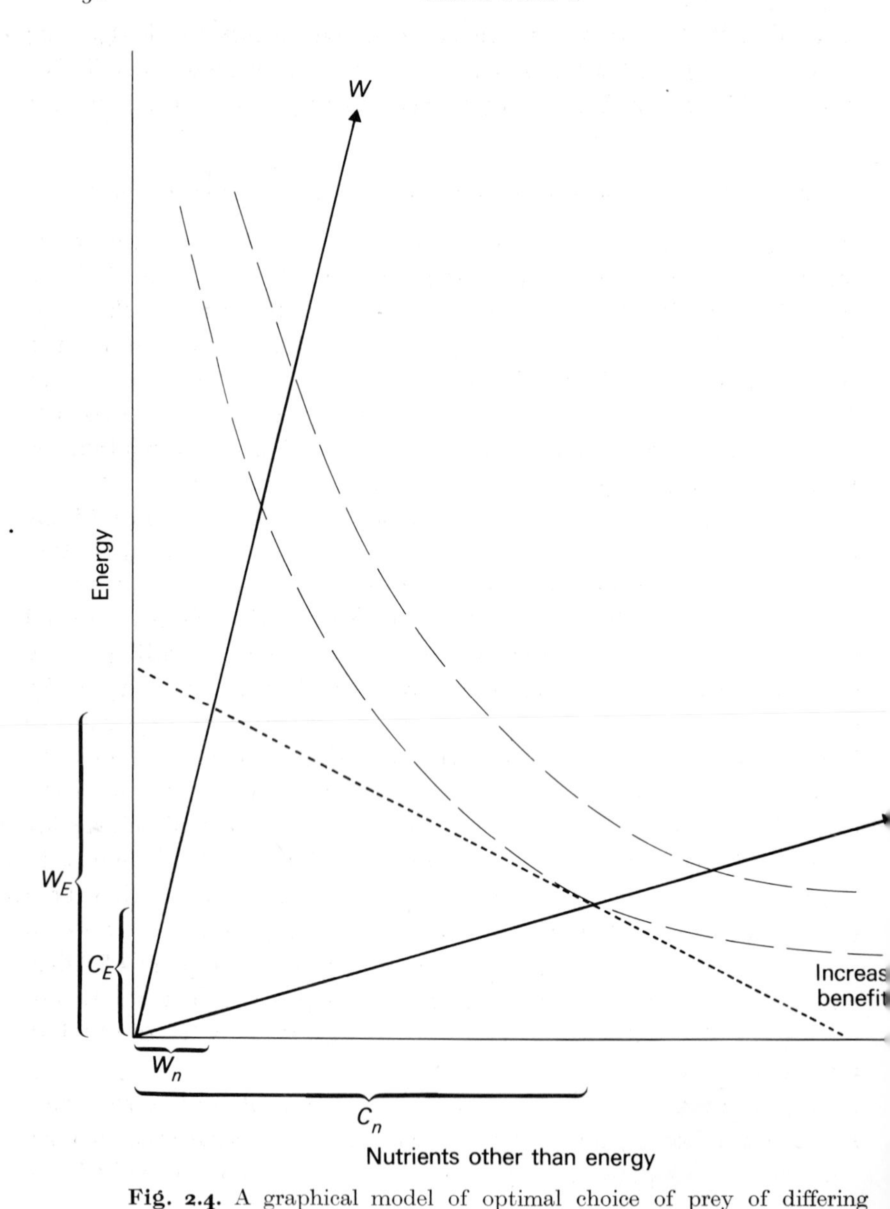

Fig. 2.4. A graphical model of optimal choice of prey of differing quality. The two prey W and C contain differing proportions of 'energy' and 'other nutrients'. If the predator eats W, its intake of energy and nutrients follows the line W, while if it eats any prey type C, its intake follows the line C. Any mixture of W and C will give an intake line between the two extremes. The broken lines are isoclines of equal benefit derived from various combinations of 'energy' and 'other nutrients'. The predator should choose its diet so as to cross the highest possible isocline of benefit, but it is constrained by the budget

nutrient (W_n:C_n). In the particular example drawn the budget line crosses the highest benefit isocline at a diet of pure *Corophium*. This is one possible explanation of why redshank prefer *Corophium* to worms.

2.2.5 *Other models of prey choice: switching and searching images*

Many predators seem to show frequency dependent preference for different prey types: they take a disproportionate number of the commonest prey. This type of predation has been termed 'switching' by Murdoch (Murdoch & Oaten 1975) and it has attracted some interest from geneticists concerned with maintenance of polymorphism (e.g. Allen & Clarke 1968) (see section 5.3.2) as well as ecologists analysing the influence of predators on prey stability. What is the relationship between switching and optimal foraging?

Switching may depend on the predator concentrating on profitable patches containing only one sort of prey (Murdoch *et al.* 1975) (section 2.3.1), but other examples of switching occur when prey are presented in a mixture. Here the functional significance of switching may depend on the fact that the predator becomes more efficient at dealing with a prey type when it is encountered frequently. Lawton *et al.* (1974) for example, showed that the attack success of *Notonecta* on mayfly larvae increases as a result of experience, and some animals are known to change their digestive physiology to cope more efficiently with the commonest food type (Moss 1972, Miller 1975). If, as a result of learning or some physiological change, the predator's efficiency at dealing with a particular prey changed drastically one could imagine that prey types could change their rank order of profitability, so that an optimal forager should switch (Hughes in press). However, one point of difference is that Murdoch's hypothesis of switching is couched in terms of relative prey density while optimal foraging models refer to absolute densities or encounter rates with prey.

One learning mechanism which can produce switching is 'searching image formation', which refers specifically to a perceptual change, whereby a predator selectively attends to one particular type of cryptic

---

line (dotted) representing the available energy or time budget for foraging. The ratio W_E/C_E shows that W gives about two and a half times as much energy per unit foraging time, while W_n/C_n shows that prey C contains more than six times as much 'nutrients'. The budget line crosses the highest isocline of benefit at a diet of pure C, so in this example the predator should eat the prey which is energetically less profitable but contains more nutrients (after McFarland in press).

prey (Dawkins 1971a, b). Hunting by searching image is not an alternative to optimal foraging, since it refers to a proximate mechanism underlying prey choice rather than an ultimate goal of prey selection. One can view the effect of searching image on optimal foraging in two ways, as a constraint or an adaptation. Firstly, prey for which the predator has no searching image are effectively not encountered (the predator does not see them), so that the relationship between encounter rate and density can be complex. Secondly, if the predator decreases its recognition time by concentrating on one prey type, then searching image formation can be viewed as an adaptation which enables the predator to increase the profitability of particular prey types.

2.3 Exploiting patchily distributed food

2.3.1 *Choice of profitable patches*

Actively searching predators usually hunt for food which is clumped or patchy in distribution. The patches might be discrete natural units such as rotting tree stumps full of insects or bushes laden with berries, or they might be statistical heterogeneities in a superficially uniform habitat, for example quadrat sampling will show that earthworms on a lawn are clumped. Just as with choice of diet, one can easily see that an optimal predator should forage preferentially in the most profitable patches and include less profitable patch types in its foraging time only when the availability of good places is low (Royama 1970). In other words, Fig. 2.2c could equally well apply to patches if we re-define 'profitability' as food intake per unit foraging time within a patch. There are numerous examples of predators and insect parasitoids which prefer to forage in patches with the highest prey density and roughly speaking rank patches in order of profitability (Fig. 2.5). This phenomenon has been termed the 'aggregative response' by Hassell and May (1974) since one of its consequences is that predators tend to aggregate in profitable patches.

Pursuing the analogy between prey types and patches, we are led to ask how many types of patch an optimal predator should visit. Since a predator could in theory spend different amounts of time in each patch the question can be phrased more precisely to ask how long an optimal predator should spend foraging in each patch. The answer depends on how the quality of patches changes with time. There are three distinct possibilities: patches might stay constant in quality with time, although

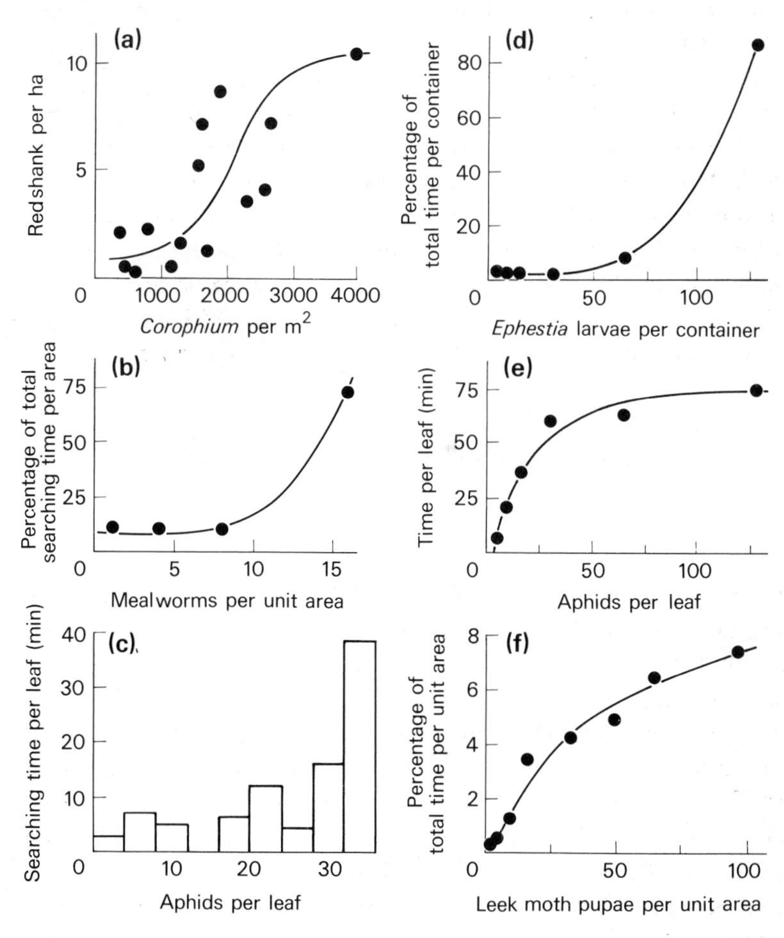

Fig. 2.5. Data from a variety of animals show that predators and insect parasites tend to aggregate in regions of high prey population density. (All curves are fitted by eye). (a) Redshank (*Tringa totanus* (L.)) density in relation to the average density of its amphipod prey (*Corophium volutator* (Pallas)) per m². (b) Percentage of total searching time by great tits (*Parus major* L.) for different densities of mealworms (*Tenebrio mollitor* L.). (c) Searching time of a coccinellid larva (*Coccinella septempunctata* L.) for different densities of its aphid prey (*Brevicoryne brassicae* L.) per cabbage leaf. (d) Percentage of total searching time by the Ichnuemonid parasite (*Nemeritis canescens*) for different densities of its host (*Ephestia cautella* (Walk.)) per container. (e) Searching time of the braconid parasite (*Diaretiella rapae*) for different densities of its aphid host (*Brevicoryne brassicae*) per cabbage leaf. (f) Percentage of total searching time by the parasite (*Diadromus pulchellus*) for different densities of leek moth pupae (*Acrolepia assectella* (Zell.)) per unit area. (After Hassell & May 1974).

this seems rather unlikely except perhaps for short lived predators; patches may change in quality as a result of the predator's activity, that is to say the predator could deplete patches; finally, patches could change over short time periods independently of the predator's activity due, for example, to diurnal or seasonal patterns of prey abundance. These three possible types of change in patch quality have different implications for an optimal predator, so they are considered in separate sections.

2.3.2 *Patches which do not change in quality*

As was emphasised above, this is not likely to be common in nature, but there are some examples of laboratory studies in which predators approach the optimal strategy. The simplest choice a predator could face is two patches of different quality with no depletion or other changes in quality with time. This is the problem facing a pigeon in a Skinner box with two keys offering food at two different reward rates (rewards per peck—a so-called variable ratio schedule). Once the optimal pigeon has detected the higher reward rate key (equivalent to the more profitable patch) it should expend all its effort in working on that key. Figure 2.6 shows two examples of such studies; both great tits and laboratory pigeons perform close to the optimal strategy by working almost exclusively at the key offering the higher reward rate. Obviously any pecks delivered to the less profitable key can only reduce the predator's average rate of food intake.

2.3.3 *Patches which are depleted by the predator*

Very often predators significantly deplete the patches in which they forage. This could be a result of direct exploitation of the prey, or because the prey became less accessible as a result of the predator's activity, for example dung flies leave a cowpat more or less as soon as a yellow wagtail (*Motacilla flava*) starts to hunt for them, and the bird is inefficient at catching the flies once they have dispersed into the grass (Davies 1977a). The consequence of depletion is that when the predator forages in a patch, the profitability declines with time, and the optimal predator has to choose the point on this curve of diminishing returns which maximises its overall rate of food intake. The optimal solution is shown graphically in Fig. 2.7a (Charnov 1976b). The diminishing capture rate in an average patch for a particular habitat is

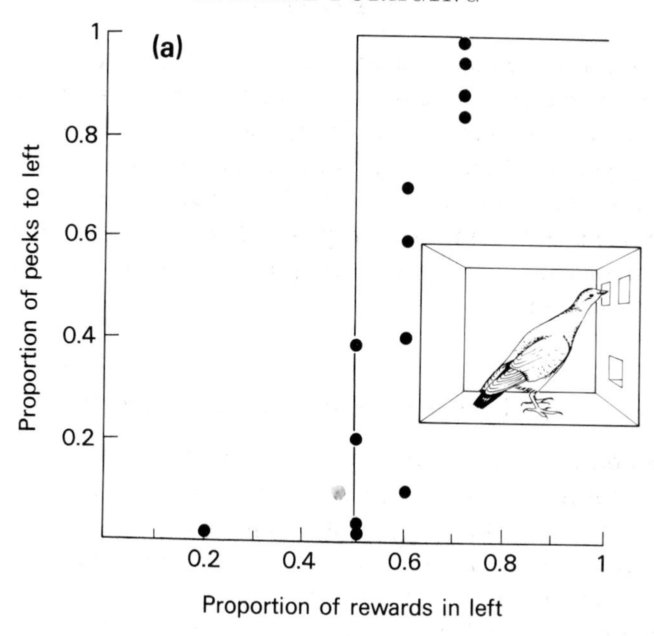

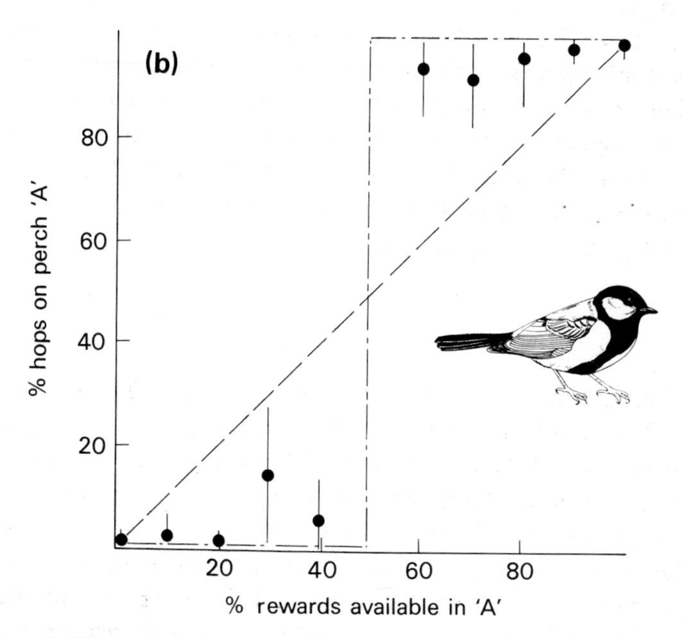

Fig. 2.6. When pigeons (a) and great tits (b) are faced with a choice of two keys or perches to obtain food, they go for the one with the higher reward rate. Ordinate shows the proportion of responses on one of the keys or perches and the abscissa is the proportion of rewards available. (After Herrnstein & Loveland 1975 and Krebs *et al.* 1978.)

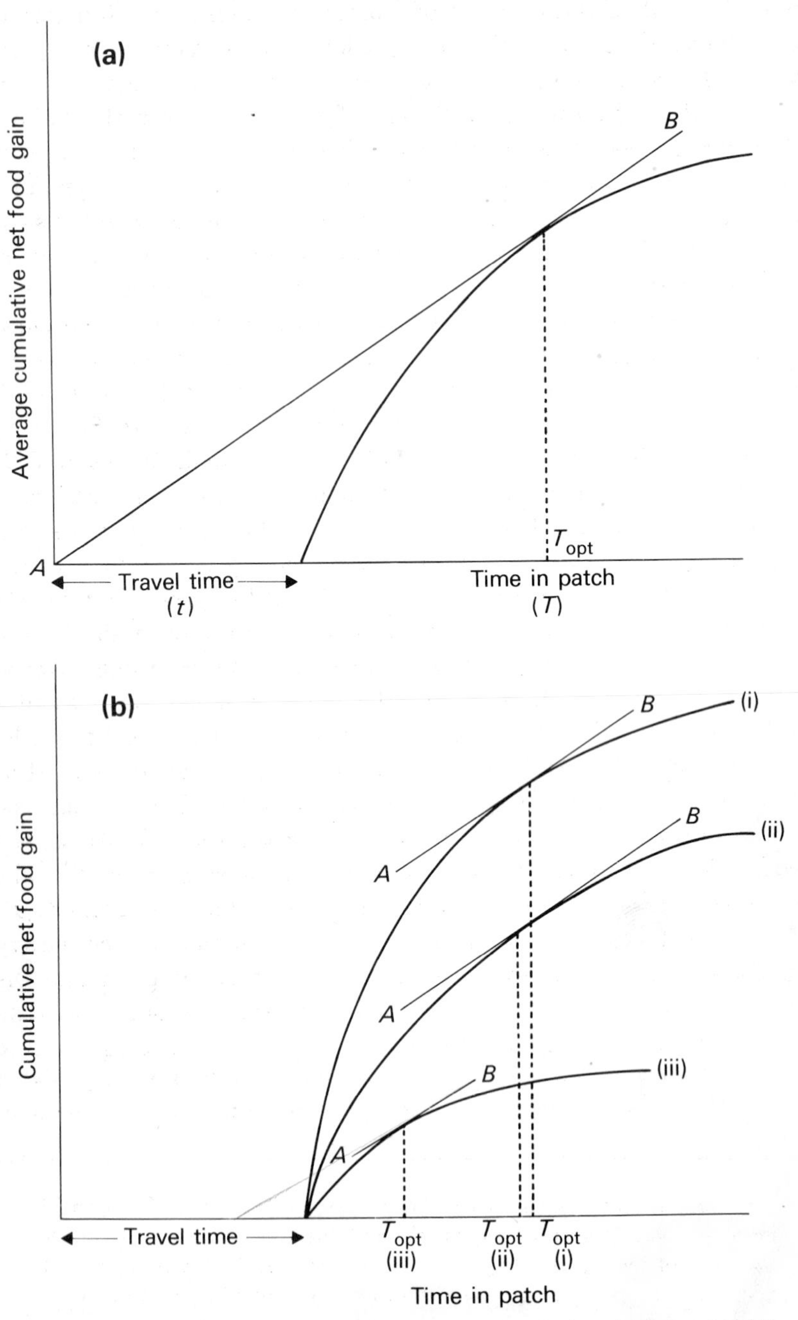

Fig. 2.7. (a) A graphical solution of the optimal time in a patch with depletion of food. The curve $f(T)$ represents the average cumulative food gain (f) as a function of time in a patch (T). The travel time spent

shown by a decelerating curve of cumulative intake as a function of time in a patch. Assume that the predator visits many patches, and spends all its foraging time travelling between or foraging within patches. The predator's overall rate of food intake for the habitat is the average food intake per patch divided by average time in a patch plus travel time. In order to maximise this quantity the predator should choose to eat in each patch just long enough to make the slope of the line AB, the net rate of food intake for the environment, as steep as possible. This line has, of course, to touch the curve of cumulative intake within a patch, and it does so where the slope of that curve is equal to E/T (the net rate of food intake). If the habitat contains a variety of different types of patches (Fig. 2.7b) the optimal solution is to stay in each patch type until the rate of intake drops to E/T. Very poor patches, in which the initial rate of intake would be below E/T, are not in the optimal set, but as with prey types, there may be a rejection time associated with suboptimal patches. [Note that there is a description of this same model in Chapter 8.] To summarise, the optimal predator should stay in each patch until its rate of intake (*the marginal value*) drops to a level equal to the average of intake for the habitat; this average includes time spent within and travelling between patches), or to put it another way, the predator should not stay in a patch when it could do better by travelling to another one. The model makes the following predictions: (1) all patches should be reduced to the same marginal value, (2) this marginal value should equal the average rate of intake for the habitat. To make these decisions, the predator has to behave as if it knows both the average rate of intake for the habitat, and the instantaneous rate within each patch. To calculate the former of these, it is nessary to know the average quality of patches and the distance between them. It is also implicit that the predator 'knows' when it moves from one habitat to another, habitats being defined rather vaguely as areas within which the average capture rate remains fairly constant. It is clearly unrealistic to expect a predator to have this sort of omniscience, except perhaps if E/T is stable over

---

travelling between patches (t) is also plotted on the x-axis. The optimal predator should choose to stay in the patch just long enough to maximise the slope of the line AB (representing the average food intake per unit time for the habitat as a whole). In order to do this the predator leaves a patch at time T_{opt}. (b) If individual patches in the habitat have different curves of $f(T)$, the predator should apply the same giving up criterion to all of them.

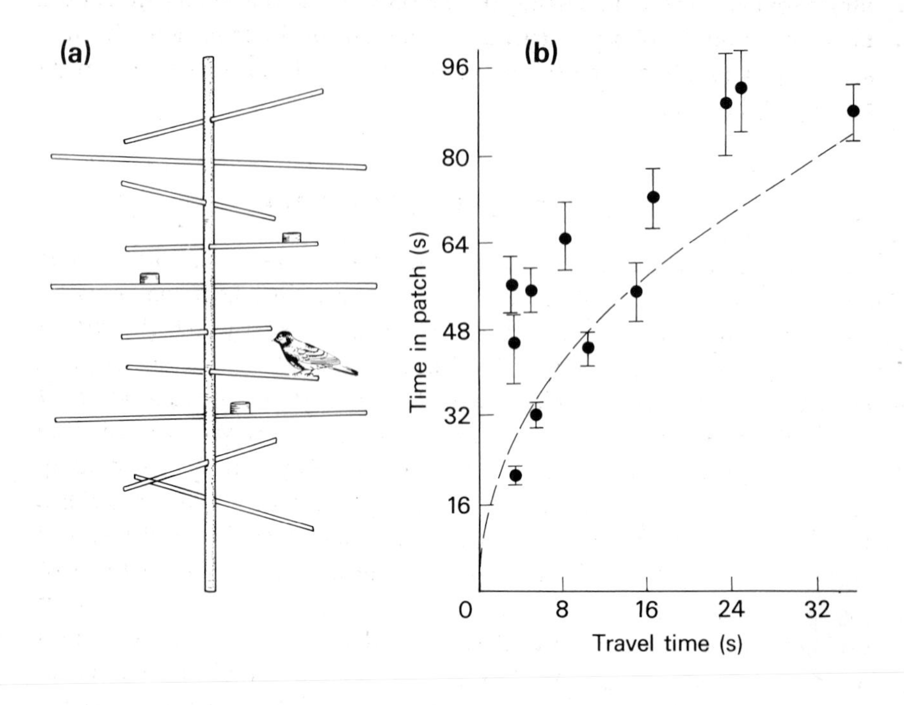
(a)
(b)
96
80
64
48
32
16
0
8
16
24
32
Time in patch (s)
Travel time (s)

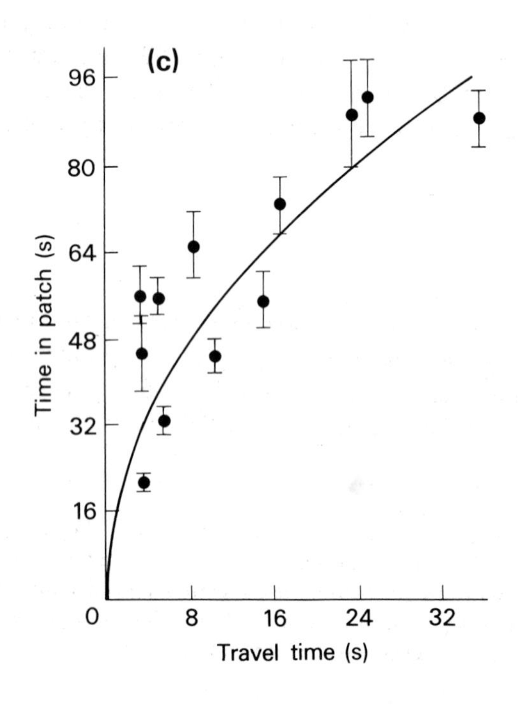
(c)
96
80
64
48
32
16
0
8
16
24
32
Time in patch (s)
Travel time (s)

long periods, so that in testing the predictions of the model we should bear in mind that the dictates of acquiring information through sampling will almost certainly cause the predator to deviate from optimal behaviour.

2.3.4 *Tests of the depletion model*

The model illustrated in Fig. 2.7 can be tested by measuring the average curve of intake within a patch and the travel time between patches. One could then predict either the time spent in each patch (by fitting the tangent A/B), amount of food eaten per patch, or the marginal capture rate for each patch. Krebs *et al.* (1974) suggested that if all prey are small and of the same size the marginal capture rate could be estimated by using the reciprocal of the giving up time (the giving up time is the interval between the last capture and leaving a patch), but as Cowie and Krebs (in press) point out, there is an element of circularity in Krebs *et al.*'s argument because if birds leave patches at random, high capture rates are likely to be associated with short giving up times. It is important to note that in testing the predictions it is not sufficient to show that all patches within a habitat are reduced to the same marginal capture rate: one also has to show that when the average quality of the habitat (E/T) changes, the marginal capture rate or time in patch changes accordingly. The significance of this point is explained in the next section. The three examples I discuss next are all concerned with predicting the optimal time or food intake from a patch.

Great Tits

Cowie (1977), in one of the most stringent tests of the optimal patch model, studied captive great tits foraging in a large indoor aviary for small pieces of mealworm hidden in sawdust-filled plastic cups on the branches of five artificial trees (Fig. 2.8a). Cowie tested six birds

---

Fig. 2.8. (a) An experimental 'tree' with 3 patches (Cowie 1977). (b) Predicted optimal time in a patch plotted against travel time (dashed line) together with the observed mean points (±S.E.) for six birds, each in two environments. (c) The same data points and predicted time taking into account energy costs of travelling.

individually, and arranged the experiments to last for a time short enough to exclude any effect of revisiting patches. Each bird was tested in two 'habitats', with a short and long 'travel time' between patches, the prediction being that birds would adjust the time spent per patch in relation to the travel time as predicted by the optimal foraging model. As can be seen from Fig. 2.7a, if the 'travel time' increases while patch quality remains constant, the slope of the line AB becomes shallower, and the point at which this line touches the curve $f(T)$ moves to the right. The predator should spend longer in each patch when the travel time is long. The 'travel time' was manipulated by making it easy or hard for the birds to start foraging in a new patch, by placing a loose or tight-fitting cardboard lid on each plastic cup. Thus the long 'travel time' between patch visits was spent by the bird in trying to prise off the tight lid from the next patch. Cowie measured, for each bird, the travel time in the hard and easy environment as well as the curve of cumulative food intake, within a patch, which was the same in both environments, as patches always contained the same number of prey. From these, he could predict the relationship between travel time and time in patch, which is shown in Fig. 2.8b, together with the observed mean for the six birds each in two environments (12 points in all). The observed relationship is quite close to the predicted, but birds tend to spend too long in each patch. Figure 2.8c shows that the fit becomes much better when the energetic costs are taken into account: Cowie calculated from standard metabolic rate equations the energetic costs of searching within and moving between patches. The fact that the net intake gives a better prediction than gross intake is an interesting indication that the birds were not simply measuring the costs in terms of time.

Bumblebees

The desert willow *Chilopsis lincaris* in the south western United States is pollinated by the bumblebee *Bombus sonorus* (Whitham 1977). When bees extract nectar from the willow flowers they first suck nectar out of a pool at the base of the corolla, and then remove nectar from five grooves which radiate out from the corolla base. They collect nectar from the pool more rapidly than from the grooves (Fig. 2.9a) so that the profitability of a flower decreases sharply when the bee uses up the pool nectar. An optimal foraging bee would be expected to leave the groove nectar when many flowers contain pool nectar, the latter arising as a result of overflow from the grooves. Whitham found that, as

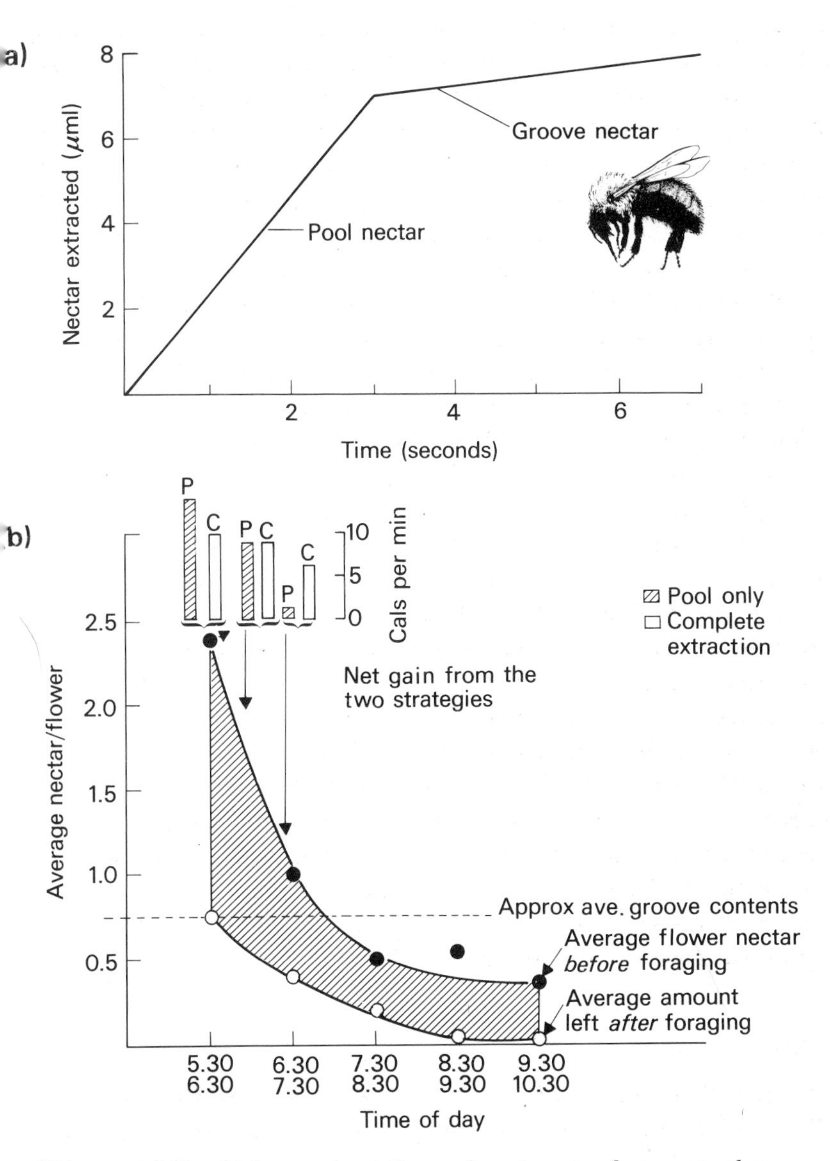

Fig. 2.9. (a) Bumblebees extract the pool nectar at a faster rate than the groove nectar of desert willow flowers. (b) As the morning progresses, flowers contain on average less and less nectar (due to depletion). The upper curve shows the nectar per flower available before a bee visits, and the lower line shows what is left after a visit. In the very early morning bees leave all the groove nectar, but by 7.30 a.m. they are collecting most of it. The histograms show the pay offs for specialising only on pool nectar (shaded) and for taking both pool and groove at three times of the morning. (After Whitham 1977.)

expected, bees tended to leave groove nectar early in the morning, when most flowers contained a lot of nectar, but not later in the morning when nectar was less abundant (Fig. 2.9b). Whitham used estimates of the energetic costs of flight and foraging and measurements of the extraction rate and calorie content of nectar extracted from a flower to calculate the net calorie rewards per minute by hypothetical bees using the 'pool-only' nectar and 'complete extraction' strategies. Early in the morning, when the flowers contain an average of 2·4 μl of nectar the pool strategy is more profitable, but later in the morning, complete extraction becomes the better strategy. Figure 2.9b shows the net caloric intake per minute from the two techniques at three stages of the morning. The bees should start to switch from 'pool-only' to complete extraction when the flowers contain just under 2 μl of nectar. By the time flowers contain 1 μl of nectar 'pool-only' foraging is much the less profitable strategy. Although the bees clearly conform to the qualitative pattern of spending longer in each flower and extracting more nectar as the morning progresses, they do not show precisely the sudden switch predicted by the model.

Waterboatmen: individual prey as patches

Many invertebrate predators feed on each individual prey for a considerable length of time, so that a single prey item can be treated as a patch. The waterboatmen *Notonecta glauca* feeds on the larvae of *Culex molestus* by slowly sucking out the internal contents before discarding the hard exoskeleton. By interrupting individual *Notonecta* after different lengths of time feeding on a prey Cook and Cockrell (1978) were able to show that the *Notonecta* at first extracts food rapidly, but that the rate subsequently diminishes (Fig. 2.10a), not because the predator is becoming satiated, but because the first bit of food is easier to extract. Cook and Cockrell varied the encounter rate of the predator with prey by presenting five different prey densities and were able to show that the time spent handling each prey varies inversely with encounter rate, as would be expected from the optimal patch model (Fig. 2.10b). As Cook and Cockrell point out, the changes in handling time per prey predicted by the optimal patch model will influence the functional response of the predator to prey density, and according to the exact pattern of change in handling time, could lead to a simple decelerating or sigmoid response.

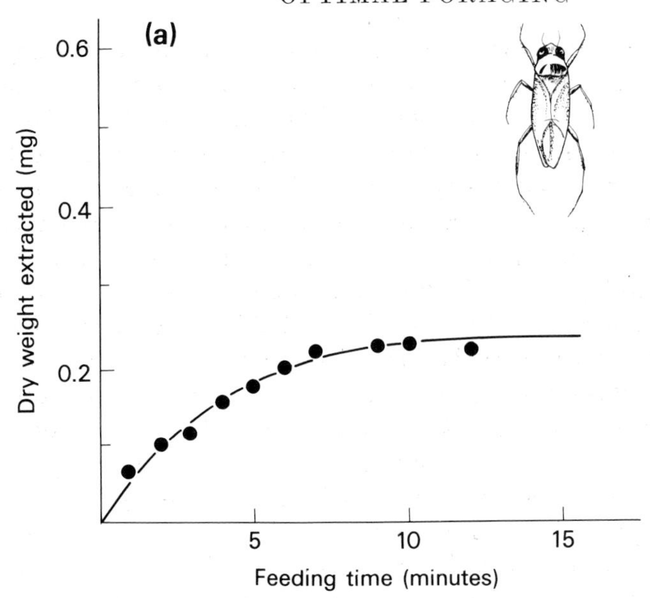

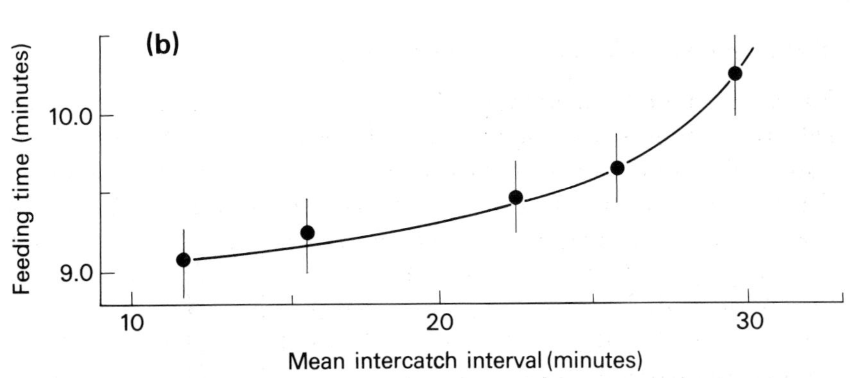

Fig. 2.10. (a) Cumulative dry weight of food extracted with time spent feeding by *Notonecta* on 2·0 mg mosquito larvae (wet weight). The curve is an exponential fitted by regression to the points obtained by interruption experiments. (b) The time spent feeding on each prey item (±S.E.) decreases as the intercatch interval decreases. (Cook & Cockrell 1978.)

2.3.5 *Behavioural mechanisms*

Giving up times and area restricted search

Results which support the optimal foraging model raise the problem of what sort of behavioural rules the predator uses to achieve the near optimal or optimal result. It is likely that a simple rule-of-thumb would bring the predator close enough to the optimal rule for the experimenter not to be able to recognise the deviation. Krebs *et al.* (1974), Hassell and May (1974) and Murdoch and Oaten (1975) independently proposed that predators might use the giving up time as a rule for leaving patches. Hassell and May, and Murdoch and Oaten both proposed that a predator might have a rigidly fixed giving up time, like a clock which is reset after each capture, and they showed that this mechanism would produce the type of time allocation shown in Fig. 2.5, which is qualitatively as expected from optimal foraging theory. Krebs *et al.* (1974) pointed out that in order for the predator to actually forage optimally, it would have to change its giving up time according to the average value of the habitat, and they were able to show that black-capped chickadees (*Parus atricapillus*) had shorter giving up times when foraging amongst rich patches than in a poor environment. However, as was mentioned above, Cowie and Krebs (in press) show that the same result could have been obtained if the chickadees were leaving patches at random, so the evidence that the birds used a variable giving up time in this experiment is not totally convincing. The difference between the fixed and variable giving up times is one of degree and it raises the important point that the 'habitat' within which an optimal predator should maintain a fixed giving up threshold and between which the threshold should change is ill defined. Probably the most realistic way to think about the concept of fluctuating 'average capture rate for the habitat' is to assume that the predator has a sliding memory window of the last n patches visited. A value of n equal to all the patches visited in a lifetime would be a fixed giving up time model. At the moment there is no evidence about the length of the hypothetical memory window.

Hassell and May also point out that predators often show an alteration of their movement path after finding a prey, for example by increasing the rate of turning and/or decreasing the speed of movement. These changes collectively are termed 'area restricted searching' and may be a simple mechanism by which predators stay in good quality patches.

Optimal foraging by Nemeritis canescens

How close can a predator approximate optimal foraging by simple behavioural rules such as fixed giving up time and area restricted search? Cook and Hubbard (1977), and Hubbard and Cook (1978) have shown that the insect parasitoid *Nemeritis canescens* forages in a patchy environment in a manner consistent with optimal foraging theory. The experimental set up within which the individual *Nemeritis* searched for prey consisted of an arena containing patches of prey (*Ephestia cantella*) concealed in plastic dishes covered with sawdust. The patches contained varying numbers of hosts. Unlike predators, parasitoids can 'recapture' the same 'prey', although *Nemeritis* does to some extent avoid superparasitism. The number of hosts parasitised in a fixed time is known to follow the 'random parasite' equation of Rogers (1972), and Hubbard and Cook used this result together with Charnov's marginal value rule to predict the optimal time spent per patch. One example of their results is shown in Fig. 2.11: the observed and predicted times per patch are very similar. However, as this experiment only involved one type of environment, the results could also have

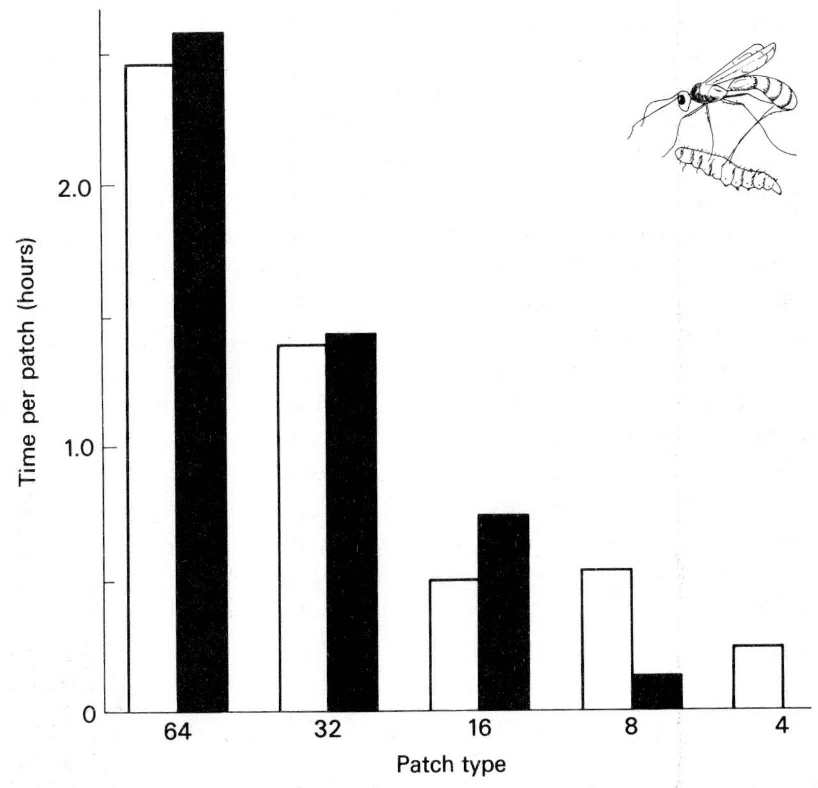

Fig. 2.11. Observed (open histograms) and predicted optimal (shaded histograms) time budgets of *Nemeritis canescens* hunting for hosts distributed in patches. (Hubbard and Cook 1978.)

been produced by the fixed giving up time model. Waage (1977) has investigated the behavioural mechanisms underlying patch use by *Nemeritis*. He showed that *Nemeritis* exhibit two responses on encountering a patch of prey: they walk more slowly, and turn back at the edge of the patch. This edge response gradually wanes with time so that the *Nemeritis* eventually leaves. Waage also showed that the decline of the edge response was due to a waning of responsiveness to the smell of the host patch, and that a successful oviposition countered the decline by increasing the olfactory sensitivity of *Nemeritis*. A comparison of ovipositions spaced and clumped in time suggested that the increment from each oviposition is independent, so that the point at which *Nemeritis* leaves a patch is probably a straight-forward interaction of the gradual waning and the independent incremental effect of each oviposition. Waage demonstrated that this simple behavioural mechanism would produce the result observed by Cook and Hubbard. He also found that the olfactory leaving threshold was not influenced by the average quality of previously visited patches which suggests that *Nemeritis* does not have the flexibility to change its giving up threshold according to the average habitat quality.

2.3.6 *Time allocation without prey depletion*

Although predators often reduce the availability of prey in a patch as a result of their own foraging activities, sometimes changes in the quality of a patch may result from changes independent of the predator's own activity. In a field study of spotted flycatchers (*Muscicapa striata*), Davies (1977b) has investigated one such situation. The flycatcher has to make the choice of whether or not to return to the same perch (analogue of a patch) after each foraging sally. Successive intercatch intervals from a perch are roughly constant, indicating that there is no gradual depletion of prey and the major factor determining the profitability of a particular perch is whether or not there happens to be a swarm of flies within striking distance, since capture success declines rapidly with length of sally made by the bird. Davies found that a flycatcher changes to another perch if it waits for longer than about 1·5 times the average intercatch interval. The optimal choice of this giving up time involves a trade off between the disadvantage of leaving while the current swarm is still within striking distance, and the cost of waiting for a long time for a next swarm to arrive if the current one has moved away. Although Davies did not measure the arrival rate of swarms, and was thus not able to quantify the second cost directly, he was able to show in a simulation model that the observed giving-up time was close to the minimum possible for achieving maximum capture rate within a swarm. Thus the flycatchers appear to be making an optimal trade off between captures within a swarm and waiting between swarms.

2.4 Search Paths

2.4.1 *Optimal search paths*

Many ecological models of parasite and predator searching behaviour assume random encounters between predator and prey, or parasite and hosts, within a patch, and models based on this assumption give a good fit to experimental results (Rogers 1972). However the assumption of random encounters does not necessarily imply random search, since if the prey are distributed at random within a patch, random encounters could result from non-random search. Random search is in fact highly inefficient, at least when the prey do not renew themselves extremely rapidly, so it seems rather unlikely that any predators or parasites do search in a truly random fashion.

There have been a number of investigations of searching efficiency using simulation models (Cody 1971, Pyke 1974 (in Pyke *et al.* 1977), Cullen cited in Smith 1974b, Jones 1977a and b). In Pyke's simulation, the predator could move from one point to any of the four neighbouring positions on a lattice grid, the total area of the grid (number of points) being varied from one simulation to another. The probabilities of moving in the four possible directions, forward, backwards, left and right (*Pf*, *Pb*, *Pl* and *Pr*) were varied in order to find the movement rules for the optimal search path. The optimal path is one in which the maximum number of grid points (which are in effect prey items) is visited during a set number of moves, or in other words the path which minimises revisiting depleted places if there is no replacement of prey. An important variable in the simulation was the rule for changing behaviour at the boundary of the grid, and Pyke considered two possibilities, a reflecting boundary, in which the rule is simply to reverse direction at the boundary ($Pb=1$), and a partially reflecting boundary in which the predator moves with approximately equal probability to the left, right and backwards when at the boundary. Pyke considered that the most realistic boundary conditions would be either an effectively infinite grid, for example a bumblebee foraging in a large meadow, or a grid with a partially reflecting boundary, this pattern of behaviour being shown by hummingbirds hunting on an artificial grid. If the searching path is symmetrical ($Pl=Pr$), then the path can be characterised by one measure, ($Ps-Pb$), which Pyke calls 'directionality'. On an infinite grid, the optimal directionality is obviously 1, and the results of Pyke's simulation model showed that with a partially reflecting boundary, the optimal directionality was somewhere between 0·8 and 1·0, depending on the size of the grid. Pyke computed the directionality for a number of real search paths (e.g. Siniff & Jessen 1969, Kleerekoper *et al.* 1970, Smith 1974a), and found that the observed directionalities lay between 0·1 and 0·8. Although it is not clear that all of the examples analysed by Pyke were in fact search paths as opposed to other types of movement, the conclusion is clear: Pyke's simulation does not adequately account for observed search patterns, unless of course the real search paths are all well away from the optimum. Pyke suggests two reasons for the discrepancy between his results and the real world, the most important being that his simulated predators did not have any sense organs, they did not detect a prey until hitting it by landing on a grid point. A second factor influencing real search paths is that predators often, as mentioned in section 2.3.5, modify their search path after a capture by decreasing the directionality.

The simulation model cited in Smith (1974b) is somewhat more realistic than Pyke's in that it includes both sensory abilities of the predator, represented as a 'domain of danger' around each prey, and a greater flexibility in the choice of directions open to the predator for each move. The model is less realistic in other respects: it assumes a very restricted prey universe and has no prey depletion, or rule for staying within the boundary of the prey universe. In the model, the predator starts at a central point surrounded by a ring of prey, and the optimal search path is one which maximises the 'hit rate' of prey by the predator. The optimal path was, as in Pyke's simulation, far from random, but had a directionality somewhat lower than the 0·8 of Pyke's model.

Many of the real search paths that were analysed by Pyke have a slight asymmetry, that is, the mean direction deviates significantly from zero (counting left turns as positive, and right as negative), and Cody (1971) in a simulation model similar to that of Pyke's, concludes that a slight asymmetry increases search efficiency. Cody also concluded that his observed directionality of 0·64 for finch flocks was optimal, but as Pyke shows, it is only optimal under the particular grid and boundary conditions chosen by Cody for his simulation.

In summary, there are no quantitative conclusions that can be drawn from these simulation studies, but they all reach the qualitative conclusion that directional search paths are more efficient than random, and that to specify an optimum more precisely one needs to know about the predator's sensory abilities, the effective grid size, boundary rules, and prey distribution.

2.4.2 *Modification of search path through experience*

Many predators probably use longer short term memory in modifying their search paths. For example both Smith (1774b), and Pyke (1974) found that blackbirds and bumblebees respectively tend to alternate left and right turns in their search paths and, as mentioned in section 2.3.5, area restricted searching is a commonly observed phenomenon.

Sticklebacks

Beukema (1968) investigated the ability of sticklebacks (*Gasterosteus aculeatus*) to improve their search efficiency by learning to exploit a particular distribution of prey, in an artificial environment consisting of a large tank divided into 18 small hexagonal cells. The fish could travel between cells through gates, and once within a cell, they could detect any prey immediately, so that the optimal search path involves visiting as few cells as possible per prey encountered. In one experiment, Beukema placed a single prey item in one of the 12 outer cells of his tank, so that a fish, starting from a randomly chosen point in the outer ring, could in theory encounter the prey after visiting an average of six cells. This would result from a strategy of swimming directly around the outer ring of cells. Buekema calculated the encounter efficiency (number of prey encountered per cell visited) of a variety of search strategies, varying from random to the optimal. The sticklebacks performed well above random, mainly because they tended to search in a directional manner, and they also gradually improved their performance over a series of trials, but did not reach anything like the optimal value of 0·17. This result shows that sticklebacks are capable of adaptively modifying their search path in response to a particular distribution of prey, and the fact that they did not reach the optimum path may be a reflection of the rather artificial task with which they were faced.

Area restricted searching

Many insect predators and parasitoids show area restricted searching. This is an adaptive mode of searching for clumped prey and there is some evidence that area restricted searching is more pronounced in species feeding on clumped food than in dispersed prey (Waage 1977). It seems, however, that in insects the tendency to perform area restricted searching is a constant species characteristic, although this is not so in birds. Smith (1974b) found that blackbirds (*Turdus merula*) perform area restricted search after a find by making successive turns in the same direction instead of the usual pattern of alternating left and right moves (a blackbird searches for prey—usually worms, on a lawn, by running a short distance, pausing to scan, turning through a certain angle, and making another short run. Smith analysed the bursts of running as 'steps'). Smith investigated the ability of blackbirds to increase or decrease the amount of area restricted searching in response to changes in the dispersion of artificial prey (pastry caterpillars) on a lawn. At low overall prey density (0·064 prey/m^2) the birds showed no area restricted searching when hunting for regularly distributed prey but they did when hunting for random or clumped prey. At high prey density (0·3 prey/m^2) the birds did not show area restricted search with random or regular distributions and only a very slight tendency to do so with clumped prey. The reason why the results were less clear cut at high prey density may be because the search path *before* a find was more tortuous, so that it was difficult to show an increase in convolution *after* a find.

2.4.3 *Optimal return times*

Finch flocks

In the discussion of search paths it has been assumed that prey are essentially non-renewing, but if one assumes that a predator might not only deplete food stocks during its foraging, but also that the food replenishes itself rapidly, one can consider how rapidly a predator ought to return to a particular place (Cody 1971, Charnov *et al.* 1976a). In theory one could imagine that a predator ought to travel around its foraging area in a pattern which maximises the harvest from a particular patch on each visit. As Charnov *et al.* (1976a) point out, such an 'optimal return time' searching strategy requires that the predator has exclusive use of an area, since otherwise there may be interference with the recovery of prey after depletion. This reduction of interference could be achieved by territorial exclusion (Charnov *et al.* 1976a, see also Chapter 11) or by group foraging (3.3.1 and 4.3.2): if all the predators in an area forage in one group, there will be no interference between individuals in their return times (Cody 1971). In his discussion of mixed flocks of finches, Cody (1971, 1974a) considers that one of the major benefits of flocking is that it enables the birds to optimise their return times, although he does not actually show that the flocks he studied have exclusive ranges, an essential prerequisite for the hypothesis. Cody studied flock movements in two areas, one close to a mountain range and one further away towards a desert. The former, wetter, area had a more abundant, but less rapidly renewing, food supply (ripening seeds), while the dry desert had sparse but rapidly renewing supplies of seed. Cody's data suggest that flocks in the desert area moved faster and took bigger turns than the birds close to the mountains. These two factors together would decrease the return time in response to the sparser, but more rapidly renewing food. It is unfortunate that in Cody's study two variables, food density and renewal rate, were confounded, so that one cannot tell which factor influences the birds' movement pattern. It would be interesting to find out how the flocks behave when both food density and renewal rate is low.

Honeycreepers

The most detailed work to date on return times by territorial individuals is that of Kamil (in press) on a Hawaiian honeycreeper, the amakihi (*Loxops virens*), pairs of which defend breeding territories around blossoming trees of the genera *Sophora*, *Chrysophylla* and *Myoporum*. By numbering individual clusters of blossoms in each of five territories, Kamil was able to record the revisiting pattern of colour marked individuals. He found that territory holders tended to pay exactly one visit to blossom clusters more than expected, and that when they visited the same place twice, the two visits were well spaced in time. Intruders are much more likely than residents to visit a cluster shortly after it has been depleted, and measurements of the rate of nectar recovery in depleted flowers showed that the intruder gets only about two thirds as much nectar as a resident. Kamil was not able to show exactly how the honeycreepers managed to avoid revisiting the same flowers, but he did note that territorial pairs tended to use different parts of the territory and hence avoided interference with each others' return time strategy.

2.5 Sampling and Optimal Foraging

None of the models discussed so far has included a provision for the predator to acquire information through sampling. They are essentially equilibrium condition models, which might apply when the predator has already spent some time in the habitat encountering prey and patches, but what does a predator do when it comes to a completely new environment, or if the environment changes? This crucial question has been little studied either theoretically or experimentally.

A study which has shown empirically that predators sample and can use the information when the environment changes, is that of Smith and Sweatman (1974) who trained captive great tits to search for hidden mealworms in six patches containing different prey densities within a large aviary. The tits soon learned to concentrate their foraging effort in the most profitable patch, but they also spent more than the optimal amount of time in the various less profitable patches (Krebs & Cowie 1976). Smith and Sweatman showed that when the best patch was suddenly reduced in quality, the tits switched to foraging primarily in the second best patch, which seems to show that the birds had sampled each place and stored up information about its relative profitability.

It is intuitively reasonable that a predator should spend some of its effort in sampling even when the environment is stable, as a hedge against possible changes (Oster & Heinrich 1976, see also 4.2.2), and one might also expect a predator to spend time exploring before it begins to exploit a totally new environment. In a study of the optimal trade off between exploration and exploitation, Krebs *et al.* (1978) measured how long a great tit will carry on sampling two unknown, non-depleting, patches of differing quality, before deciding to exploit the more profitable patch. (The two patches consisted of two perches at opposite ends of an aviary on which the birds were trained to deliver a food reward.) They found close agreement between the birds' behaviour and a theoretical optimum calculated by a dynamic programming algorithm. The theoretical optimum is one which maximises the number of rewards obtained during the foraging period, and intuitively one can see a trade off between deciding too soon and perhaps making the wrong choice, and sampling for too long, missing the opportunity to exploit the more profitable patch.

2.6 The Energetic Cost of Foraging

Any model which tries to predict the behaviour of an efficient predator ought to take into account the costs as well as the benefits of a particular decision rule. In optimal foraging models the costs are measured as time and energy spent foraging, but most of the tests of these models have not attempted to measure energy costs. We saw in section 2.3.4 how the energy costs of travelling between patches altered the predicted optimum time spent in each patch by great tits. The observed results were closer to the optimum calculated by taking into account the energy costs than to an optimum based only on time costs.

The energy costs of travelling are also important in predicting the optimum meal size for hummingbirds (DeBenedictis *et al.* 1978). When a hummingbird forages by extracting nectar it does not fill its crop to capacity within one bout of feeding before switching to another activity such as territorial defence or resting. If hummingbirds were either maximising the energy intake during a single visit to the food source, or minimising the time spent travelling to and from the flowers one would expect them to fill the crop to capacity during each visit. However, a crop full of nectar is an appreciable load for a hummingbird to carry, and there is a trade off between the cost of carrying nectar in the crop between foraging bouts and the benefit of saving travel time by eating as much as possible at each visit. DeBenedictis *et al.* calculated the optimal meal size resulting from this trade off, assuming that the goal is to maximise the net rate of energy intake over the total foraging time. The observed results were close to the predicted optimum, which suggests that hummingbirds take into account the energetic costs of foraging.

A more general model of search costs and benefits is shown in Fig. 2.12. The model considers two types of feeding behaviour. One has a low cost per unit searching time and results in a low capture rate (for example walking very slowly), the other has a high cost per unit search time and yields a high capture rate (for example running after prey). The model predicts that when prey density is low, a predator should use a cheap, low return method of foraging, but when prey are abundant it should switch to the high cost, high yield method (Norberg 1977b, Evans 1976). Although there is no quantitative evidence in support of this prediction, Evans (1976) noted that bar-tailed godwits (*Limosa lapponica*) use less costly foraging methods (slower walking rate) when

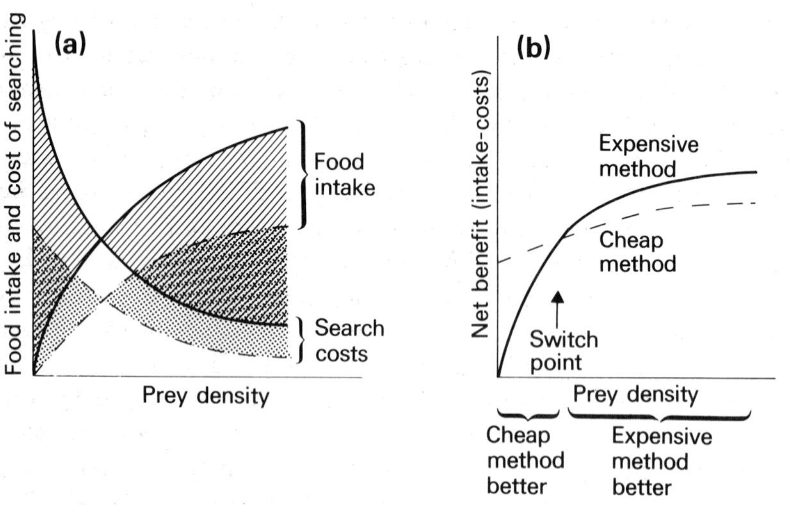

Fig. 2.12. (a) Food intake and search costs in relation to prey density for two foraging methods. The solid lines are for an expensive search mode, such as running, which yields more food at a particular prey density than does the cheap method of hunting, for example walking, shown by the broken lines. The shapes of the curves are based on simple assumptions: food intake in relation to prey density is a standard type II functional response, and search costs are inversely proportional to prey density. The cross-hatched area and the stippled areas respectively show the net benefit from the expensive and cheap foraging methods. (b) A plot of the net benefit for the two foraging methods, which corresponds to the height of the cross-hatched and stipple areas at each prey density in (a). Where the food intake lines are below the cost lines the benefit is negative.

prey (*Arenicola marina*) are scarce and may stop foraging altogether under adverse conditions of prey availability.

2.7 Ecological Consequences of Optimal Foraging: Resource Partitioning

Although it is not often explicitly stated, the hypothesis that community structure is determined largely by competition contains implications about differences in foraging efficiency between species. The hypothesis assumes that the species in a community can be arranged along a series of continuously distributed resource axes (which are usually assumed to be independent) and that competition between two species can be measured by the degree of overlap of their 'utilisation functions' (niche overlap) on one or more of these axes, food being the most commonly studied axis. These basic assumptions can be used to calculate in theory how much overlap is possible without one species excluding the other and hence how many species can be packed into a community along a particular axis. Pulliam (1975b), for example, predicted with some degree of success the type of finch species occurring in three different communities on the basis of estimates of utilisation functions and availability of different sizes of seeds. Other most widely such as Pulliam's, to predict community structure, the most widely accepted types of evidence for competition as a determinant of community structure are firstly the rather constant ratio of sizes of the feeding apparatus of adjacent pairs when species are ranked by size, indicating that they are regularly spaced on the food resource axis, and secondly changes in the food utilisation function of remaining species when similar species are removed ('competitive release'). This is taken to suggest that normally the range of food eaten by a species is limited by competition.

The arguments briefly outlined above contain the implicit assumption that each species is most efficient at utilising its own favoured food resources, and that during evolution tightly packed species have become restricted to a narrow utilisation function because a generalist is less efficient on any one resource than a specialist. The first of these assumptions is, not surprisingly, borne out by good evidence: for example, when closely related finches (Kear 1962), and sunfish (Werner 1977) differ in size, there is a relationship between size of feeding apparatus and size of prey dealt with most efficiently. Partridge (1976a) measured the ability of blue and coal tits in utilising different types of micro-habitat, and showed that differences in efficiency correlated well with differences in habitat preference (see Chapter 12). These differences between species are, on the whole, a consequence of morphological adaptations and hence cannot tell us much about the role of competition between specialists and generalists in the evolutionary origin of niche separation and niche width. More relevant are comparisons within one species; for example Grant *et al.* (1976) found that individual *Geospiza fortis* (one of Darwin's finches) with larger bills prefer large *Opuntia* seeds and they are more efficient at handling them than are individuals with smaller bills. Partridge (1976b) showed that individual great tits differ in their efficiency at performing different foraging tasks and that they prefer to forage in the microhabitat where they are most efficient. These studies, in showing how fine scale differences between individuals can lead to segregation along a resource axis based on foraging efficiency, suggest that a similar process may have led to the evolution of species differences.

Turning to the dynamics of existing communities, we can ask whether the presence of a competitor x increases or decreases the optimal diet breadth of species y or optimal number of patch types visited by y. The competitor does not change the profitability ranking of prey types to species y, so if overlap is sufficient for x to eat any prey which fall into the optimal set of y, the effect will be to decrease y's encounter rate with these prey and force it to expand its diet to include lower ranking prey (MacArthur 1972). With patches, the effect of a competitor may be different. If the patches preferred by x are reduced in profitability so that y no longer includes them in its optimal set, the presence of a competitor should cause y to shrink the range of patch types visited, or possibly switch to new patches (Cody 1974a, Werner 1976). If the different patch types contain different prey, then decreasing the number of patch types visited may as a side effect decrease the breadth of diet, so that a competitor could cause y to become more of a diet specialist.

2.8 Conclusions

Optimal foraging models are often successful in predicting decision rules for predators in simple laboratory or field conditions, and when the environment is relatively stable. In a fluctuating environment the predator's problem is to continually update its estimates of capture rates and availability in order to make optimal decisions, and since most real environments change with time, this important but neglected aspect of optimal foraging must be the next major goal for theoretical and experimental work.

A common criticism of optimal foraging theory is that it is essentially a circular argument. The basic premise is that natural selection is an optimising process, and the models and experiments end up showing that predators are at least sometimes optimal. Although this reasoning may seem circular, it is not, because in formulating an optimal foraging model we have to make a guess about a cost function which is being maximised. The models discussed in this chapter have been based on the guess that optimal predators are designed to maximise net rate of food intake while foraging, but this could be an incorrect guess in many instances. Predators might be designed to minimise the variance of food intake (Thompson *et al.* 1974), minimise the risk of overheating while foraging or a host of other possibilities, so that any one hypothesis about the nature of cost function could be refuted and replaced by another.

There is however a problem when the models' predictions do not work. Some tests of optimal diet models have produced negative results (e.g. Emlen & Emlen 1975) but these are not discussed here because it is impossible to distinguish between three possible reasons for the discrepancy between predicted and observed behaviour. The hypothesised cost function could be wrong, the premise of optimal behaviour could be wrong, or both of these could be right, but the animal could have been tested in an environment to which it is not adapted. For example, a hummingbird successfully avoids revisiting artificial flowers on a vertical spike when the flowers are arranged in their natural spiral pattern, but it fails to do so when the flowers are in a simple linear arrangement up the spike (G. Pyke, pers. comm.). The premise that most animals are in some sense optimal could be wrong for at least three reasons, evolutionary lags in response to a changing environment, frequency dependent selection (in which the commonest genotype is always at a disadvantage), and local versus global optima. I take the optimistic view that in spite of all these difficulties, optimal foraging theory is by no means a failure in predicting how predators make decisions.

Chapter 3
Living in Groups: Predators and Prey

BRIAN C. R. BERTRAM

3.1 Introduction

Many of the most striking characteristics of animals' structure and behaviour are adaptations concerned with predation—with obtaining prey and with avoiding being preyed upon. The form and speed of the cheetah, the tense alertness of an antelope, the quills of a porcupine, the sensitive ears of an owl, and the camouflaging colours of a ground-nesting bird are but a few examples of the various aids used in the struggle to find food but not become food. In some circumstances, but by no means always, companions can provide a further measure of protection from predators and of assistance in procuring prey. In this chapter I shall consider (1) the variety of possible ways in which group living may assist animals in avoiding being preyed upon, (2) the ways in which predators may benefit in their hunting from living in groups, and (3) some of the other attendant advantages and disadvantages of a social way of life by both predators and prey.

For the purposes of this chapter, a 'group' consists of two or more animals together, either temporarily or permanently. It is worth bearing in mind at this stage that the costs and benefits of being a member of a group alter with the size of the group, and the topic of group size will be discussed later.

Before outlining the relationship between grouping and predation, however, it must be put into perspective. Obviously, living in groups is only one of the many and possibly conflicting ways of reducing predation; for example, in Chapter 5 Harvey and Greenwood deal with camouflage and warning colouration, which are others from among the armoury of defence methods used by potential non-victims. In some circumstances, too, grouping may lead *not* to a reduction of predator pressure but to an increase. Similarly, predators need to use a variety of different strategies to procure their food, such as those described by

Krebs in Chapter 2; their problems are not solved but may sometimes be reduced by being in groups. On the other hand, it should be stressed that there are often other advantages of living in groups, quite unconnected with predation; for example in Chapter 9 Emlen reviews co-operative breeding in birds, and in 3.5 some of the other consequences, beneficial and detrimental, of group living are outlined.

Any species is subject to an array of different kinds of selective pressures operating on it; some are stronger than others, and they push in opposite directions, some tending to make the species more social in the course of its evolution, and some tending to make it more solitary. Predation, whether by or on the species, is only one of these selective pressures. Sometimes it is an extremely important pressure, while sometimes it is dwarfed by others. In addition, the relative strengths of different selective pressures probably change as the species evolves. For example, I suggest that the advantages of communal hunting were probably the major pressure in tending to make lions (*Panthera leo*) evolve into social animals, but that these advantages may well have been *relatively* reduced in importance now by the many other advantages of their social way of life, especially their co-operative reproduction which has been described elsewhere (Bertram 1975).

As the above example illustrates, I shall concentrate particularly on the social mammalian predators, partly because my own experience lies in that field, and partly because they provide such good examples of highly developed predatory behaviour and social organisation. This paper is not intended to be a comprehensive review of social grouping, as may be found in Alexander (1974) or Wilson (1975). Nor is it an attempt to compare different kinds of social systems—see, for example, Kleiman and Eisenberg (1973) for discussion of the probable evolutionary history of dog and cat social groups. It is not an attempt to provide a general functional explanation of why some animal species gather or live in groups while most do not, and even less is it an attempt to attribute grouping mainly or generally to predation, which in some cases probably has negligible effects. My aim is merely to disentangle and consider separately the variety of selective pressures, particularly those concerned with predation, which operate on animals in groups. For example, I want to examine the many different possible ways in which being part of a group may help animals of different species to survive the attentions of predators.

Four general points should be borne in mind: first, that prey animals have to cope with the attentions of a number of different types of predators (as well as parasites) which use different hunting strategies;

second, that the predators feed on a variety of different types of prey with even more diverse defence strategies; third, that the predators themselves are also the potential prey of some other predator; and fourth, that the distribution and availability of food must play an enormous part in favouring group living in some circumstances and in precluding it in others. Thus deciding what is the best grouping strategy for a species to use is bound to be a complicated process, for the strategy is most unlikely to be the response to only a single selective pressure.

3.2 Advantages to Prey by Reducing Predation

Potential prey animals may benefit in a number of different ways from being in a group as will be outlined. Not all or indeed any of these ways apply to any particular species; on the other hand the different ways are not necessarily or even usually alternatives to one another.

3.2.1 *Avoiding detection by the predator*

In general one would expect that a group of prey animals would be more conspicuous than a single animal, and so more easily detected by a predator; but because groups are scarce a predator may find it more difficult to find one. Vine (1973) and Treisman (1975) have considered theoretical cases of detection and capture of camouflaged prey with different distributions. Under certain circumstances, a predator may have a lowered success rate when searching for grouped prey; this applies if he is able to capture only one member of any group he discovers, and particularly if he is hunting by vision. Nonetheless, aggregating with others usually makes prey animals in general more rather than less easy to detect.

However, not all prey animals are equally susceptible to capture. In groups above a certain size, a particularly vulnerable prey individual may be able to escape detection by a predator by shielding itself from view behind its companions. Kruuk (1972) and Estes (1976) described the way in which vulnerable wildebeest (*Connochaetes taurinus*) calves could be led by their mothers to the opposite side of the herd from a spotted hyaena (*Crocuta crocuta*); as a result the hyaena did not realise that the herd contained a possible meal, and consequently did not make an attempt to kill any of them. The other animals in the herd did not benefit (or lose) from being in a group, in this instance, because healthy adults are largely immune from the attentions of single hyaenas resting in the daytime (Kruuk 1972).

3.2.2 *Detecting the predator*

For most mammals, dependent for their safety on fleeing rather than on camouflage, early detection of the predator usually means the escape of the prey. It seems obvious, first, that the more detectors the

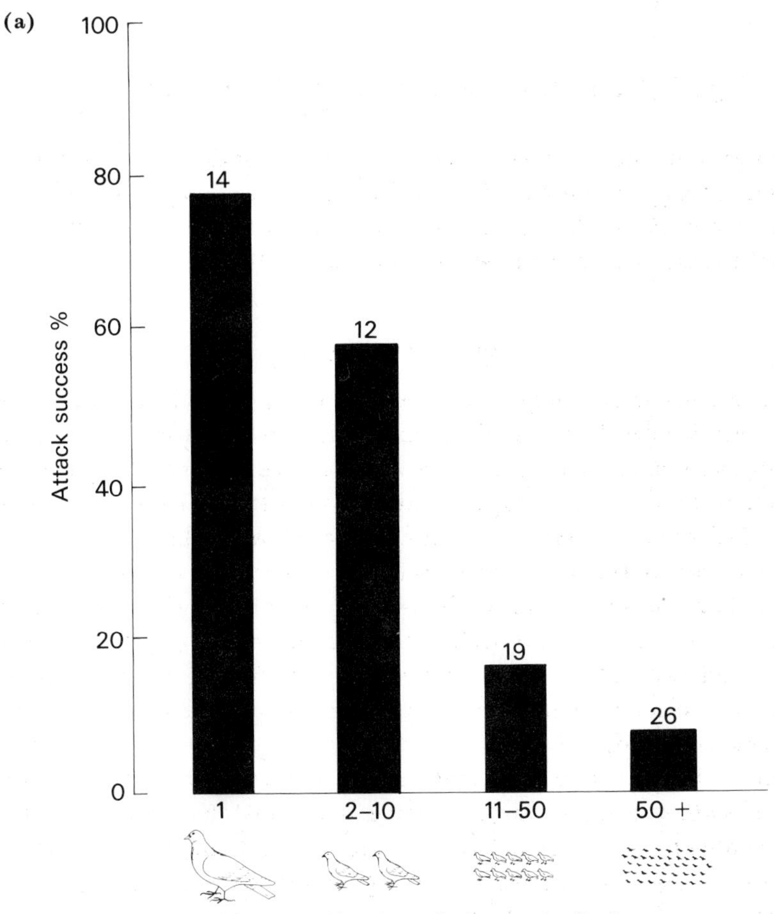

Fig. 3.1. The value of flocks of woodpigeons in detecting the approach of an avian predator, a goshawk. See text. (a) shows the dependence of the goshawk's attack success on the size of the flock. (b) Shows the distances at which different numbers of woodpigeons detected the predator's approach. (Kenward 1978.)

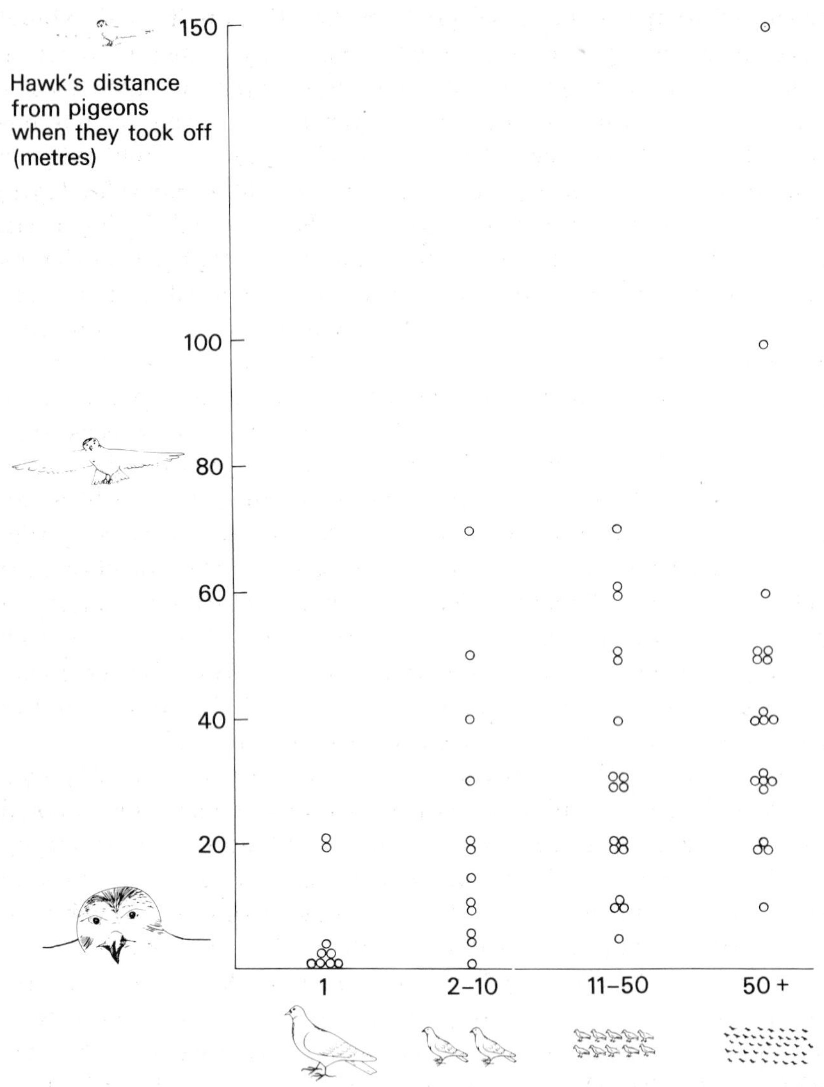

Fig. 3.1.(b)

higher the probability of such early detections, and second, that provided an animal can perceive that one of its companions has detected a predator, the animal will benefit from the presence of companions. Pulliam (1973) produced a theoretical model in which birds could spare a fixed proportion of their time not feeding but with their heads up and looking for predators; he showed that the probability that an approaching predator would be detected increased rapidly with group size at small group sizes, but quickly levelled off. The proposed double benefit of grouping—increased predator-detecting ability and reduced time spent on vigilance—has been occasionally tested in practice. Powell (1974) and Siegfried and Underhill (1975) 'flew' hawk models over bird groups of varying size. In general the larger groups responded sooner to the 'predator'; in addition they were able to spend more of their time feeding and less being watchful. Kenward (1978) used a live trained goshawk (*Accipiter gentilis*) to attack feeding flocks of woodpigeons (*Columba palumbus*), and found that the predator's flights at the pigeons became less successful as the number of attacked birds increased (Fig. 3.1a). The effect was partly due to the poor condition of the single pigeons, but was still significant between flocks of different sizes. Enhancement of detection by flocks appears to be involved too, because the distance of the hawk from the pigeons when they took flight also increased with the number present (Fig. 3.1b), even if flights at single birds were excluded. Hoogland and Sherman (1976) showed that larger colonies of bank swallows (*Riparia riparia*) detected the presence of a (stuffed) predator sooner than small colonies did. Schaller (1972) found that lions were more successful when hunting prey animals which were on their own than those which were in small groups, probably in part because of improved detection by the group, but also perhaps because solitary individuals of the social ungulate species are in some way defective and therefore vulnerable.

The improved-detection benefit of being in groups probably does not go on increasing with group size, for a thousand pairs of eyes would see little if any more than a hundred would. Indeed, Crisler (1956), watching wolves (*Canis lupus*) hunting caribou (*Rangifer arcticus*) calves, considered that the latter were more easily caught when in large herds because they could not see the wolf's approach.

Benefitting from the vigilance of companions has sometimes been taken a stage further: in some species of primates, an adult male often remains on a vantage point and may thus act as a lookout while the others are feeding (Hall 1965, Gartlan 1968). Even if, as is likely, he is mainly looking for male rivals, he is nonetheless in a better position than the rest of his group to detect an approaching predator. Another example of an adult male acting as a lookout is provided by klipspringers (*Oreotragus oreotragus*) (Dunbar & Dunbar 1974). In dwarf mongooses (*Helogale undulata*) subordinate males do a disproportionate amount of the watching (Rasa 1977). Lookouts do not benefit as much as the other group members do, and the evolution of this possibly altruistic behaviour is probably helped in some of these cases by the high levels of relatedness among group members, as discussed later (see 3.5.3).

3.2.3 *Deterring the predator*

If the prey species is not a great deal smaller than the predator, or has particularly effective weapons, several of them acting in concert can sometimes deter or thwart the predator's attack, whereas a single prey animal would be unable to do so or unwise to risk attempting it. Wasps, of course, defend their colonies very effectively against casual predators thousands of times their size. It is possible that the mobbing of a predator by many birds sometimes deters it from trying to catch one of them. Certainly both Kruuk (1964) and Hoogland and Sherman (1976) have showed that mobbing by black-headed gulls (*Larus ridibundus*) and bank swallows respectively deterred at least some of the predators on their eggs and young, and that more mobbing birds were more effective than fewer. The evolution of mobbing behaviour is discussed by Harvey and Greenwood in Chapter 5.

Mobbing is not confined to birds. Mammals such as dwarf mongooses will also cluster together and mob some of their predators (Rasa 1977), while banded mongooses (*Mungos mungo*) will bunch with their young in the middle, face, deter, and even attack large avian predators (Rood 1975).

Musk oxen (*Ovibus moschatus*) threatened by wolves gather into a defensive formation with an array of powerful horns facing the predators and with vulnerable animals in the middle (Mech 1970). Eland (*Taurotragus oryx*) co-operate in a similar way in preventing attack on themselves and their calves by hyaenas (Kruuk 1972). In the primates, group defence against, and attack on, predators has been reported quite often (Kortlandt & Kooij 1963, Crook & Gartlan 1966, Stoltz & Saayman 1970, Eisenberg *et al.* 1972).

It is possible that a predator may also be deterred from attacking a group of prey by the risk of injuries incurred indirectly. For example, Tinbergen (1951) pointed out that a peregrine falcon (*Falco peregrinus*) stooping at extremely high speed at its starling (*Sturnus vulgaris*) prey would be severely injured if it struck any but its intended victim, as might well happen if the falcon were to plunge into the tight flocks which starlings form in response to its presence above them.

3.2.4 *Confusing the predator*

If the predator has been neither detected nor deterred, it still has to select one prey animal and rush at or pursue it. Again, it seems likely that if there are several potential prey animals which flee in unpredictable directions when he attacks, a predator may be confused and less able to concentrate on or follow any one of them. Neill and Cullen (1974) have shown clearly in experimental conditions that such confusion of the predator caused lower success rates by squid (*Loligo vulgaris*), cuttlefish (*Sepia officinalis*), pike (*Esox lucius*) and perch (*Perca fluviatilis*) attempting to prey on shoals of small fish; all four predators did best when hunting single fish, and worse when hunting fish which were in large shoals as compared with small shoals. Otherwise there is a dearth of good evidence for such a confusion effect. It is probable that the 'explosion' in all directions, including upwards, of a herd of impala (*Aepyceros melampus*) when startled has a confusing effect on the predator trying to catch one (Jarman 1974). Wyman (1967) and Kruuk (1972) described the way in which a female Thomson's gazelle (*Gazella thomsoni*), whose fawn is being pursued by a jackal (*Canis mesomelas*) or hyaena, will run back and forth across the path of the predator; how effective this behaviour is was not clear, but it presumably has some survival value. The other side of the coin here is that the prey animals themselves can also be confused, collide with one another and find their escape impeded: the size of the prey group is clearly important. Schaller (1972) considered that this was one of the reasons why lion attacks on large herds of ungulates tended to be more successful than attacks on medium-sized groups. However, in addition there is the point that larger herds are statistically more likely to contain an animal which is particularly vulnerable to predators through sickness, injury or old age.

3.2.5 *Diluting the predator's effects*

Assume that the predator has approached a group of its intended prey undetected, and catches one of them. If the prey are capable of taking evasive action, the rest of them are likely to escape. For any one predator attack, the larger the group of prey animals, the smaller is the chance that any particular individual is the unfortunate victim. Hamilton (1971) showed that gregariousness can be selected for in this way even if the predation pressure on the prey species is higher than if the prey were better dispersed. Thus an individual can protect itself by buffering itself with companions who can become substitute victims.

This form of protection by 'dilution' is a mathematically verifiable one. The degree of protection afforded by companions is reduced as the proportion of any prey group who are captured increases, and is reduced if the rate of predation attempts on a group increases with the size of the group. Nevertheless it is clear that with most mobile animals the majority of prey individuals in a group do escape from any single predator attack. For example Schaller (1972), Kruuk (1972), Mech (1970) and Malcolm and Van Lawick (1975) all showed that even the four social large predators—lions, hyaenas, wolves and wild dogs (*Lycaon pictus*)—only occasionally caught more than a single animal from the prey group they were hunting. Equally, it is most unlikely that the number of attacks on a group is proportional to its size: a herd of, say, a hundred antelopes does not have to withstand a hundred times as many predator attacks on it as a lone animal. Thus companions usually help to protect an individual animal, by diluting his chances of being caught if the group is attacked.

3.2.6 *Avoiding being the victim*

It is likely that an animal can protect itself better the closer it is to its companions. Hamilton's (1971) geometrical model showed that each individual could reduce the size of its own 'domain of danger' by positioning itself nearer to another animal, and thus making itself less likely to be the prey individual nearest to the predator. Clearly, not all the animals in a group can be in the safer middle of the group, but their tight clumping when a predator approaches represents their attempts to get as close to it as possible. This may result in a compact group from which it is more difficult for the predator to obtain a victim (e.g. Tinbergen 1951 already quoted), or it may result in a milling unmanoeuvrable mass which makes the predator's task easier (Schaller 1972); but in either case each prey individual is generally safer than if he were on his own, and the central ones are much safer.

Other than from observations in captivity (e.g. Neill & Cullen *op. cit.*) there is a shortage of data showing that it *is* particularly the animals at the edge of groups which are taken by predators. Exceptions are the work on colonially nesting birds by Kruuk (1964) and Patterson (1965) where in general, the nests at the edges of the colonies were more likely to be preyed upon than those at the centre. However, for animals more mobile than birds' eggs, information is scanty. Partly, there are practical problems in obtaining it in the wild, and partly there are problems in the interpretation of it. If edge animals are taken more, to what extent is it due solely to their position, and to what extent due to their being more vulnerable (slow, unalert or uncompetitive), which is why they are to be found at the undesirable edge positions? For example, Coulson (1968) showed that kittiwake (*Rissa tridactyla*) nests at the edge of a colony were less successful than more central ones, and that this was not caused by predation (which was

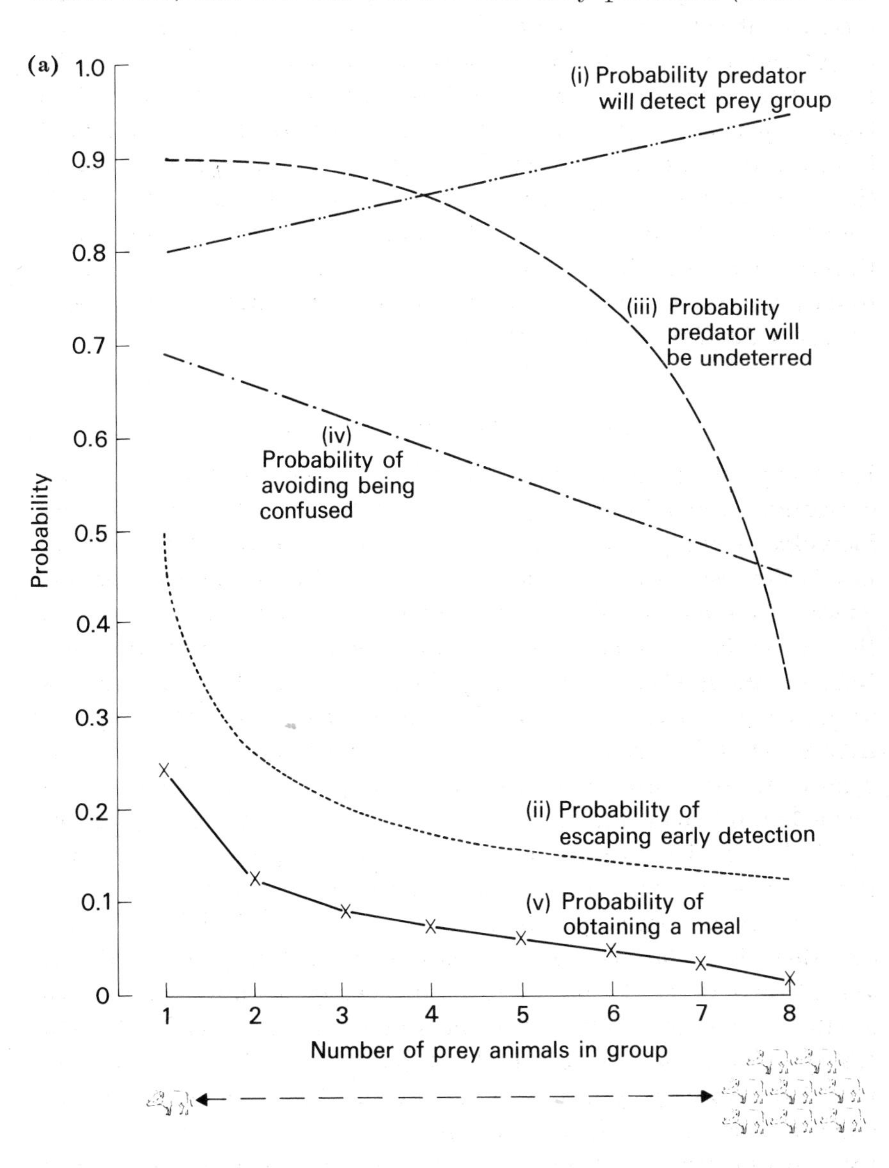

Fig. 3.2. The effects of prey group size on the probabilities of predator success (a) and of prey capture (b). See text for full explanation. I am envisaging a solitary predator such as a leopard approaching prey such as warthogs. The values of the predator's chances and how they change with group size are purely hypothetical.

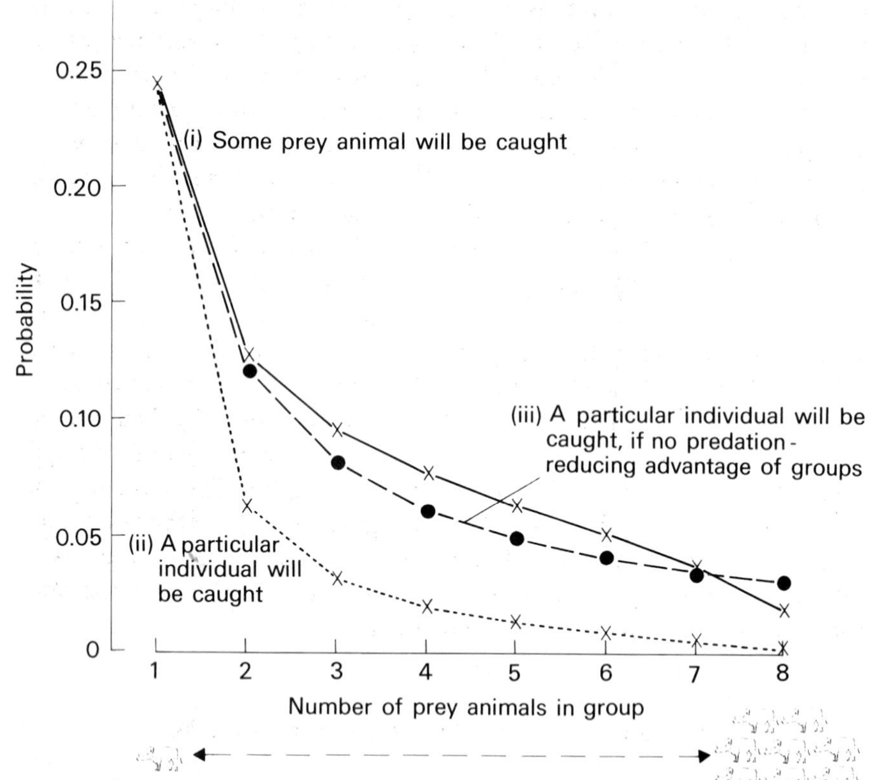

Fig. 3.2.(b)

negligible) but by the fact that peripheral birds were of poorer quality. On the other hand it is possible that predation effects may ultimately have been one of the factors causing these birds to compete for the central nests.

3.2.7 *Effect of group size on individual safety from predators*

We have seen that there are various possible advantages of being in groups in surviving against predators. Some of these advantages apply more to small groups and some to larger ones, as I have suggested in Fig. 3.2a. This figure shows the various factors influencing the chances of success of an imaginary predator stalking towards small groups of prey animals, and suggests tentatively how these factors might vary with group size; obviously a large number of variables and assumptions are involved. The predator's probability of being able to detect the prey group (line i) may increase steadily with group size.

His chances of remaining undetected (line ii) during his approach decrease rapidly at first as groups get larger (Pulliam 1973). The likelihood that he will be undeterred (line iii) by the imaginary prey's co-operative defence is probably high until there are more than a threshold number of them. His ability to concentrate on and follow one particular victim and not be confused by the others (line iv) is less when tackling a larger group (Neill & Cullen 1974). His overall probability of success (line v) is the product of each of these.

Figure 3.2b looks at the position from the point of view of the prey animals. Line i is the same as line v in Fig. 3.2a, the probability that the group will end up with one member killed by the predator. Line ii shows the probability that any particular individual will be the victim; it is simply line i divided by the number of animals in the group. It is clear that an individual protects himself enormously by being in a group. The curves in Fig. 3.2a were based on a large number of assumptions. If instead we make the conservative assumption that a predator's chance of succeeding does not alter with group size, nonetheless it can be seen (line iii) that each individual's chances of dying are still considerably reduced by his being with companions, and that the first few companions are by far the most important.

3.3 Advantages to Predators in Obtaining Food

As already mentioned, predators themselves may often be the potential prey of other predators. For example, nearly 10% of the diet of leopards (*Panthera pardus*) in the Serengeti was composed of other carnivorous species (unpublished observations). Insectivorous birds are preyed upon by raptors, which also kill mongooses which also prey on snakes which eat rodents. In considering the advantages to predators of living in groups, we should remember that they may also enjoy the benefits already outlined as accruing to prey animals. In addition there are a number of other advantages concerned with obtaining their own food. As will be apparent in what follows, I am interpreting the term 'predator' in a very wide and rather loose sense.

3.3.1 *Improved location of food*

Ward and Zahavi (1973) suggested that the behaviour of a number of bird species in gathering into large flocks might enable them to exchange information about food sources. It would be beneficial as well as possible for a bird which had had a relatively unsuccessful day's foraging to observe and follow other individuals which appeared to be either better fed or to have knowledge of the whereabouts of food. The successful bird, on the other hand, would probably be unable to prevent such observation and following without wasting considerable time and effort; depending on the kind of food source he might not suffer from any depletion of it by his followers; and by himself observing other birds he could gain useful information about the relative merits of his own feeding place and of others in the vicinity. Ward and Zahavi argued that the value of such a system of information exchange would be greatest for species feeding on large, clumped, widely dispersed and unpredictable food sources, and that information exchange provided the most likely functional explanation of the large communal roosts of birds such as starlings, weavers, herons and seabirds.

Whether the system does in fact operate—whether animals do observe others and so gain information about the distant whereabouts of food and assistance in finding it—is difficult to test. Menzel (1971) showed that captive chimpanzees (*Pan troglodytes*) could determine by observation of the behaviour of a companion whether it had previously been shown a source of food, and to some extent how much food there was there.

Krebs (1974) observed great blue herons (*Ardea herodias*) at a nesting colony and at their varying feeding grounds. He found that birds from neighbouring nests tended to leave at about the same time to seek food, and to go to somewhat the same places. It was not clear, though, whether or not the less successful birds were observing or following the more successful. The herons were also clearly attracted to places where one or more others (or dummies) were already feeding. So, of course, are many other species—for example vultures (Houston 1974), woodpigeons (Murton 1971), and hyaenas (Kruuk 1972)—but there is no need to live in groups in order to respond to the sight or sound of other individuals feeding.

It has been shown that information transfer can occur on a smaller scale and over short time periods. Krebs *et al.* (1972) showed experimentally that captive great tits (*Parus major*) in groups of four obtained more hidden food than when alone; success by one bird attracted the others, and also caused them to search more in the proximity of the food source and to investigate similar hiding places. They tended also to modify their searching behaviour in a similar but weaker way in response to a bird's unsuccessful examination of a hiding place. Murton (1971) showed that the food taken by woodpigeons depended on, and was the same as, the food which their companions were eating; subordinate pigeons observed the food choices made by the well-fed dominant birds, and selected the same food types.

Exchange of information about the location of feeding areas or types of food has been little tested or demonstrated in mammal groups. The first might occur in species which gather and disperse frequently in search of unpredictable food: bats and badgers are some of the most likely candidates. The second probably occurs widely in carnivore and perhaps in primate groups.

Another possible advantage, connected with locating food, of being in groups has been proposed by Cody (1974a). It may be that birds can exploit food more efficiently (in terms of quantity of food eaten per unit distance travelled) in an area if they forage over it as a single large flock, because they can then avoid going to a part where they have recently fed (see also 4.3.2 and 2.4.3). Single birds foraging independently, on the other hand, would often be searching in places which previous feeders had denuded.

Competition will be discussed later.

3.3.2 *Improved chance of catching prey*

Once the predator has located potential prey, it faces the difficult task of capturing animals which are adapted to avoiding capture. As a result, most predator hunts generally fail, at any rate when hunting prey which can take rapid evasive action. For some capture methods, two or more predators may do better than one alone. For example Schaller (1972) found that lions' success rate when hunting gazelles, zebras (*Equus burchelli*) and wildebeest was doubled when two or more lions hunted together. Their co-operative hunting methods mean that a prey animal escaping from a lion which it has detected is liable to run within range of another which it has not. For the lions' method of hunting and type of prey, more hunters generally do better than one alone; but for leopards and cheetahs (*Acinonyx jubatus*) the same advantages of greater numbers do not apply, for reasons discussed by Bertram (in press).

Other predators may improve their hunting success through numbers for different reasons. Wyman (1967) found that jackals hunting gazelle fawns generally failed when hunting alone but succeeded when hunting in pairs. Kruuk (1972) showed a similar phenomenon with hyaenas hunting wildebeest calves. In both cases a second predator was able to counter the attacks of the mother, enabling the first to capture the offspring unimpeded. Martinez and Klinghammer (1970) described co-operative hunting by killer whales (*Orcinus orca*) which surrounded and hemmed in their porpoise prey in a way which would clearly be impossible for a single predator.

There are relatively few studies which have measured the success rates of solitary versus social hunts, and there are problems besetting the conclusions. First, for example, Krebs (1974) showed clearly that herons feeding in flocks took in food at a much faster rate than those feeding singly; yet this was not due to any kind of co-operation or to the number of birds *per se*, but resulted from the fact that flocks only built up where feeding conditions were good. Second, a lone predator of a normally social species such as a lion may well be different in skill or motivation as well as in number from lions hunting communally. And third, a failure to find a differential success rate may be due to the predators' allowing for it when assessing the chances of success before embarking on a hunt; if a lion has companions, for instance, it is conceivable that it may start trying to hunt prey which it would ignore as being impossibly difficult to catch alone.

Few predatory birds hunt in a co-ordinated way, although a number of insectivorous birds forage in groups. Pelicans dip their heads in synchrony in one of their methods of fishing, bunched close together, and presumably benefit from such co-ordination. Many plunging seabirds such as gannets fish at the same place; it is possible that in such cases there is a confusion effect, rendering the prey easier to catch, but it is likely that the 'heron difficulty' I have referred to above applies. Confusion-causing, driving, and chasing and wearing down would be methods of hunting which would generally benefit from co-operation among hunters, while ambushing or overtaking would be methods where single animals would be expected to do as well.

3.3.3 *Catching larger prey*

Even if a social predator is no more successful in its hunting than a solitary one, it can generally capture larger prey. This has been well documented for the large mammalian carnivores: Bourlière (1963), Kruuk (1972, 1975), Schaller (1972) and Schaller and Lowther (1969) have all drawn attention to the fact that the victims of the socially-hunting carnivores tend to be big. The social species —lions, spotted hyaenas, wild dogs and wolves—habitually kill prey animals which are at least their equal in weight, and often several times as large. By contrast the solitary cat, dog and hyaena species rarely do, but generally feed on prey considerably smaller than themselves.

Among some of the smallest mammalian carnivores—stoats (*Mustela erminea*) and weasels (*Mustela nivalis*)—this generalisation does not hold. Both hunt alone, and regularly kill prey animals which are much larger than themselves (Hewson & Healing 1971), their victims consisting mainly of lagomorphs, rodents and birds (Day 1968). Birds, being inactive at night and having very slender necks which a small carnivore can seize, are particularly vulnerable. It is likely too that the scale effects of small size are important. To try to illustrate the point, consider a stoat which attacks a rabbit six times its own size; if the rabbit falls over onto the stoat, the latter is far less likely to be injured than a single lion would be which landed up under the huge weight of a falling buffalo.

Few within-species comparisons are available. Malcolm and van Lawick (1975) found that whereas small packs of wild dogs killed mainly wildebeest calves and gazelles, some larger packs specialised in taking adult zebras, sometimes with a division of labour in the killing process. Solitary wolves live largely on carrion or on small food items whereas packs take adult moose and deer which can defend themselves (Mech 1970). Single lions rarely kill adult bull buffaloes (*Syncerus caffer*) while groups of lions do (Schaller 1972).

There are probably two reasons why social carnivores take larger prey species. The first and obvious one is that the collective strength of several predators helps in pulling down a large victim, especially if the latter is defending itself. This is particularly the case with wild dogs and hyaenas which lack specialised methods of killing their prey. The second reason is that a social predator can afford to run greater risks of personal injury when tackling large food animals which may resist violently. I (pers. obs.) and Schaller (1972) have recorded instances of seriously injured lions surviving for months on food killed by their companions. By contrast it is likely that a lame cheetah, needing to hunt by speed and alone, would stand a much higher chance of dying of starvation. It is likely that the problem of potential injury is one of the main reasons why no avian predator has gone in for killing large prey co-operatively. If they were to attempt to capture large birds, they would risk severe injury when crashing with their prey; against large mammals they would lack weight; and against large fish they would be out of their element.

Being able to take larger prey as well as small should theoretically increase the range of different prey species which a predator can feed on. In practice, however, the number of prey species which it actually does feed on is reduced (Bourlière 1963, Kruuk & Turner 1967, Kruuk 1975). A group of predators not only can but needs to take larger prey animals to feed all the members of the group, and in most habitats there are fewer large species than small ones.

3.3.4 *Competing better for food with other competing species*

Being in a group may enable a predator to hold its own better against other predator species, both directly and indirectly. Direct competition is generally stronger between predator species than between herbivore species because their food (meat) is in discrete, small, high quality packages which can be fought for. Direct competition between lions and hyaenas, for example, takes place over carcasses which have been killed by either of the two predators or which have died of natural causes. Kruuk (1972) and Schaller (1972) described the importance of numbers in such encounters. A single hyaena is no match for a lion; a pack of them can drive a lion from a carcass, especially if she is a lioness; but a group of lions can displace a pack of hyaenas. Similarly a leopard is unlikely to lose its kill to a single hyaena, but is certain to do so to two or more. Estes and Goddard (1967) considered that a small pack of wild dogs would have much greater difficulty than a larger one in preventing spotted hyaenas from appropriating prey caught by the dogs.

Among fish, Robertson *et al.* (1976) have examined the effects of being in a shoal on competitive ability against other individuals and other species feeding on the same food. Striped parrotfish (*Scarus croicensis*) are subordinate to damselfish (*Eupomacentrus planifrons*), and are inhibited in their feeding by the attacks of territorial members of the latter species. Schools of parrotfish can feed at higher rates than single individuals because the damselfish's attacks are in a sense 'diluted' by their numbers; he cannot drive a whole shoal out of his territory. Just the same advantages accrue to shoal members against territorial conspecifics who similarly try to drive them off.

There are few data on indirect or ecological competition between social predator species. However it seems likely that if lions are more efficient predators as a result of being social (for the reasons given above), then they will be more effective in whatever indirect competition with other species occurs. They will presumably take a greater proportion of the vulnerable section of the prey population, leaving fewer for their competitors. The importance of indirect competition of this sort is not clear; it is likely to be greatest between predator species which use similar hunting methods and so take victims which are vulnerable in similar ways. Thus hyaenas probably offer much more indirect competition to wild dogs than lions do, because the first two species both run down weaker prey individuals, while lions capture a higher proportion of healthy adults (Kruuk 1972, Schaller 1972).

3.4 Other Advantages

Many of the advantages described so far apply whether or not the members of the group are of the same species. Mixed-species groups occur, although not as widely as single-species ones. For instance, small groups of single territorial male antelopes of more than one species are common; it is likely that each member of the group benefits by their increased ability to detect predators, and each runs less individual risk of being the victim if they fail to do so. Baboons (*Papio anubis*) and impala are often found in large mixed herds (Altmann & Altmann 1970); part of the reason for this may be their improved detection of predators by combining their respective abilities in the field of colour vision and high vantage points with extremely sensitive noses and ears.

Among predators, multi-species groups are common in the mixed feeding parties of insectivorous birds (Croxall 1976, Macdonald & Henderson 1977). Whether and how much each species benefits is not clear. It is possible that not all do: some may act involuntarily as 'beaters', flushing insects which other species take advantage of, but reaping no benefit themelves. Heatwole (1965) showed that cattle egrets (*Bubulcus ibis*) obtained more food with less expenditure of energy by associating with cattle and feeding on the insects which the cattle disturbed. The 'nuclear' species in many mixed-species flocks may act in a way similar to the cattle. Birds in such flocks may benefit from one another's presence in a more subtle way. Krebs (1973b) showed experimentally that captive chickadees (tits) could learn about food sources from the behaviour of other species of tits with which they were foraging. Mixed species groups are not found among the raptors nor the mammalian carnivores; although some may have attendant scavengers, the relationship is one-sided.

Although all the above advantages from being in groups relate to aspects of the living environment—to food and predators—in certain cases animals may benefit from grouping by an improved ability to resist the inanimate environment. The familiar example of woodlice huddling and thereby being better able to resist desiccation is a good illustration (Allee 1926). For the warm-blooded vertebrates, resisting heat loss is a more significant benefit, and probably explains the communal roosting of some of the smaller birds, such as wrens (*Troglodytes troglodytes*) (Armstrong 1955). Sealander (1952) found that deer mice (*Peromyscus*) could survive for longer at low temperatures if they were allowed to huddle together. Trune and Slobodchikoff (1976) showed experimentally that pallid bats (*Antrozous pallidus*) roosting in clusters conserved metabolic energy more effectively than if they were isolated, partly because single bats were more restless. Incubating emperor penguins (*Aptenodytes forsteri*) form dense huddles, which presumably helps them to withstand blizzards and extremely low temperatures (Stonehouse 1953).

Another aspect of the inanimate environment to be overcome is the resistance to locomotion. The shoal formation of some species of fish is such that each individual benefits from the turbulence of the others, and thus the drag on all of them is reduced (Weihs 1973). In addition, some fish species improve their own swimming efficiency by producing a drag-reducing mucus, and the washing-off of this might help reduce the drag on other individuals swimming along behind (Breder 1976). Aerodynamic advantages of formation flying have been posulated for large birds such as geese, pelicans and flamingoes which fly in V-formations, although the relative advantages for them are probably much smaller (Gould & Heppner 1974).

At this stage I should stress once more that only certain of these advantages of group living apply to any particular species. Sheep do not benefit from their grouping by being able to kill larger prey, and mussels do not escape from a predator by confusing it during a chase! Which species benefit, how much, and in which ways depends on their abilities, their food and their foes. And most species would not experience a net benefit at all—witness the many species which do not live socially and so do not suffer from some of its detrimental consequences, particularly increased intraspecific competition for food.

3.5 Intra-Specific Consequences of Group Living

The advantageous consequences of being in groups have been considered so far from the point of view of finding food or avoiding becoming food, and mainly at the species level. Some of the intra-specific consequences which result from group living by predators and prey alike will now be examined. As before these will be classified for convenience under a number of separate headings, but it must be borne in mind that the various consequences are not independent but interact with one another in many ways.

3.5.1 *Permanent groups*

Animals which are habitually in groups, particularly in small groups, are often in permanent groups. Wolf packs, for example, are not loose temporary collections of different animals but long-lasting associations of the same individuals (Mech 1970). The same applies to lion prides (Schaller 1972, Bertram 1975), wild dog packs (Frame & Frame 1976), hyaena clans (Kruuk 1972), mongoose packs (Rood 1975, Rasa 1977), many ungulate groups (references in Jarman 1974), most primate groups (references in Crook 1970, and Clutton-Brock & Harvey 1976), and some rodents (e.g. Barash 1974), and marsupials (Kaufmann 1974). Wilson (1975) gives many other mammalian examples.

The same is true among a number of bird species—see Chapter 9 by Emlen. Birds' nesting, being static, necessarily restricts the movements of individuals, and militates against flexibility in the composition of breeding groups, more than is the case with many mammals whose young do not go through such an immobile stage.

One result of this stability of composition of a group is that the members of it are able to recognize one another individually and can learn a great deal about one another. Both allow considerably more complex social relationships to develop than if the group had a changing and unknown membership. Some of these relationships—such as dominance hierarchies and co-operative behaviour—are described below; it is important to remember that they depend on constancy of group membership, and usually on individual recognition.

3.5.2 *Inbreeding*

Another result of group stability is that animals in groups are likely to find themselves in the proximity of potential mates who are close relatives. Inbreeding tends to have deleterious consequences. Greenwood *et al.* (1978) examined 12 years of breeding records of great tits in Wytham wood near Oxford. They showed that matings between close relatives resulted in a nesting mortality 71% higher than matings between unrelated birds, but the subsequent survival and breeding performance of their inbred progeny did not differ from those of outbred birds. Packer (1977b) found that matings between closely related baboons resulted in a 40% drop in the viability of their offspring.

The number of instances in both these studies was small, as is the number of such studies, for it is striking how rarely animals in general breed with their near relatives. In great tits, Greenwood *et al.* (*op. cit.*) showed that this was largely because of the different dispersal distances of the two sexes: females tended to move further from their natal areas than their brothers did. In species such as wildebeest, weaver birds or starlings, living in huge associations apparently almost devoid of social structure, the chances of a female meeting and mating with her father or brother must be minute; incest avoidance in such species could be almost entirely statistical. In small groups such as primate troops and carnivore packs the chances of such matings would be high. In most of these species incest takes place much less often than one might expect on statistical grounds, because the offspring of one sex or the other (or sometimes both) leave their natal group and try to join or found another group.

Which sex stays and which emigrates varies according to the species and its social system. Female wild dogs usually leave their natal pack and its males (Frame & Frame 1976); so do female chimpanzees (Pusey in press), Arabian babblers (*Turdoides squamiceps*) (Zahavi 1974), and both sexes of gorillas (*Gorilla gorilla*) (Harcourt *et al.* 1976). It is the males which leave in hyaenas (Kruuk 1972) and lions (Schaller 1972); in baboons (Packer 1975), rhesus monkeys (*Macaca mulatta*) (Sade 1967), and most other primates; and in most group-living ungulates (Geist 1974b, Jarman 1974).

As a result of these emigrations, incest seldom occurs. It would be unwise, however, to assume that these emigrations have necessarily evolved primarily as incest-avoiding mechanisms. It is clear that in lions, for instance, subadult males leave largely because of expulsion by adult males (who also monopolize mating opportunities and get earlier access to food); their own pride holds little future for them at that stage, and they leave in search of another pride which can be taken over and where they will do better (Schaller 1972, Bertram 1975). Incest in lions is usually avoided not through an incest avoidance 'mechanism', but as a by-product of the social system. In the one case I know where young males have been able to remain, they have mated freely with their mother and other relatives, as they do in captivity. On the other hand, Zimen (1976) reported that among wolves littermates failed to show sexual interest in one another; here there did appear to be a fairly specific incest avoidance mechanism. A similar mechanism appears to operate among other ones in humans: children reared in close proximity in some Israeli kibbutzim never subsequently marry other members of their peer group (Bischof 1975 and 6.2).

3.5.3 *Relatedness*

If animals are spending all their lives in permanent groups, their companions and not only their mates will often be relatives. Many field studies have followed individuals over long periods of time and demonstrated that this is so. Examples are to be found among carnivores (Mech 1970, Kruuk 1972, Schaller 1972, Bertram 1975, Rood 1975, Frame & Frame 1976); among primates (Sade 1967, Missakian 1973, Harcourt *et al.* 1976, and references in Clutton-Brock & Harvey 1976); among herbivores (Klingel 1965, Douglas-Hamilton 1973, and references in Jarman 1974); among rodents (Sherman 1977); and among birds (Watts & Stokes 1971, Brown 1972, Ridpath 1972b, Zahavi 1974, and references given by Emlen in Chapter 9).

I have dealt with the special problem of being related to possible mates. Being surrounded by related companions produces no such problems; indeed it allows the evolution of greater complexity in social systems through the operation of kin selection, a concept introduced by Hamilton in 1964. Because replicas of an individual's genes are likely to be present not only in an animal's own offspring but also in its close relatives, one would expect natural selection operating at the gene level to favour behaviour which benefits relatives. Parental behaviour, which benefits offspring, is of course the most widespread and best known relative-helping behaviour, because offspring are generally both the closest relatives and those most in need of assistance. But other types of assistance can be given too, and to other relatives. Such assistance can take the form of co-operation, or of refraining from competition, or of helping to bring up the offspring of relatives: all raise the 'inclusive fitness' of the helper (Hamilton 1964, see 1.5).

Not all relatives can be assisted to the fullest extent, but rather the total finite available quantity of assistance has to be distributed among them. Ideally, each recipient should receive a quantity proportional to his degree of relatedness to the donor, who should make allowance for the needs of each potential recipient as well as for any uncertainty about his relatedness (Dawkins 1976). Depending on the social system, there may be a considerable amount of this uncertainty, particularly in animals which mate promiscuously or which mix up their eggs or

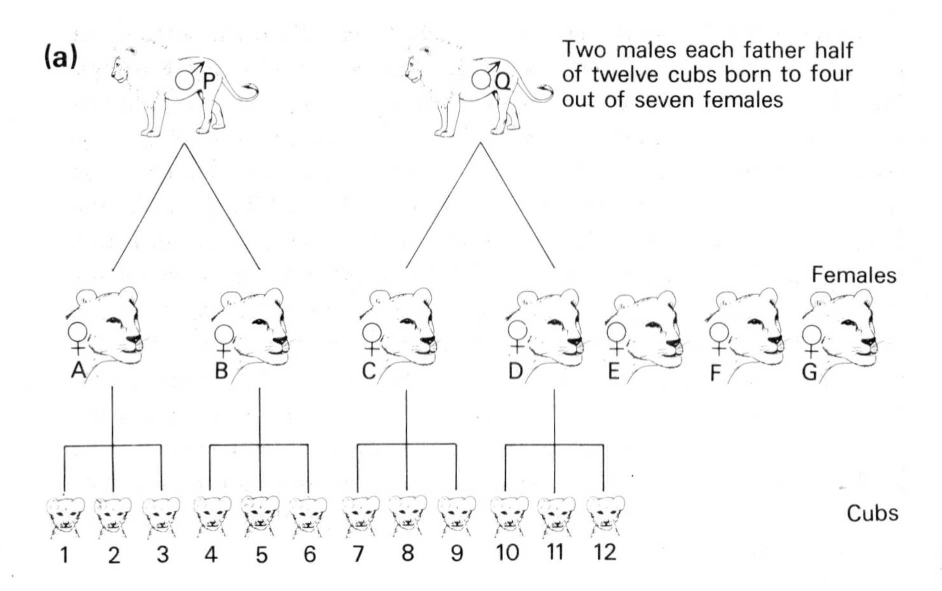

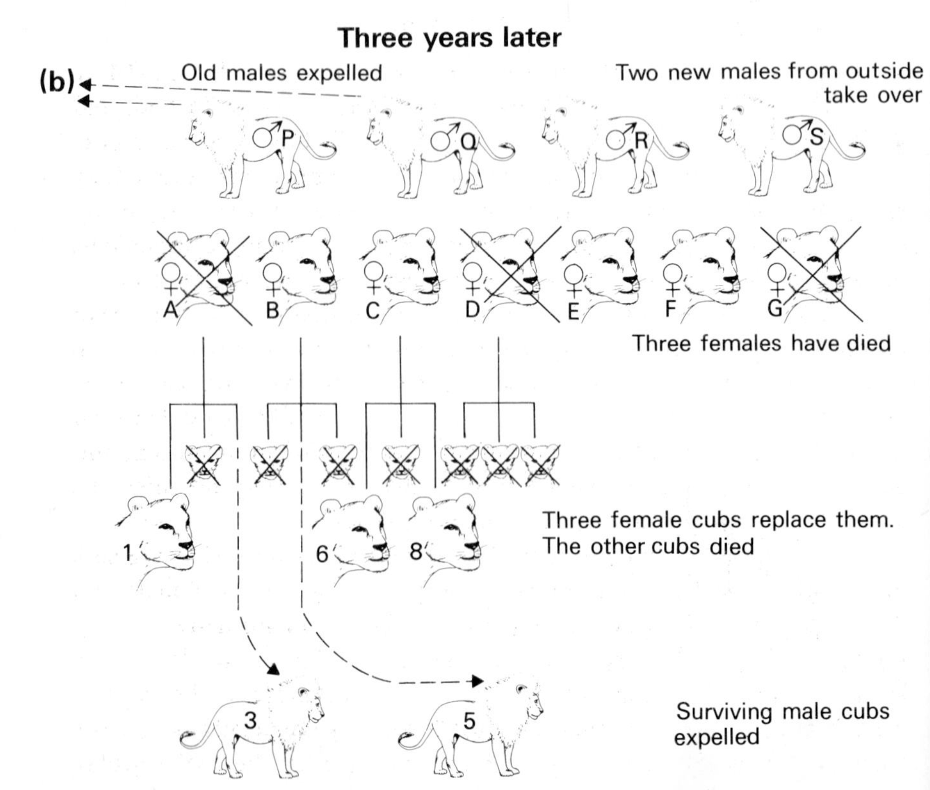

Fig. 3.3. The reproduction of a 'typical' lion pride.

offspring. Nonetheless it may be possible to calculate the average or probable degrees of relatedness among animals within social groups. For lions I made the assumptions that a typical lion pride reproduces as shown in Fig. 3.3, and calculated the resultant degrees of relatedness (Bertram 1976). I showed that the breeding adult males are on average related to one another about as closely as half-siblings, while the lionesses are approximately one another's full cousins but unrelated to the males. These degrees of relatedness offer the key to understanding much of the co-operation observed within lion prides, as described later. Maynard Smith and Ridpath (1972) showed that in the Tasmanian native hen (*Tribonyx mortierii*) the relatedness among the males in breeding trios was the basis of their reproductive co-operation. In many other social groups of animals, whether mammals, birds, insects or others, kin selection similarly provides a strong selective pressure favouring the development of the co-operation and altruism which are discussed in later sections.

3.5.4 *Who breeds*

Within a group of animals, potential mates and potential competitors for them are all close at hand. Mates are therefore easier to find, and also easier for a dominant to monopolize. Within a social group, however, potential competitors are not only competitors: often they are also relatives with genes in common and partners with many objectives in common. If an animal benefits from having companions, he or she must not so dominate those companions that the latter either die or leave. The extent and the methods of competition within social groups depend on the stakes, the weapons, the costs and benefits to victor and loser, the alternative strategies open to the loser, the degree of relatedness of the competitors, the amount of difference between them in competing ability, the extent of the need for co-operation, and the social system; all of them in turn influence the way in which the social system evolves in any particular species.

Males generally compete more than females, because by doing so a male can increase his reproductive output more than a female can (Trivers 1972) (see Chapter 7). I shall consider the males first. In many social species, competition among males results in a dominance hierarchy, with the males at the top of it doing most of the mating. This is true in most mammal and bird social groups containing more than one male. In general, the tighter and smaller the social group, the more exclusively are mating rights confined to a single male, as is the case in some mongooses (Rasa 1977), wolves (Mech 1970, Zimen 1976), wild dogs (Frame & Frame 1976), turkeys (*Meleagris gallopavo*) (Watts & Stokes 1971), and Arabian babblers (Zahavi 1974).

In many species, of course, competition among males is so strong that all the others are driven from the group by the dominant male, as, for example, in langurs (*Presbytis* spp.) (Sugiyama 1967), many other one-male troops of primates (Crook & Gartlan 1966), zebras (Klingel 1965), vicuna (*Vicugna vicugna*) (Franklin 1974), seals, and a number of antelopes and rodents.

At the other extreme, in a few species—lions (Schaller 1972) and Tasmanian native hens (Ridpath 1972b)—two or more males have equal mating rights and there is no dominance hierarchy among them. At any particular mating opportunity, first possession determines who mates. I have argued (Bertram 1976) that factors which contribute to this lack of competition between companion male lions are (a) the very high cost of fighting, (b) the need for a male to have companions in order to be able to retain ownership of a pride of lionesses, (c) the improbability that any particular mating will lead to offspring, and (d) the fact that the male companions are closely related. By contrast, competition with an unrelated rival group of males is intense, and is probably the ultimate cause of the high degree of sexual dimorphism. This competition manifests itself in fierce fighting, and in the killing of some of the predecessors' offspring. By killing the cubs to which he is not related, a male which has newly taken over a pride of lionesses manages to raise his own reproductive output, for two reasons (Bertram 1975, 1976). First, the lionesses come into oestrus again sooner, thus offering the new male an earlier opportunity of fathering offspring. And second, his own offspring when born will suffer less competition from older cubs in the pride. Infanticide occurs in similar contexts in langur monkeys (Mohnot 1971, Rudran 1973, Hrdy 1974).

Competition among males naturally leads to the evolution of ways of winning. As well as having weapons, large size is generally an aid to victory, and therefore male-male competition tends to result in large males and thus sexual dimorphism. Large size in males may then lead to other consequences. Food-parasitism by male lions is an example: the males can rob the females of food which the latter have caught. Crook and Gartlan (1966) argued that group living in savannah primates, having evolved primarily because of improved detection of predators, led to an increase in male competition within these groups; this in turn led to larger and better armed males who were then capable of defending the troop against predators. Another way of improving one's chances in a fight is by having the support of allies, as can evolve through reciprocal altruism (see 1.6).

Females compete less than males, but they compete nonetheless—not for matings, which are not usually in short supply, but for reproductive opportunities. In packs of dwarf mongooses, wild dogs and wolves only one dominant female usually reproduces, managing either to prevent subordinate females from coming into oestrus or to inhibit them from mating; exactly how she achieves this is not known. By contrast, all lionesses in a pride reproduce, and so do any of the adult females in most primate troops and ungulate groups. The evolution of a system of inhibition of the reproduction of subordinates is probably helped by the combination of several factors, none of which is either a necessary or a sufficient condition. First is if the dominant female is able to produce herself as many offspring as the group could support; a wild dog bitch for example can produce up to 16 puppies in a litter, but a lioness only up to 6 (Schaller 1972, Kleiman & Eisenberg 1973). Second is if subordinates are unable to breed successfully on their own, so they do better to remain and perhaps outlive the dominant. Third is if the group is a compact one so the dominant is able to control the subordinates. And fourth is if those subordinates are closely related to the dominant so that the selective pressure to compete is reduced through kin selection. There is a positive feedback system at work here. Any monopolizing of reproduction means that the offspring tend to be more closely related to one another. This in turn probably favours further co-operation through kin selection, and reduces still further the selective pressure on subordinates to compete for reproductive opportunities. These factors apply to most of the social carnivores, and to the social insects (see 1.5), and to some birds (see Chapter 9), but not to most other mammals.

Reproductive competition does not exclude reproductive co-operation. Male and female nonbreeding wolves, wild dogs and mongooses all bring food to others' young, and generally aid in rearing them, probably another instance of kin selection at work. Lionesses allow their companions' offspring to suckle from them and help to rear them communally, and so do banded mongooses (Rood 1975). Communal suckling is very rare in other species. For example no primates or ungulates consistently suckle one another's young, although a lost or orphaned infant may occasionally be adopted (Hrdy 1976). Some primates and ungulates live in female kin groups, and we might expect some of these species to go in for communal suckling, if kin selection is important in its evolution, as I have argued is the case for lions (Bertram

1976). However in lions there are other selective pressures also favouring the evolution of communal suckling. First, it appears that cubs are more likely to survive if they are reared together; second, as subadults they do better as a group; and third, adult males have a much longer reproductive period if they have companions (Bertram 1975). Therefore it is to the advantage of a lioness for her own cubs' companions to survive, and by allowing them to suckle from her she may be able to help them to do so.

3.5.5 *Allocation of food*

Being in a group means that there is potential competition for food among the members, even if that food was only captured because of the participation of the group. The nature of the food supply and how the animals use it obviously influence the extent and the seriousness of this competition. An errant grazing herbivore such as a wildebeest, zebra or bison (*Bos bison*), travelling extensively and feeding unselectively wherever it goes, probably does not experience any significant food competition from a few companions, although when these latter become very numerous all will have to travel even further. The presence of even a few companions would make resident herbivores, even unselective ones such as rabbits (*Oryctolagus cuniculus*) or hippopotami (*Hippopotamus amphibius*) (Olivier & Laurie 1974) have to travel further from their burrow or base to feed. On the other hand there would be little direct competition in the sense of fighting. A blade of grass is not worth fighting for because that is usually much more effort than finding another.

Direct food competition is likely to be similarly slight for small mobile food such as insects or small fish. A dominant animal cannot appropriate a subordinate's catch before the latter has consumed it, although he may be able to choose the best hunting position. The extent to which the group members of social insectivorous or piscivorous species, such as babblers, mongooses, coatis or lemurs, tend to reduce one another's food intake is not clear; it probably depends on the distribution of the food supply and the catching methods of the hunters.

Many fruit-eating primates and the social carnivores are subject to much greater possibilities of direct as well as indirect food competition, since their food tends to be in large scarce portions. I shall deal with primates first. Those species which live in large troops use larger areas than do those which live alone or in small troops (Clutton-Brock &

Harvey 1977), which suggests indirect food competition. Within troops, the subordinate animals may be displaced by a dominant one before they have been able to consume the food item, such as a cluster of fruit, which they have found.

Food-sharing, as well as food-robbing, occurs in primates. Baboons and chimpanzees which have possession of a piece of meat will sometimes allow others a share of it (Strum 1975, Teleki 1973). Another form of food-sharing is by calling or leading companions to a patch of food. Wrangham (1977) described the phenomenon in male chimpanzees, for example. He argued that in this case the motivation behind the calling was to attract companions for other purposes such as impressing females or threatening neighbours; the caller uses the food source as a lure, and by advertising his presence there he warns others that the food will no longer be available unless they come quickly. No vertebrates other than Man have been shown to be able to achieve the feat which honeybees regularly perform with their waggle-dance—namely to give information to their companions about the location of a remote source of food (see section 4.2.3), as opposed to simply calling or leading them to it.

Food competition is very intense among the social carnivores. A companion at a carcass means a 50% reduction in the amount of food available to the eater, in a way which does not apply to a grazer with a companion. Hyaenas and wild dogs all gobble food as fast as they can from a carcass, with some squabbling but without a clear rank order among the adult feeders. The same applies with lions when a carcass is large and the animals not too hungry; with smaller pieces, the first in possession is temporarily dominant over the others, who keep clear. But size is important, and despite not having first possession a larger lion can rob a smaller one; thus lionesses rob cubs, and male lions rob females and cubs. This may result in the starvation of the cubs at times of food shortage and small kills (Schaller 1972). Among wild dogs on the other hand, the young are dominant over the adults, and thus are able to feed first.

We can guess at some of the evolutionary reasons for the differences between these two species. The average adult wild dog is much more closely related to a puppy in its pack (from the data given by Frame & Frame 1976) than is an average lioness to the cubs in her pride (Bertram 1976), and partly for this reason through kin selection she is more likely to allow the young to assert their dominance. In addition, most adult dogs will never reproduce directly at all themselves, only indirectly by helping to rear their relative's offspring; thus an adult raises rather than lowers its inclusive fitness by allowing young to feed first.

Food sharing as opposed to food competition is also frequent among the social carnivores. In lions it is slight. In some circumstances a male lion is surprisingly more tolerant towards cubs than lionesses are, and will allow them to feed from a carcass from which he is excluding all other lions. A male lion is likely to be much more closely related to cubs in the pride than a lioness is to cubs which are not her own (Bertram 1976). Other social carnivores are more active in their food sharing. Wild dogs regurgitate food to other members of the pack, both to adults and to young. Any wolf or mongoose regurgitates or brings back respectively food for the young in the groups, whether or not it has offspring of its own. A banded mongoose which has found a pile of elephant dung full of dung beetles will twitter and churr and thus attract other members of its pack (Rood 1975). The degree of food-sharing as well as of reproductive co-operation is highly developed in these carnivore packs, most of whose members are closely related to one another: the social insects exceed them in all three characteristics, where it is likely that kin selection has weighed even more strongly in favour of co-operation instead of competition (Wilson 1971).

3.5.6 *Other co-operation*

In a hostile world, companions often help against enemies, many of which are members of the same species. Being in a group helps animals to deal with unfriendly conspecifics, as it does in combating predators. With both nomadic lions and hyaenas, for example, there may be competition between groups for carcasses; again numbers help in the same way as described earlier for competition *between* these two carnivore species—a large group can appropriate a carcass from a smaller one (Kruuk 1972, Schaller 1972). Rood (1975) found that larger packs of banded mongooses usually won in clashes with smaller packs. The same is probably true of many primate species. The advantages of being a larger group do not necessarily outweigh the advantages of being on home territory, but they may well be important in deciding the results of disputes over territorial boundaries.

All members of a dwarf mongoose pack take part in territorial defence (Rasa 1977), both directly and by scent-marking. Co-operative defence of the territory against rivals is found in most group-living territorial animals. Not all individuals participate equally, however. For example, male lions do more patrolling, marking and defence of the territory than females (Schaller 1972); this is probably both because they are more at risk from rival males and because they are larger and so better at fighting. Male lions fight co-operatively, as do male langurs taking over a new troop (Hrdy 1974), and male turkeys (Watts & Stokes 1971) against rival male fraternities.

Sexual dimorphism in size can lead to a sexual division of labour within the group, although not necessarily through any kind of co-operation among members. Thus male lions do less hunting than females, perhaps partly because they are probably less good at it or more preoccupied with territorial defence, but mainly because they do not need to do so because they can plunder the females of food which the latter have caught. While waiting for the lionesses to capture a meal, males may remain with and merely by their presence guard the cubs which are likewise waiting, and all may therefore benefit somewhat, if unequally. Thus what may at first sight appear to be co-operative division of labour may be solely the result of each individual doing what is best for itself. The lookouts in dwarf mongooses are generally subordinate males (Rasa 1977). It is possible that this division of labour tco is an indirect result of the social behaviour of the species. If, as in some ungulate species, subordinate males tend to be driven to the edge of the group where the risk is higher, they are forced to be more alert for their own safety; and the other central group members can then benefit from that increased vigilance both in terms of their own safety and because they need to spend less time being vigilant themselves.

The guards in banded mongooses do appear to be more altruistic, in that the returning foragers do not bring back food for them (Rood 1974). Dwarf mongooses have been reported (Rasa 1976) to show considerable altruism in their care of a sick member of the pack, modifying their behaviour in such a way as to assist the invalid. There is the remote possibility, though, that in addition pack members may benefit more selfishly from the presence among them of a sick member who is likely to be the individual which is picked off by predators. The other social carnivores also show a mild form of altruism in allowing (although not helping) non-productive animals (old, sick or lame ones) to feed, but otherwise they do not assist them.

3.5.7 *Further social development*

Long-lasting stable societies of related individuals allow the development of many complex aspects of the social system, some of which have already been touched on. Another concerns individual animals' preferences, choices and 'personalities', and other individuals' responses to them. This is an important topic, but an exceedingly difficult one to study, precisely because the sample size tends to be only one. Even in our own species we do not know why John is fond of Jane, jelly, jazz, boasting, beer or bullying Bob (although we may be able to find a few factors which contribute); nonetheless, the behaviour, preferences and personality of Jane, Bob and the brewer are all influenced to some extent by John's, but it is difficult to determine how much or in what ways. Doubtless many of the same kinds of interactions apply within animal communities. For example, I do not know why lioness A prefers the company of B to that of C or D, and it is probable that many different factors contribute to it. Her preference may well exert an influence on her pride's organisation and even success: females tend to synchronise their reproduction (Bertram 1975), especially with close companions, and this synchrony is followed by improved cub survival. In addition, companionship could perhaps also influence the success of hunting attempts and of territorial confrontations.

Living in a society vastly increases the scope of cultural transmission of behaviour, because it facilitates close contact with a wide range of different individuals. There are more animals to learn from, or to respond to more often. In addition living in a group frequently means that offspring remain with their parents for a much longer period; at least some of the young of the social species of dogs, cats, and mongooses probably stay in close contact with their parents for at least three times as long as the young of comparable solitary species do. During this time a great deal of learning no doubt takes place. Some is of particular feeding skills or methods, such as the 'fishing' for termites by chimpanzees (van Lawick-Goodall 1968), or the washing of food by Japanese macaque monkeys (*Macaca fuscata*) (Kawai 1965), but much is less obvious. Nonetheless, young animals doubtless benefit by observing the hunting methods of adult lions (Schaller 1972), baboons (Strum 1975), and chimpanzees (Teleki 1973); by observing where are likely places to search for food; by observing what foods are suitable and how they should be tackled; by observing the characteristics of predators and how they should be avoided; and by observing the characters of companions and how they should be treated. Having more companions to observe, and for longer, presumably increases enormously the amount which can be learnt directly instead of by individual trial and error. And if those companions are relatives and therefore favourably disposed towards the learners, active teaching can help the latter to learn more and faster. For example Rasa (1977) described an instance of an adult dwarf mongoose apparently teaching a juvenile during an enounter with a snake.

The greater the number of interactions with other individuals, the more signals are needed in those interactions. Both Kleiman (1967) and Kruuk (1975) stressed that there is a tendency for the social carnivores to have a larger range of signals and of signalling characteristics than the solitary species, although there are striking exceptions. With the graded signals characteristic of mammals, the amount of information transferred by a signal is extremely difficult to determine (Kleiman & Eisenberg 1973) (see also Chapter 10). It may be that the social species are better at interpreting rather than at sending signals. Equally, misinterpreting a signal from a social companion may be much less serious than misreading a signal from a hostile neighbour.

3.6 Optimal Group Size

I have referred throughout this chapter only to being and living 'in groups', without specifying the number of individuals in a group. We have seen that there may be certain advantages to an individual in being a member of a group, but these advantages are not necessarily greater for larger groups. For example, as the group's size grows, the competition for food among its members may increase linearly or exponentially, while its ability to withstand predator attacks probably reaches a plateau. If these two selective pressures were the only ones operating on the species, there would be some optimal group size at which the average individual does best, enjoying moderately good protection by detection of predators yet not suffering from too much food competition. What this optimal group size is depends on the relative strengths of the different selective pressures. In practice, however, there are dozens of different selective pressures operating, of differing strengths and in different directions. We cannot yet measure all these selective pressures for any particular species, even if they remained constant, which they do not, so it is still impossible to calculate what is the optimal group size which we would ideally expect natural selection would lead the species to adopt. All we can do is measure the strengths of the selective pressures which we think are likely to be the most important, calculate what would be the optimal group size based on those pressures only, and see to what extent naturally occurring groups are of this size. For example Caraco and

Wolf (1975) found that according to their analysis of Schaller's (1972) data, hunting lions were often in groups of optimal size when hunting gazelles (for maximum meat per lion per hunt), but in supra-optimally sized groups when hunting zebra and wildebeest. In practice natural groups of any species are of a range of different sizes, but we would feel pleased if the mean group size corresponded to our calculated optimal size.

Our pleasure should be shortlived, however. We must remember that the various animals in a group may be subject to different selective pressures; for example males and females may favour very different group size. On the other hand, even within the same sex not all individuals are equal. A dominant male animal may favour a larger group size than would a subordinate, since he suffers less from food competition, benefits more from an abundance of mates, and can better buffer himself against predation by being surrounded by more vulnerable companions: he is also in a much better position than the subordinate to control the size of the group. A certain group size may be favourable for him but not for any of the other members of the group; but if those others' only alternative option is to leave on their own, they may well do better to remain.

If groups are permanent, as we have seen they often are, and are composed of known companions and perhaps relatives, optimal size may be impossible to maintain, even if all individuals in the group did favour the same size. As the animals reproduce, the group grows, and thus is above optimal. If it were to split in two, the halves would each be far below optimal, and all individuals might do less well than in a single group of supra-optimal size. To expel only the excess portion and thus bring the main group down to optimal size would be an alternative possibility; but if the animals expelled are offspring or other close relatives, the remaining animals' inclusive fitness may well be reduced by condemning them to the hardships of life in a group which is far below optimal size.

Another complication, besetting small groups particularly, is chance fluctuations in size due to accidental causes. If it is much more harmful to the group members for the group to be below than above optimal size, the best safe strategy would be to have the group usually above what would otherwise be the optimal size. This of course merely adds yet another stage of complexity and refinement to the definition of 'optimal'.

The concept of optimal group size thus has most relevance in species where groups are temporary associations, probably for feeding. In truly social species, the size of the group is probably determined less by ecological factors and more by within-group interactions.

Chapter 4
The Economics of Insect Sociality

BERND HEINRICH

4.1 Introduction

When compared with vertebrate systems there are certain peculiarities that must be kept in mind when considering the role of economics in the functioning and possible evolution of insect societies. In vertebrate animals co-operation and sociality are based in large part on economics of group living and resource defence. In insects, however, relatedness and possibly 'parental manipulation' are additionally important factors in the evolution of sociality that presumably differ in relative importance from one instance to another (see discussion by Evans 1977).* A female's reproductive output could be enormously enhanced when she is able to harness the aid of co-workers or a portion of her offspring to help rear later offspring to sexual adulthood. However, there are several pre-suppositions before this route(s) to sociality becomes feasible: (1) there must be food for the young; (2) the young (or co-workers) must remain close enough to the mother to be manipulated, as they do after parental care is established; (3) the young (or co-workers) must tolerate or encourage the manipulation, presumably for some benefit they gain from it.

It is not intuitively obvious why offspring or other individuals would allow themselves to be manipulated at the expense of their own reproductive output. However, a genetic predisposition that could ease resistance to manipulation, provided the advantages of sociality are great enough, is that the offspring share many genes with each other. Thus, by aiding siblings to reproduce (aiding their mother's reproductive output) they are passing on many of the same genes indirectly rather than through themselves. As pointed out rigorously by Hamilton (1972) the genetic predisposition of aiding siblings should be particularly great in the Hymenoptera where, because of haploidy in males,

*This is discussed in more detail in section 1.5.1.

sisters share on the average $\frac{3}{4}$ of their genes, whereas they would share on the average only $\frac{1}{2}$ with possible offspring. It has therefore been suggested that sisters could gain more inclusive fitness (genetic investment to future generations) by care of their younger sisters than by equal amount of investment to produce offspring of their own (see 1.5). (In Hymenoptera the males leave the nest and do not take part in aiding the nest economy.) If this is not sufficient cause for sociality then it should at least reduce the evolutionary resistance to it.

In termites, with diploid males, colony mates share on the average only $\frac{1}{2}$ their genes with each other, as they do with potential offspring. Thus there is no genetic predisposition for an offspring to sacrifice its reproductive output for that of a sibling. However, this has obviously not been a barrier to sociality in this group of organisms. Possibly the opportunities for parental manipulation have been greater than in Hymenoptera, or economics and defence have assumed more significance. Being able to utilize wood, the roach-like ancestors of termites were presumably restricted to localized and nearly unlimited amounts of food, where dispersal of the young to find new food was not necessary, thus providing more opportunity for parental manipulation.

There are both economic costs and benefits associated with sociality. In the aggregation of many individuals at one locality, the need for resources increases linearly with group size. However, the availability of resources in any one area is finite. As a consequence, in an insect colony as well as in any other social aggregation, there are limits on the rate of resource availability, on group size, or both. These limits have been subject to natural selection through mechanisms of resource acquisition and processing.

Insect societies have evolved to specialize on a wide variety of resources, and they have evolved various 'strategies' of harvesting them. (The term 'strategy' implies purpose, but as here used it will denote only a set of evolved adaptations that acts in concert and aids survival.) Indeed, when examined closely, a wide variety of behaviour of the social insects appears to have direct bearing on their foraging strategies, and hence their energy economy.

The purpose of this chapter is to review selected aspects of behaviours of some social insects relative to foraging and energy balance. Each of the possible insect groups that could be examined has, in reality, multiple adaptations related to foraging and energy balance. However, in order to show some of the diversity within the group as a whole I have arbitrarily restricted the discussion to those features that appeared conspicuous, and possibly characteristic, for selected taxa, rather than discussing the relative contribution of various strategies found in any one particular species.

4.2 Bees

4.2.1 *Stingless bees—the aggressive strategy*

One of the means of securing limited resources is to fight for them. But fighting for ungathered food resources is by no means a universal phenomenon. Most social bees are well known to be highly aggressive in defending their nests containing stored resources and young, but they generally do not interfere with other foragers at most natural food sources. However, a well-documented exception is found in some species of stingless bees (*Trigona* spp.) in Costa Rica (Johnson 1974, Johnson & Hubbell 1974). This work provides a good example of the possible selective pressures for and against aggression. The prime requisite for aggression is that it is successful, and the basic question then becomes: Under what circumstances can aggression be successful, and what characteristics does the social insect require to carry it out?

At a site in Turrialba, Costa Rica, five species of stingless bees were found to co-exist. Interactions of the bees were observed at artificial baits (sucrose solutions) and at natural food sources (flowers). The largest of the five bees, *Trigona silvestriana*, was the most aggressive. These bees always displaced the other species from bait when it was held by fewer than approximately 50–60 bees. One pair of grappling bees could cause 40–50 bees to suddenly leave the bait, possibly in response to alarm pheromone release.

The aggression was most intense between colonies of the same species. After a 2-day battle between three colonies of *T. corvina*, at an experimental grid of 25 baits, 1,812 dead bees were found. Most of the dead bees were found near the baits of highest sucrose concentration (2·5 molar). After two days each colony exclusively controlled a set of quality baits (high sucrose) from the grid of 25 with varying rewards.

Aggression occurred also at natural food sources. Under one pollen-laden *Bactrix* palm inflorescence open for 2 hours, 44 dead bees were found: 5 *T. silvestriana*, 26 *T. corvina*, two pairs of interlocked *T. corvina*, and 3 *T. silvestriana* attached to a total of 6 dead *T. corvina*. *Trigona silvestriana* occupied the same banana inflorescences in August 1971 and in February 1972, defending them against attacking bees, wasps, and other insects.

The presence of a more aggressive species significantly reduced the feeding durations of the less aggressive. In one contest a 2·4 molar bait between *T. silvestriana* and *T. testacea*, the *T. testacea* spent on the average 5 seconds hovering and 11 seconds feeding in the absence of *T. silvestriana*, but after *T. silvestriana* had started to utilize the same food source, the *T. testacea* spent about 11 seconds hovering and only 3 seconds feeding. In an encounter lasting 11 days between *T. silvestriana* and the less aggressive *T. latitarsus*, *T. silvestriana* partially excluded *T. latitarsus* from the preferred 0·8 and 2·4 molar baits. Since the more aggressive species harvest higher quality resources and excluded less aggressive species from them, it could be asked why do the less aggressive persist through evolutionary time?

Coexistence between the aggressive and non-aggressive species could involve several mechanisms. First, the small less aggressive species may specialize on small widely-spaced plants providing relatively slight amounts of pollen and nectar. Presumably they could make an energy profit at less rewarding food that is inadequate to large bees (Heinrich & Raven 1972). The aggression of the large bees is presumably only profitable at high quality food sources (compact and in high concentrations). With the availability of such resources expected in the tropics because of many high-energy pollinators, the exploitation by way of aggression should, in turn, select far greater body size in aggression.

Rather than utilizing only low-quality food resources, some bees, like *T. fulviventris*, employ an opportunistic strategy. *Trigona fulviventris* are non-specialists, utilizing many kinds of flowers. They were always the first bees to discover baits and they were very investigative of novel objects in their environment. By foraging in dispersed rather than in a grouped manner, and by having a large worker force 'uncommitted' to any particular resource, many workers of these bees are available for discovery of new food resources, and for recruitment to them. They find and gather the resource before an aggressive species finds it and drives them away. *Trigona silvestriana* had a similar opportunistic strategy, except that they were also aggressive.

Group vs. individual foraging is of prime importance in food partitioning based on aggression. It has been shown, for example, that *T. fulviventris*, which forages as individuals or in small groups, is restricted to small clumps of flowers of *Cassia* in an area where a group forager, *T. fuscipennis*, occurs (Johnson & Hubbell 1975). The *T. fuscipennis* controls the Cassia plants with large numbers of flowers apparently by driving the single foragers of other species away. The *T. fuscipennis* probably recruits and marks its exclusive foraging areas with pheromone. The pheromone marking by *T. fuscipennis* greatly increases the likelihood of *T. fuscipennis* visitation. It appears that the pheromone marks create the 'boundaries' of resources with 'fuzzy' edges, marking the resource space to be defended.

The aggressive species owe their success not only to large body and colony size, but also to efficient recruitment. The less aggressive species probably owe their competitive edge to a greater ability to find new food sources. Possibly the ability to bring large numbers of recruits to monopolize a good food source relies on having a reserve of 'unoccupied' bees in the hive. But the ability to utilize many food resources simultaneously may depend on having many bees searching in the field, at the expense of potential recruits remaining in the hive.

The potential advantage of efficient recruitment is great. For example, at high quality baits occupied by many potential competitors, lone bees, such as scouts, are frequently excluded by being repelled or killed. But successful scouts of *T. mexicana*, for example, may return with large numbers of recruits arriving 'in a small cloud' (Fig. 4.1), and by their combined efforts dislodge the competitors from superior bait.

Clearly, the aggressive strategy of the *Trigona* bees is advantageous at rich and highly-clumped food, but it is generally only possible in bees that can efficiently recruit large numbers of nestmates that can inflict costly damage on their competitors.

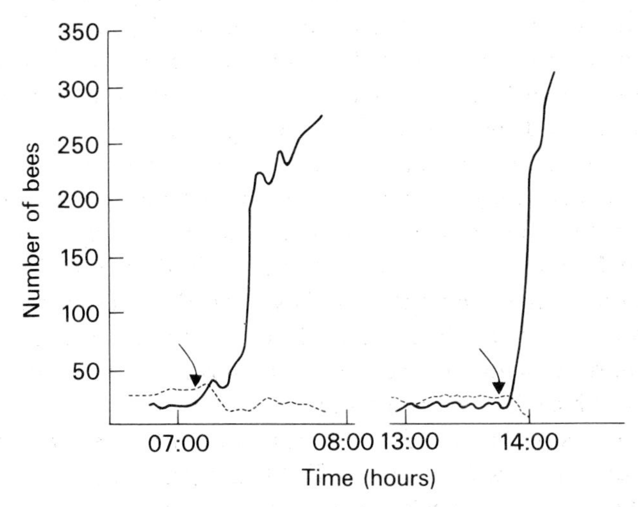

Fig. 4.1. Two examples of group arrival of *Trigona mexicana* (———) to baits occupied by *T. fulviventris* (. . . . .) (from Johnson 1974).

4.2.2 *Bumblebees—individual specialization*

In the north-temperate regions most plants bloom in synchronous bursts. Flowers tend to be distributed over large areas, and the food rewards per flower are generally minute. What are some of the strategies for harvesting these resources?

An alternate strategy to fighting is to allocate as much energy as possible directly into the acquisition of resources, without attempting to repel competitors. The example we will explore here is that of social bumblebees, *Bombus* spp. In these bees the emphasis in competition for limiting resources is exerted at the level of foraging skill, acquired through learning and specialization. Genetically fixed flower specialization found in many solitary bees (Linsley 1958) would obviously not work for social bees that require a continued food input to the hive. Social bees rear successive batches of young throughout the growing season, and since a succession of different plants will have bloomed the colony cannot base its entire food economy on any one of them.

Bumblebees are unable to recruit foragers to a point source of food. As a consequence of this lack in recruitment ability they cannot over-

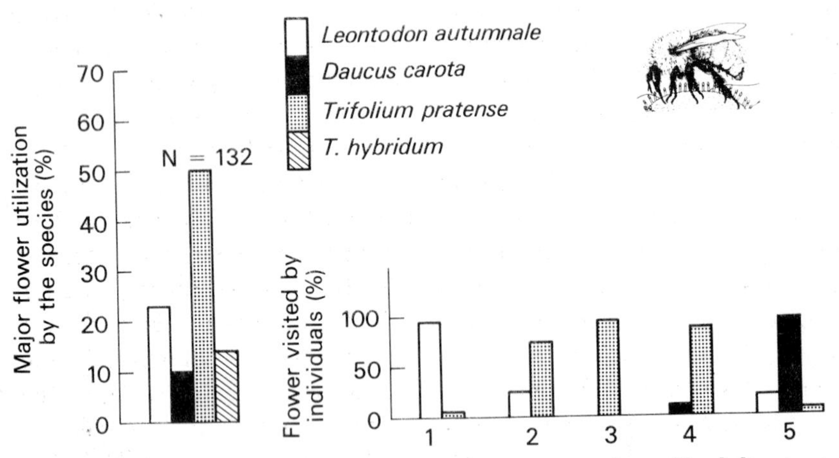

Fig. 4.2. Flower specialization by *Bombus vagans* workers. The left set of bars shows the percent of workers of the species specializing on four different intermingled and concurrently blooming flowers, indicating apparent species-preferences. The right set of bars shows the percent of the different kinds of flowers visited by 5 individuals, indicating individual preferences independent of species preferences (from Heinrich 1976).

power other bees and rob their nests, nor can they hold and defend feeding territories like stingless bees. As far as we know each forager must individually determine which are the best flowers at any one time. But for the colony as a whole it is also important to be able to track the changing resources through time.

Individual bumblebees specialize (Fig. 4.2). Any one colony usually has individuals specializing from different kinds of flowers (Brian 1952). As a result, a colony is simultaneously tapping the energy from a wide range of sources. Specialization aids foraging success, particularly at many of the morphologically complex flowers. For example, to collect pollen from *Solanum dulcamara*, the bees grasp the flowers with mandibles, vibrate them by contractions of the thoracic muscles, and shake the pollen from the tubular anthers onto the ventral surface of the abdomen, from where it is transferred into the corbiculae. While collecting pollen from wild rose, the bees grasp groups of anthers in the flat cup-shaped flowers, shake loose pollen, grasp another group of anthers, etc. and scrape the pollen from the body hairs. Collecting nectar from *Aconitum* requires crawling into the interior of the flower past the anthers and taking the nectar from the tips of two modified petals modified as nectar receptacles. There is a wide variety of flower morphology requiring specialized movement for nectar/pollen collecting (Fig. 4.3). Individuals utilizing the complex flowers for the first time are often clumsy and inefficient, or they may not find the food rewards at all. Foraging efficiency may range from 0%—such as in bees attempting to collect nectar from flowers that provide only pollen—to 100%, an arbitrary assigned efficiency of experienced bees that gather the appropriate reward in the minimum time (Heinrich 1976).

Young 2–3 days old *Bombus vagans*, initiating foraging, have been observed visiting on the average 5 different kinds of flowers on their first 2–3 foraging trips, but within a day or two they become restricted to those areas and to those flowers offering the most net profit (Heinrich unpublished). However, following the depletion of these 'best' flowers they again sampled and sometimes switched to other flowers. As the original high-reward flowers are being depleted by many bees, the rewards *available* become roughly comparable to normally less rewarding but previously unutilized flowers (Fig. 4.4). Thus, at high bee populations the low food reward flowers are also utilized by some bees that have switched from the depleted flowers, and by new bees which have sampled the reward spectrum of the available flowers for the first time.

The kind of flowers visited are in part related also to the bee's

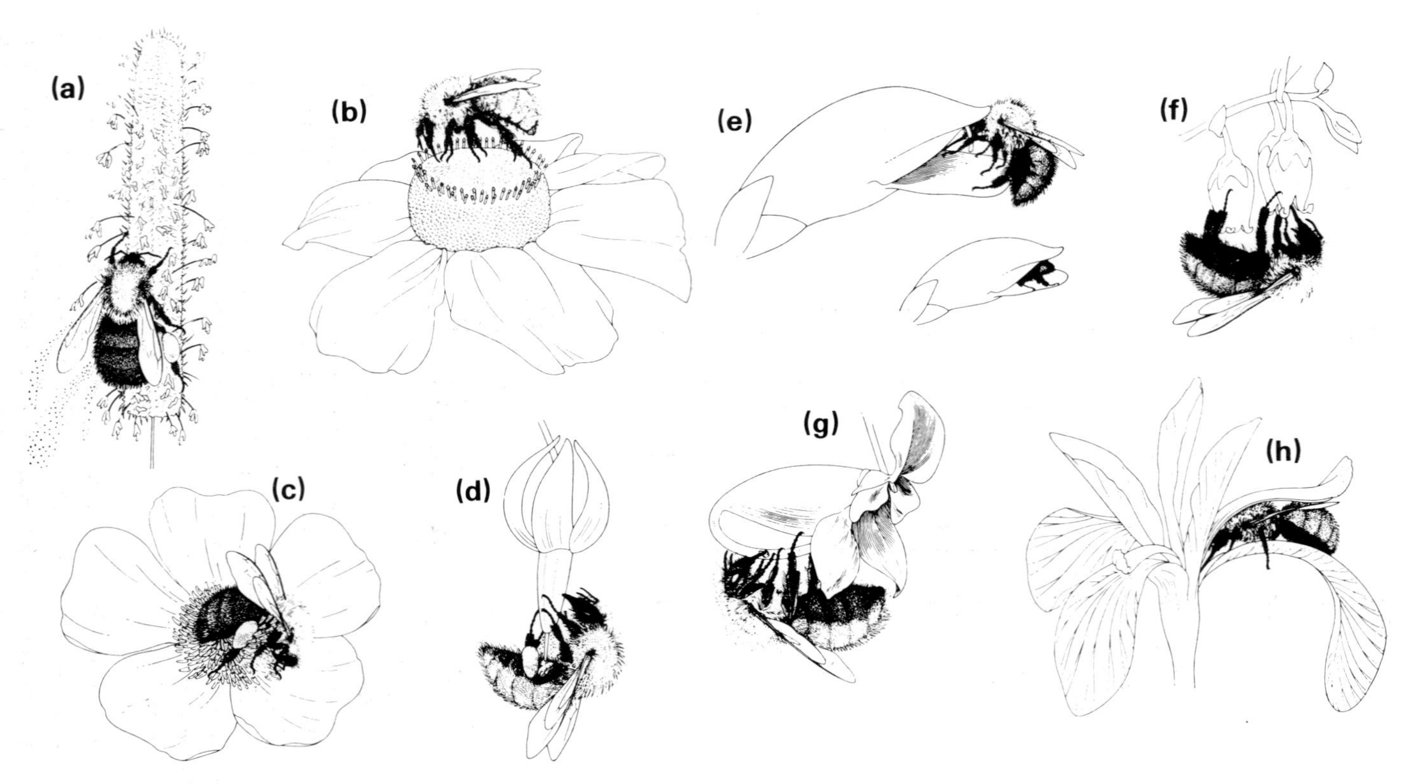

Fig. 4.3. Behavioural differences of bumblebees specializing to collect pollen (A, B, C, D, E, F) and/or nectar (B, E, G, F, H) from morphologically different kinds of flowers (from Heinrich 1976).

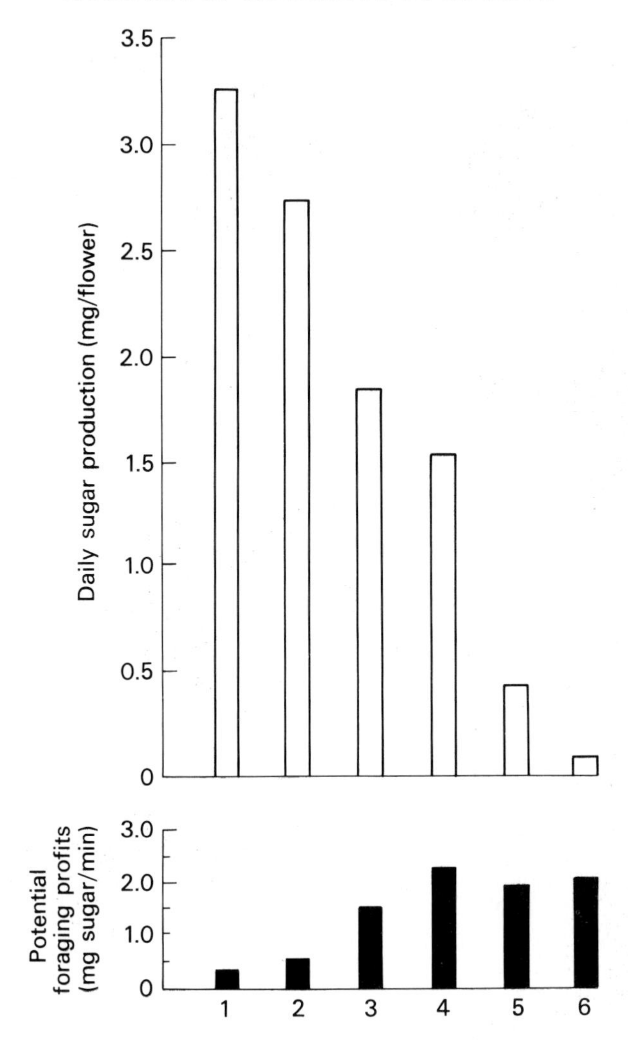

Fig. 4.4. Relationship between daily sugar production of individual flowers (open bars) and net calculated sugar rewards available per unit time in flowers left open to the foragers, which were primarily bumblebees (solid bars). In part due to equilibration of nectar supplies by depletion, some of the low-reward flowers come to yield as much or more as the high nectar producers (from Heinrich 1976). 1 = *Chelone glabra*; 2 = *Impatiens biflora*; 3 = *Epilobium angustifolium*; 4 = *Robinia pseudo-acacia*; 5 = *Rhododendron canadense*; 6 = *Prunella vulgaris*.

genetic pre-conditioning. For example, in laboratory studies the bumblebees *Bombus terricola* became $>90\%$ flower-constant to blue flowers within 50 flower visits, each rewarded with 1 μl 50% sucrose, while the same constancy to white required at least 250 flower visits (Heinrich *et al.* 1977). Genetic preconditioning to blue might be adaptive since many flowers with appropriate food rewards (adapted for bee pollination) are coloured blue.

Both in the field (Heinrich 1976) and in the laboratory (Heinrich *et al.* 1977) conditioned individuals tend to become resistant, but not incapable, of switching to new flowers. Conditioned individuals continue to visit flowers that offer relatively low rewards in comparison to other more highly rewarding available flowers. Possibly they improve in foraging at their specialties while making little profit if forced to switch to a new morphologically complex flower. At the present time no direct quantitative data are available on the net foraging gains per unit time, so we can only speculate and make inferences.

One possible mechanism of keeping track of several resources is multiple specialization. Most bumblebee foragers eventually specialize and utilize primarily one kind of flower, their 'major'. Besides having a 'major' specialty, however, they often have one or more 'minor' specialties, at least where several species are flowering concurrently. When the minor flowers become more remunerative, or when the major flowers become scarce, the bees readily switch and forage primarily from their second specialty flower (Heinrich 1976). By mathematical modelling (Oster & Heinrich 1976) we have shown that the pure strategy of 'majoring' is always better than random foraging if the reward structure remains constant through time. 'Minoring' is a necessary compromise required to track resources changing through time. However, recent experiments (Heinrich unpublished) show that the amount of 'sampling', and the number of 'minors' are functions of the duration that the bees have been foragers, as well as of the relative food rewards between major and minor flowers. As a consequence, the bumblebee's foraging behaviour allows for both the benefits of specialization and the rapid response to changing resources on the basis of individual forager initiative (see also 2.5).

Competition in bumblebees for food resources is primarily an exploitation (scramble) rather than an interference (contest) competition. It is exerted by way of foraging 'skill'. In part, foraging skill at different flowers depends on the main foraging 'tool'—the tongue. For example, the longer the average tongue-length of workers of different species, the more clover flowers per unit time they visit (Holm 1966), and the more rewards are presumably collected. Similarly, long-tongued bees require much less time at each flower of *Delphinium barbeyi*, having hidden nectar, than short-tongued bees (Inouye 1976).

Thus they visit more flowers per unit time and achieve a greater rate of food harvesting on these particular flowers than their short-tongued competitors.

The bees seldom if ever have highly clumped and rich food resources available in the natural environment. They often visit hundreds of flowers on each foraging trip, and these flowers are distributed over areas often over 500 m^2 (Heinrich 1976) that are also utilized by many other individuals. Not being able to recruit, no colony can control a foraging area, and large numbers of bees from many different colonies may forage from the same flowers. In such a situation it is disadvantageous to relinquish foraging activity to chase and attack other individuals. Indeed, bumblebees at flowers appear to be totally absorbed in foraging, never taking time and energy to chase other individuals.

Despite the fact that bumblebees forage earlier as well as late in the day, and at a much more rapid rate than honeybees, they accumulate generally only enough food reserves to tide them overnight or over a few days of bad weather. Their food intake is 'immediately' converted to offspring. In contrast to honeybees, the colonies are temporary, ending in the autumn. Thus no food stores for the winter need to be collected, and all of the resources are channelled instead to produce a large crop of new queens and males at the end of the colony cycle.

In summary, in bumblebees we observe a foraging strategy whose success is based on individual initiative of the foragers. Foragers assess the state of nature individually, and by specializing on those flowers where they make the most profits the different individuals tap the most rewarding food resources available, which promotes the hive economy.

4.2.3 *Honeybees—communication*

Honeybees, like stingless bees, cannot be easily pigeonholed as using any one foraging strategy. They rely on communication as well as individual specialization, and they are able to efficiently exploit point sources of rich food, as well as diffusely distributed resources. I shall briefly discuss some aspects of their flower specialization, and then relate their foraging behaviour to their well-known capacity to communicate.

Although flower generalists as species, the individual honeybee foragers, as well as workers of most other social bees, are relatively flower constant. As already recorded by Aristotle, individual bees tend to restrict themselves to specific kinds of flowers. Charles Darwin presumed the adaptive significance of this behaviour is probably in improving foraging skill at different flowers that vary widely in morphology. This idea still holds. Weaver (1957) noted large individual differences in the foraging speed and in the method of flower manipulation of individual honeybees foraging from vetch and concluded that these differences were in part due to experience. Honeybees are, of course, well known to learn to associate food with specific signals such as scents, colours and geometric patterns (reviewed by von Frisch 1967).

The learning of specific signals can be very rapid. Learning speed is partially signal-dependent, and the bees become 85% flower-constant to violet signals after being rewarded with only one honey-stomach load of 2 molar sucrose solution (Menzel 1968). Bees in the field generally restrict their foraging to one specific area where they forage from one kind of flower.

Although individual honeybees specialize the colony as a whole can respond rapidly to changing resources. The colony can also mobilize most of its worker force to work on one crop, provided it yields the best food rewards at the time. The colony's flexibility to assess the resource environment and to specialize at the most rewarding food sources is due to its scouts and its communication system (see review by Gould 1976). Scout bees are that small percentage of investigative 'non-conformist' individuals that explore new potential resources. Having discovered a profitable food source they become foragers and begin to dance in the hive.

As observed by Oettingen-Spielberg (1949) and reviewed by von Frisch (1967) a normal colony in the spring has about 1000 new young bees ready to assume foraging duty each day. Of these only a vanishingly few fly forth as scouts to discover new sources of food (Oettingen-Spielberg 1949). By far the majority wait in the hive. After completing 'Innendienst' or hive duties, bees generally aged a month or more (Oettingen-Spielberg 1949, Lindauer 1952) are ready to assume foraging and assemble on the 'dance floor' where they are solicited to forage at some definite goal. These new bees follow all dances that come in their way, but the experienced foragers are recruited either by scent of their specialty flowers alone, or they assemble on the dance floor to be recruited by hive mates at those times when their specialty is yielding food (Körner 1939). They are generally not recruited by competing guilds foraging from other flowers. Lindauer (1952), observing individually marked young bees, found that in 79 out of 91 successful first foraging trips the recruitees returned with the same kind of harvest that the recruiters had collected, and 41 of them danced, indicating approximately the same distance and direction as the recruiters they had followed previously.

Ordinarily different scouts are soliciting foragers for different food resources. How are decisions made regarding the hive's foraging effort? First, some confusion is avoided in that the number of scouts is small. Secondly, scouts generally do not recruit until after making several foraging trips, unless the food source is obviously of high quality. Only a small percentage of the foragers dance. Thirdly, the dances themselves by their vigour indicate overall foraging profitability, in addition to distance and direction, thus providing a threshold that may or may not be sufficient to alert given new bee recruits. Fourth, the recruitees are given food samples by the dancer from which they can presumably gauge not only the kind of food source by its scent, but also its quality by sugar concentration. Fifth, as shown by Boch (1956), readiness to dance of one group of bees depends on whether or not another one is finding something the bees judge better. 'Tanzlust' depends on hive demand. Foragers get rid of their nectar (but not pollen) by regurgitating it to receiving bees in the hive. The receiving bees indicate hive demand by taking nectar from those returning foragers that have the most concentrated (high quality) nectar rather than from those that bring dilute nectar (unless the colony is in need of water to prevent overheating at high air temperatures). When returning foragers find no takers for their collected nectar they stop foraging and further dancing, thus it occurs that one colony may end up specializing on one kind of flower. Undoubtedly, however, the actual processes are somewhat more complicated than the above generalizations might indicate.

In summary, the strategy of the honeybee permits all the advantages of individual specialization, combined with the ability to respond quickly to temporally changing reward availability. The hive acts as an infomation centre (see 3.3.1). Rapid recruitment by the aid of scouts facilitates the utilization of new floral resources when old ones become depleted or when new and better ones become available.

4.3 Ants

4.3.1 *Army ants—group predation*

Foraging in ants involves a diversity of apparently co-operative actions, varying both with the type of resource and with its distribution. For example, in some honeydew collecting ants small individuals may act as gatherers that regurgitate their crop contents to larger 'tankers' that carry the honeydew back to the nest (Sudd 1967). In *Formica rufa* two workers carry house-fly size prey significantly faster back to the nest than individuals (Chauvin 1950), and in *Pheidole crassinoda* if individuals cannot move large prey they recruit helpers from the nest that augment their dragging power (Sudd 1960). In ants, the evolution of a chemical language—the recruitment pheromones—allows for considerable flexibility in foraging response to variations in the amount and distribution of potential food.

A highly evolved strategy of foraging in some carnivorous ants is that of raiding in large groups with the aid of pheromones affecting group cohesion and group response. The most well-known of these include 'driver ants' of the genus *Anomma* in Africa. A colony may

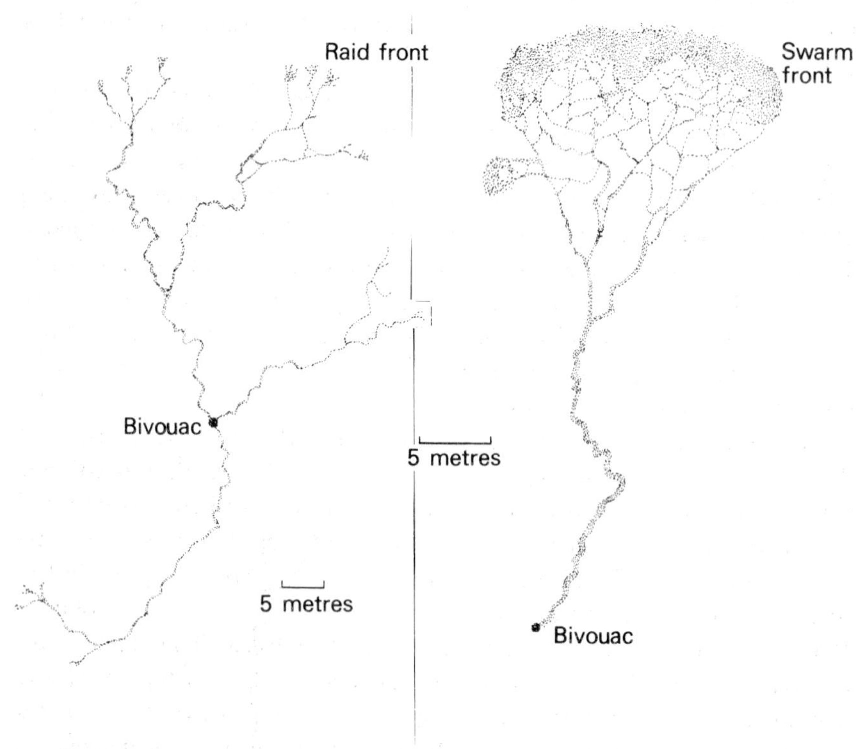

Fig. 4.5. Raiding patterns of army ants. Left: Column raiding by *Eciton hamatum*. Right: Swarm raiding by *Eciton burchelli* (from Rettenmeyer 1963).

consist of 20 million individuals, and daily raids of several million workers issue from the nest pseudopodium-like, surrounding and engulfing prey, and then draining back to the nest. Because of sheer numbers, and sharp cutting mouthparts, the ants can overpower and dismember not only large insects and insect colonies, but also stationary vertebrate prey, like nestling birds and young mammals.

The New World counterparts, the 'army ants' of Central and South America, principally *Eciton*, similarly overpower prey not normally available to small groups. Chadab and Rettenmeyer (1975) report that in *E. hamatum* the most common prey are colonies of other insects and arthropods larger than themselves. Small numbers of attacking ants were thrown off wasp nests, but the wasps surrendered their brood while being attacked by large numbers of ants. Consequently, massive and swift attacks are advantageous, and *Eciton hamatum* has one of the most efficient recruitment capabilities of any ants so far studied. Thirty seconds after an ant returning from a wasp nest bait traversed the 20–30 cm to a nearby raiding column, 50–100 ants were recruited and a continuous stream of ants was established almost immediately (Chadab & Rettenmeyer 1975).

Two types of raiding have been recognized in *Eciton*, 'swarm raiding' exemplified by *E. burchelli*, and 'column raiding' by *E. hamatum* (Fig. 4.5). In the raiding 'swarms' of *E. burchelli* masses of ants 15 m to 25 m wide and 1–2 m in depth typically advance steadily broadside, covering the forest floor as a great net, depleting the arthropods in their path. Individuals advance a few cm at a time and turn back, laying down scent, while others advance the front still further. The 'swarm' may advance at about 20 m/hr. The more typical hunting mode of many species is that of column raiding, where the raid advances simultaneously along many odour trails.

What are the advantages of group as opposed to individual raiding? There can be no doubt that in the first line both swarm and column raiding aids the ants in overpowering large prey and in utilizing the brood of wasp and other insect colonies that would otherwise not be available to them. Column raiding is presumably most adaptive in the utilization of highly clumped resources such as other insect colonies that can only be taken by massive and swift attacks. The assaults are greatly enhanced by massive and rapid recruitment, as shown by the column raider *E. hamatum* utilizing clumped prey (Fig. 4.6). Swarm raiding, on the other hand, should be a particularly efficient mechanism for capturing small highly mobile prey. A small running beetle, for example, may easily outrun most individual ants, but if confronted

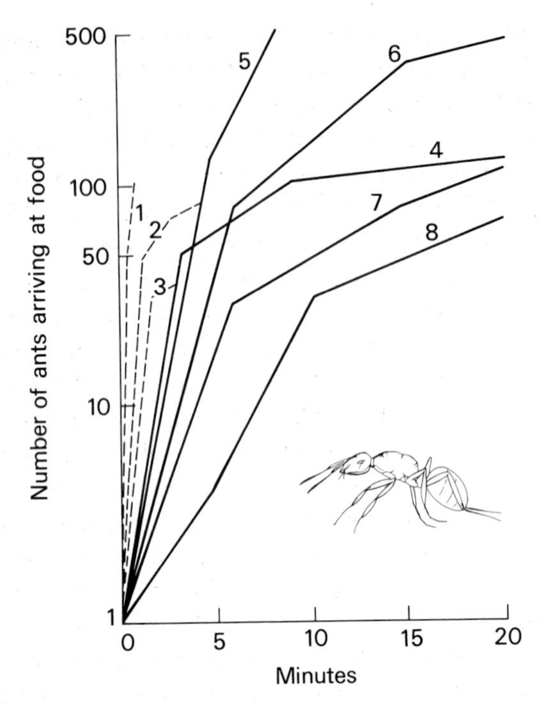

Fig. 4.6. Comparison of recruitment responses of six species of ant mass recruiters (those orienting primarily to scent of trail). Dashed lines (1, 2, 3) *Eciton hamatum*; 4, *Formica fusca*; 5, *Solonopsis invicta*; 6, *Myrmica americana*; 7, *Pogonomyrmex badius*; 8, *Crematogaster ashmeadi*. (Adapted from Chadab & Rettenmeyer 1975).

with a swarm of raiders, the insect attempting to escape from one ant would have a good chance of running into the receptive mandibles of others. Swarm raiders, being less reliant on recruitment and odour trails, take a much wider range of booty (Schneirla & Topoff 1971). Their foraging bands appear to constitute a 'dragnet' that indiscriminately rounds up all stray arthropods in the path of the steadily advancing front.

Army ant colonies are thinly distributed in their environment (Scheirla & Topoff 1971), and almost nothing is known about inter-colony interactions. There is no direct evidence that they avoid going into areas that have just been depleted by other raids. However, their mechanisms of raiding should reduce time and energy investment for foraging in such areas. Ants recruited by columns that move into areas where food is being found, would automatically tend to be deflected from poorly rewarding areas. Nevertheless, repetitive hunting raids by millions of individuals in any one area, on probably relatively thinly distributed prey (relative to colony food requirements) would expectedly lead to the local depletion of food. For example, it may take many months for new wasp nests to become established and to grow to large size. However, the ants have evolved a unique set of behavioural and physiological traits that permit periodic emigration, or nomadism. This has been studied in greatest detail in *E. burchelli* (Schneirla & Topoff 1971, Rettenmeyer 1963, reviewed by Wilson 1971).

The basic pattern in nomadism involves an endogenous cycling of 'statary' and 'nomadic' phases closely tied with the reproductive cycle. During the statary phase the ants cluster or camp in the same bivouac site for 2–3 weeks from which trails issue periodically. During this time the ant cluster contains the pupae, and near the middle of the statary period the queen lays within the cluster in several days, a batch of 100,000 to 300,000 eggs. Some short while after the larvae emerge the pupae from the previous synchronously produced brood eclose. The sudden presence of the many thousand of callow new workers stimulates the colony to emigrate and the nomadic phase begins. This phase is also of 2–3 weeks duration, and it terminates when the larvae stop feeding and pupate. During the nomadic phase the ants bivouac at a different site after each day's raid, thus their forays each day cover new territory.

4.3.2 *Harvester ants—seed foraging strategies*

Numerous species of ants (*Pogonomyrmex*, *Veromessor*, and others), commonly called 'harvester ants', are highly specialized behaviourally for the harvesting and utilization of seeds in arid regions. Their foraging strategies are of interest because the temporal and spatial abundance of the seeds varies greatly. Do the ants have different strategies adapted to different seed densities?

A recent detailed study of *Pogonomyrmex barbatus*, *P. rugosus*, and *P. maricopa* (Hölldobler 1976b) in the mesquite-acacia desert of Arizona and New Mexico explored species differences. It leaves no doubt that, contrary to casual observations in the past, *Pogonomyrmex* ants can recruit. Scout ants, after discovering a new rich food source such as a new seed-fall or a freshly killed coakroach, return to the nest dragging their extruded sting on the ground and depositing a recruitment pheromone from the poison gland (Hölldobler & Wilson 1970, Hölldobler 1976b).

Two of the species, *P. barbatus* and *P. rugosus*, utilize very similar food resources. These highly territorial species partition the foraging ground between competing inter- and intra-specific colonies by trunk trails extending to 40 m or more from the nest (colonies are generally only 18–19 m apart). Foragers leave the trunk trails on individual foraging sorties at some distance from the nest, and return to the trunk trail at almost the same spot where they left it. The relatively permanent trunk trails survive heavy rainfalls, and they are derived from recruitment trails. Because of the trunk trails the ants spatially partition their food resources, and mass confrontations between neighbouring colonies are avoided. The trunk trails allow the ants to respond rapidly to patchy food resources at long distances from the nest. When two colonies simultaneously utilize the same patch of seed, the colony that is closest to its nest, or to one of its trunk trails, tends to recruit more foragers and displace its competitors. Each colony has several trunk trails, and although given foragers generally remain associated with given trunk trails, they can be directed from one to another. The trunk trail strategy clearly extends the foraging range in these ants, and it improves the harvesting ability on rich and patchy resources.

One of the other sympatric *Pogonomyrmex* studies by Hölldobler (1976b) in Arizona, *P. maricopa*, exhibited a different foraging strategy. Unlike *P. barbatus* and *P. rugosus*, this species did not leave the nest in well-established trails before diverging on individual excursions. Individual foragers generally left the nest in all directions and *P. maricopa* showed relatively low recruitment intensity. The nests of these ants were dispersed relatively widely, possibly because foraging areas in this species were not partitioned by trunk trails.

The significance of the divergent foraging strategy of *P. maricopa* relative to *P. barbatus* and *P. rugosus* is apparently related to its food resources. *P. maricopa* is a food specialist, preferring small seeds that were little contested by the other two species. Individual *P. maricopa* foragers range widely, being guided by landmarks and a well-developed time-compensated astromenotaxis. These ants probably utilize food that is dispersed rather than clumped. There is little advantage to recruit when individual initiative to locate and harvest the small dispersed seeds is at a premium.

A previous study of three species of seed-eating harvester ants in the Mojave Desert of the southwestern United States addressed itself to behavioural plasticity and evolved responses to varying seed densities and distributions (Bernstein 1971, 1975). *Pogonomyrmex californicus* and *P. rugosus* (both from high elevations in the Mojave where vegetation and presumably seed densities are relatively high) were found to use exclusively the individual foraging method. These ants were reported to wander about until they find a seed, to then orient visually and return straight to the nest where they deposited the seed and then returned directly to the vicinity of where the seed was found to search for another seed. Thus individuals would be generally consistent in the direction and distance they travelled from the nest, but every individual would forage independently of the others.

The individual strategy of *Pogonomyrmex* at relatively high (presumed) seed densities is probably adaptive if the ant would *always* find seed no matter in what direction it travelled, as when the seeds are relatively *evenly* distributed about the nest. Bernstein's observations on *P. rugosus* conflict with those subsequently made by Hölldobler (1976b) on the same species. Hölldobler observed trunk trail foraging and recruitment to piles of seed. Since recruitment in this species is now solidly established it must be concluded that the lack of recruitment seen in the first study does not imply the lack of this behaviour under appropriate circumstances, such as amply clumped resources.

The third seed-eating species, *Veromessor pergandei*, was observed by Bernstein (1975) to be both a group and an individual forager. The ants used the individual foraging method during the brief time of high seed density. However, at low elevations (where it is realtively dry and vegetation and seeds are sparse), except at time of maximum seed abundance, all foragers left the nest in one direction in one continuous column. The ants fanned out at the distal end of the column, where they foraged individually. The ants appeared to forage over the area in a systematic fashion: on consecutive days the foraging columns rotated about the nest (either clockwise or counter-clockwise), presumably in the direction of increasing food density. The average amount of rotation per day varied from 14° to 71°, and the amount of rotation appeared to increase with decreasing food density. Unfortunately, the precise mechanics of column rotation were not reported.

Why should the column-sweep foraging method be advantageous at low seed densities? As mentioned previously, *P. maricopa*, a food specialist, forages singly for dispersed specialized seeds and generally gains little advantage from recruitment. But *V. pergandei* forages cooperatively at low seed densities (presumed). Possibly the column-sweep method is a still more effective method than individual (as opposed to recruit) foraging because it reduces the incidence of individuals traversing ground recently cleared by others (see 3.3.1). If so then *V. pergandei* should be able to outcompete *P. maricopa* for food.

In another study on the foraging behaviour of *Veromessor pergandei* in Death Valley, California, Rissing and Wheeler (1976) concluded, like Bernstein, that these ants have a highly flexible foraging behaviour.

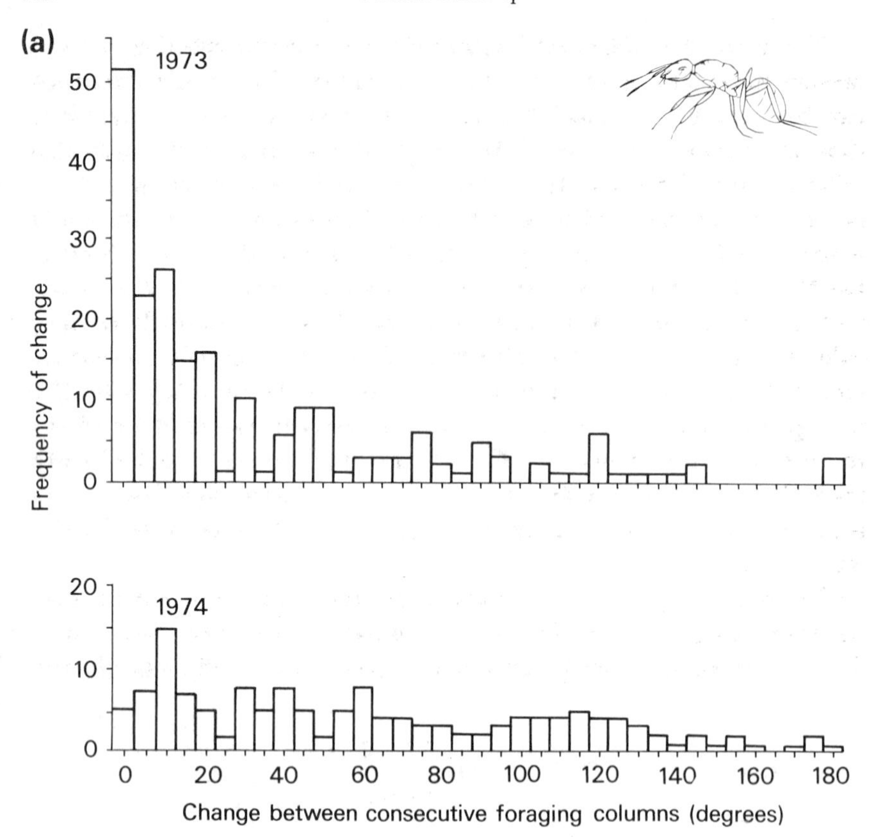

Fig. 4.7. (a) The frequency of changes in degrees between consectuive foraging columns of *Veromessor pergandei* during 1973 (combined data from 4 nests) when ample rains resulted in large seed-falls, and in 1974 (combined data from 3 nests), when seed production was much less (from Rissing & Wheeler 1976).

In response to diminished annual seed production the ants decreased collecting the scarcer seeds, and collected 'less desirable' seed as well as non-seed plant material. In contrast to Bernstein, however, Rissing and Wheeler observed column foraging even when seed production was unusually high (Fig. 4.7a). During 1973, when there had been rain and much seed production, about $\frac{1}{2}$ of the foraging columns on consecutive days varied by 10° or less. On the other hand, in 1974 when seed production was less, consecutive foraging columns appeared to strike off at random directions to one another (Fig. 4.7a), although successive columns from selected nests did show a general tendency to rotate about their entrance in a characteristic clockwise or counter-clockwise fashion (Fig. 4.7b).

What are the behavioural mechanisms of column rotation? At the present time, before column-rotating behaviour in specific directions can be accepted as established fact, the studies raise more questions than they answer because of the lack of precise data on the food distribution and abundance that presumably affects the behaviour. It can be inferred, however, that as individual *Veromessor* deplete the seeds near the end of an expanding column, they necessarily move laterally. Do the individuals that have moved laterally return directly to the nest, laying down a recruitment trail, and thus automatically causing column rotation? Does the direction of the foraging column remain constant throughout the day, and only change the next day after the trail pheromone has evaporated and the foragers returned by direct route to the area where they left off the previous day? Undoubtedly results would vary greatly from one area and from one time to the next depending on local distributions of food and its discovery by the ants.

Bernstein (1971) experimentally tested the ant's foraging behaviour by attempting to manipulate seed abundance by adding seeds about the nests of both *P. californicus* and *V. pergandei*. (Unfortunately the

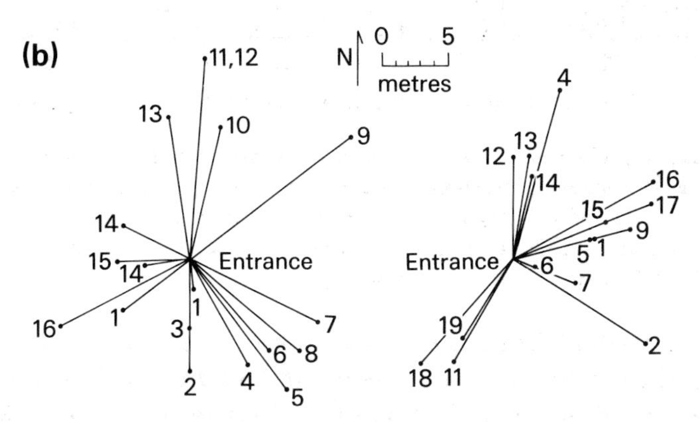

Fig. 4.7. (b) Column rotation during consecutive foraging periods of two nests of *Veromessor pergandei* in 1974 (each day had two main foraging periods—early morning, and afternoon or evening after temperatures had fallen). Lines indicate length of foraging columns, and numbers indicate consecutive foraging periods. In the nest illustrated at left the columns generally rotated counterclockwise. In the one shown at right the pattern was more complicated, but tended to be generally clockwise. During periods 3, 8, and 10 the ants left the nest in all directions. The ants did not forage to the west and northwest of the nest entrance of this nest possibly due to lack of plants in these areas (from Rissing & Wheeler 1976).

initial abundance of seeds in the nest vicinity which might have affected the response to the added seed, was not known.) Both species exhibited roughly equal success in retrieving coloured seeds spread evenly (with a mechanical spreader) in a 3 m radius of the nest entrance. Both adopted the individual foraging method. In contrast, when she placed similar amounts of seed into 4 piles (seeds of each pile dyed a different colour), each pile 1·5 m from the nest entrance, the foraging success between the two species differed. *Veromessor* generally only found those seed-piles which coincided with the direction of the existing foraging column. *Pogonomyrmex*, on the other hand, rapidly found all 4 piles. It appeared, therefore, that the colonies with individually foraging *Pogonomyrmex* are more efficient than *Veromessor* colonies with pre-existing columns in locating and exploiting new clumped resources, unless the *Veromessor* can counteract with more efficient recruitment to those sources they do locate. The ants probably do not recruit when resources are thinly spread, thus they forage individually. However, during individual foraging, one ant necessarily forages over at least some of the territory recently depleted by a colony-member, and column-rotation is a mechanism that should reduce this source of foraging inefficiency.

The above studies have set a groundwork for possible future work on seed-foraging strategies in ants at different food densities. Unfortunately there is much confusion due to lack of knowledge on actual food densities in the nest and in the field, as background to experimental manipulations.* Secondly, there is confusion due to terminology. 'Group recruitment' conventionally refers to orientation provided by a single recruiter to a group of individuals by tactile, chemical, or both types of cues, to contrast it from 'tandem running' (tactile cues only) and 'mass recruitment' (chemical trail rather than individual ant as primary orientator). But Bernstein refers to 'group', 'recruit' and 'individual' foraging, although little or nothing of the recruitment mechanisms was known.

4.3.3 *Leaf-cutter ants—farming*

The cultivation of micro-organisms, fungi, plants, and possibly other insects is widespread in social insects. Attention to the phenomenon was already drawn by Linnaeus who referred to aphids as 'ant's cows'. Many groups of ants collect the sugary anal excretions (essentially

*Editors' footnote: Davidson (1977) has recently done an experimental study of seed harvesting ants.

phloem filtrate) from a variety of homopteran plant-sucking insects. The 'cows', in turn, receive the ant's protection, and close symbiotic relationships have evolved including some where species of root-feeding aphids are found only in ants' nests (see discussion in Sudd 1967). The 'granaries' of harvester ants, as well as the enormously distended repletes of other desert species that store sweet fluids including the 'honeydew' secretions of aphids, indicate that some ants have evolved mechanisms for long-term energy economy.

Long-term energy economy may in some cases involve not only the storage, but also the harvesting of resources. A good example that has recently come under renewed investigation is the foraging of the leaf-cutter ants *Atta*. These ants may practice 'prudence' in harvesting leaves, thereby ensuring a long-term supply.

Leaf-cutter ants, *Atta* and *Acromyrmex*, base their economy on fungi grown on masticated leaves in subterranean gardens. Colonies can contain up to several million workers, and their large food and energy demands make them serious agricultural pests in parts of Central and South America where there are plant monocultures. The ants have been reported to strip a garden bare in a day or so.

Undoubtedly the *Atta* 'technology' of using fungi to extract energy from a rather common resource—fresh leaves—contributes to their large colony size. Some termites have evolved a similar 'solution' to the energy problem—in addition to utilizing endosymbiont bacteria or flagellate protozoa to digest cellulose some species culture fungi. There is at least one important difference in the ant's and the termite's use of plants. The termites utilize only dead material, and have no potential to exert control over its long-term supply. But repeated defoliation of plants might kill them, thus removing them as future food sources to the 'permanent' *Atta* colonies. (*Atta* queens live as long as 20 years.) It has been calculated that a single *Atta* colony harvests several tons of leaves in its first 6 years of existence when it grows in population from a single queen to over several million individuals (see Weber 1966). Despite massive leaf harvests, ants in the natural environment with many tree species available rarely kill trees because of over-defoliation in the nest vicinity (Rockwood 1975). For *immediate* optimization ('exploitation') the ants should first strip trees near the nest, moving out gradually at ever-greater distances when the local supplies become depleted. It is well known, however, that nests are often surrounded by rich vegetation, and the ants travel long distances to cut plants similar to those near the nest (Eidmann 1935). Do the ants have 'strategies' of harvesting that optimize long-term energy balance?

Some recent work gives strong evidence for the idea that the ants practice 'prudence' in harvesting, as would be expected of any efficient farmer provided he had exclusive use of the land. Cherrett (1968) has concluded that *Atta cephalotes* in rain forests of Guyana have a conservative, rather than an exploitative grazing system, and that this results in an evening out of the grazing pressure around the nest, thus preventing over-exploitation of the animal's resources. He found that the ants foraged far from the nest, but not because appropriate species were not sampled nearby. In one extreme example, *Terminaila amazonica* was being cut 65·4 m from the nest, while an intact specimen was overhanging the nest entrance. The mean distance of 101 nearest potential food trees was 15·2 m, while the mean distance of 101 trees being utilized was 46·7 m. The zone of exploitation was constant up to 31·2 m, but exploitation increased from 31–47 m, falling off thereafter.

It has also been observed that the ants can be highly selective in the types of leaves that they use, and the selectivity varies with seasonal abundance (Rockwood 1975). In Costa Rica at the end of the dry season, when many trees are leafing out, the ants utilize specific species. In the dry season they are less selective and utilize a number of evergreens that are not utilized in the wet season. These results need not be interpreted as indicating any type of 'prudence' in harvesting, since a decreased selectivity at diminishing resources is a general phenomenon observed in most animals where it has been studied (see 2.2). However, the effect of the different 'preferences' is to conserve food supplies in the nest vicinity.

A recent paper by Rockwood (1976) provides new insights into the question of whether or not *Atta* foraging is conservative. The observations showed that if trees near the nest were not utilized it was not because the ants could not find them, but because they were actively discriminating. Both *A. cephalotes* and *A. colombica* concentrated their foraging efforts on a much smaller subset of plant species than those actually available. The plant species utilized varied with the seasons. However, the different food plants were harvested at similar rates at the same time of year by *different* colonies, even though each colony had its own foraging areas with *unique* plant species composition. Many of the rarer plant species were the most frequently attacked, and the selectivity, as well as the absolute number of trees visited, was independent of foraging distance to at least 50 m from the nest. When tree trunks were banded with ant repellent the ants in most cases still discovered the leafy crowns and utilized the banded trees by crossing over from nearby trees.

Another line of evidence indicating selectivity, and possible 'conservative' foraging, was the degree of defoliation. An ant colony is capable of gathering several kilograms of leaves per day. Despite the ants' obvious potential to rapidly defoliate trees, they did not necessarily cut large amounts of leaves from trees near the nest. But large-scale defoliation was observed in some trees up to 60 m from the nest, and some ants were foraging up to 140 m from the nest. Generally one trail continued to lead from a given harvesting area for a long period of time, while other trails switched almost daily from one source to another. Rockwood (1976) concluded that the ants were sometimes foraging exploitatively by taking advantage of seasonal changes in plant availability, but the spatial patterns of harvesting near the nest were not consistent with pure exploitation.

The above results are interesting from the functional standpoint, but they give no indication about the possible behavioural mechanisms that could produce 'prudence'. At the present time the mechanisms of leaf-cutter foraging are not well understood. Selectivity could, in part, be related to palatability changes based on nutrients and secondary plant substances. However, the ants are essentially monophagous on the fungi of their gardens, even though they are 'polyphagous' in their foraging habits. It seems improbable, however, that the ants or fungi have specific nutritional needs that require a large heterogeneity of leaf types since laboratory colonies can be maintained for years on as few as one plant species. Rockwood observed that the ants cut leaves from a palatable source at high rates the first 2 or 3 days after locating it, and then the rates steadily declined or leveled off. Previous laboratory experiments (Cherrett 1972) suggest that the leaf-cutters are attracted to *novel* sources of palatable leaves. In one experiment one laboratory colony, colony A, of *Atta cephalotes*, was given a daily excess of privet leaves (*Ligustrum ovalifolium*). Colony B was given a daily excess of veronia leaves (*Veronia angustifolia*). Every 10 days the colonies were given a choice between the two leaves, and then the leaves were reversed. In 9 out of 10 occasions foraging increased on 'changeover' days, and in 9 out of 10 cases the ants preferred the leaves they had not been given the previous 10 days. Cherrett concludes that the apparent preference for novelty has the ecological implication of spreading the grazing pressure more evenly over the available resources. By frequent changing of foraging sites the ants could also, in effect, be patrolling their area and keeping away competing ants, although Eidmann (1935) claims that *A. sexdens* is not aggressive towards conspecifics.

4.4 Termites—energy efficiency

Termites are a large and behaviourally varied group, but they all utilize plant fiber. They employ conspicuously different economic strategies from most of the social Hymenoptera. The inflow of energy to their nest economies is less related to harvesting strategy than to efficient food processing and utilization.

In general, the termites' most obvious feature from the standpoint of comparative energetics of sociality is the tendency for some species to live directly inside an essentially unlimited food source, and their evolution of extreme efficiency in the use of this food having a very low value of utilizeable calories. Although the quantity of food available to them is large, the rate at which energy becomes available is very low and the animals practice drastic economy in their use of energy. Colonies may grow to tremendous size with millions of occupants, but the energy turnover per unit time is slow. For example, at the end of the second year, a colony of *Incistermes minor* has eaten only 3 cm^3 of wood, and the colony contains only the royal pair, one soldier, and about a dozen nymphs (Light 1934).

The primitive termites Kalotermitidae ('dry wood termites') that live directly inside their food source can dispense not only with the energetic cost of foraging, but also that of nest-building. Feeding and nest-building functions are combined. Living inside their food that also provides a hard protective shell and is 'permanently' available, they escape seasonal time constraints in colony build-up that would require high rates of energy utilization. In any case, the low-energy strategy is a necessity because of the low rate of energy extraction from wood.

In contrast to the wood termites, some of the higher termites build huge mounds, and their populations of millions of individuals subsist on plant material often gathered some distance from the nest. *Mastotermes darwiniensis*, for example, may make foraging galleries 100 m in length.

The tremendously powerful adaptive step that has made possible the utilization of cellulose from ubiquitous plant fiber has been the use of microbial symbionts. Kalotermitidae rely on an intestinal flagellate fauna to digest the wood (see review by Honigberg 1970).

Some termites also have an intestinal bacterial flora, and utilize externally cultivated fungi for cellulose breakdown of their food. The fungus-growing termites (subfamily Macrotermitidae) have an intestinal bacterial flora similar to cockroaches, and their imperfectly digested fecal pellets are built into fungus combs where symbiotic fungi of the genus *Termitomyces* degrade them further. The fungus gardens are also found in those Macrotermitiade that build large nests (principally out of clay and saliva) and store food material. Possibly the use of fungus gardens evolved from accidental contamination of these stores and of fecal material.

Some termites, principally the Macrotermitidae, build nests of amazing size and complex architecture (Noirot 1970). Some of the construction features involving building materials, internal organisation of space, nest shape, and nest orientation, are of more than passing interest in an analysis of the animal's overall energy strategy. The use of fecal pellets in *Coptotermes*, *Cubitermes*, *Microcerotermes*, *Nasutitermes*, and probably others in the construction of nest walls demonstrates efficient use of undigested food materials. In addition, rather than expending energy for nest-heating, as in social bees and wasps, it has been proposed that the fungus gardens with decaying vegetation in nests of Macrotermitidae serve as a source of heat in addition to a source of food. Lüscher (1951) observed in *Macrotermes subhyalinus* (formerly *M. bellicosus*) that a volume of about 3 gallons of fungus garden with brood and workers increased to 40·5°C overnight at an air temperature of 21°C, presumably by a fermentation process of the bacteria in the 'fungus garden'. The cavities within the nest appear to be so arranged in *Bellicositermes natalensis* as to allow for the circulation of the internal atmosphere on the basis of a thermosiphon system where warm air rises to heat occupied nest cavities, and cool fresh air enters ventrally to replace the warm air (Lüscher 1955, 1961). In nests from the Ivory Coast warm air from the fungus combs in the centre of the nest rises upward and is driven through 6–12 outward running channels into ribs that run downward on the outside of the nest. In these thin-walled ribs gas exchange and cooling take place with the outside air. The cooled air enters a 'cellar' at the bottom of the nest and is siphoned upward again. In Uganda termites that are morphologically identical have a one-way circulation system in their mounds. Air enters through open channels into the 'cellar' and escapes through thin walls at the top.

Large nests take up and store heat from solar radiation, and the inhabitants migrate within the nest to suitable temperatures. Skaife (1955, p. 10) writes: 'On a hot afternoon the outer cells of the mound are empty and all the inhabitants are crowded into the lower parts of the nest, where it is cooler. During the winter, on the other hand, they will congregate in the superficial cells on a sunny day for the sake of the warmth'. In the Australian mound-building termites *Amitermes meridionalis* the tall flattened nests are oriented with their long axes in the north-south direction, thus maximizing the area for uptake of solar radiation in the morning and evening, and minimizing it at noon. Thus, by a variety of features, various termites appear to minimize their energy investment for both heating and cooling of the nest microclimate.

The termites practice a strict self-centered economy in use of energy. As already mentioned, partially digested faecal material is used for nest construction, for nest heating, and for further breakdown and retrieval of additional food energy using fungal symbiants. In addition, the termites recycle materials within the colony. Cadavers are consumed, and injured individuals are cannibalized.

Few data are available that quantify and isolate the various features of energy flow in the termite societies. Lüscher (1955) reported that a colony of about 2 million *Macrotermes natalensis*, each weighing on the average 10 mg, consumed about 0·5 ml O_2 g^{-1} hr^{-1}, or 240 l O_2 per colony per day. Hébrant (1967) found that a colony of *Cubitermes exiguus* before swarming contained a biomass of 28·6 g (excluding gut contents) and consumed 156 ml O_2/day. After swarming the biomass dropped to 19·5 g, and energy expenditure to 81·4 ml O_2/day. The metabolic rate of the colony without reproductives corresponds to an equivalent energy utilization of 40 gm glucose/yr (81·4 ml O_2/day × 365 days/yr × 5·0 cal/ml O_2 × 1 mg glucose/3·7 cal=40 gm glucose/yr). For comparison, based on a metabolic rate of 36 ml O_2/g body weight/hr (Heinrich 1974) a similar biomass of bumblebees has an energy expenditure at least 200 times as great while regulating nest temperature at 5°C.

4·5 Other economic benefits of sociality

A major factor in the economic success of an insect colony is the efficiency with which the collected resources are utilized and converted to new queens and males, or swarms. This topic is too huge to be more than brought to the attention in this presentation. I shall only point out some of the obvious areas where it applies. In broadest terms it applies to the construction of the nest as it affects the movements of the colony members within it, caste ratios and their effect on labour distribution (Wilson 1968, 1971), the economical utilization of building materials, and the efficient use and conservation of metabolically produced heat for temperature regulation. It applies also to the distribution of energy and materials between individuals within the nest so as to streamline the production of the colony's 'product'—its offspring.

Like in a factory, an important factor in efficiency of production involves division of labour. In honeybees, where all of the workers are reared in identical hexagonal cells, the products are essentially morphologically identical. Division of labour results, however, as the young bees start at hive duties and end up as foragers when they become older. In bumblebees, on the other hand, where larvae are reared in communal cells, some individuals are crowded, receive less food, and grow to smaller size. Such small individuals often never forage, restricting their entire activities to house duties. Larger bees may forage at 2 days of age. Specialization reaches the extreme in ants and termites where there are not only size- and age-dependent tasks, but where there is also distinct polymorphism with sometimes one or more sodier castes. The ability to perform the various labours efficiently, however, depends not only on skill of the individual members, but also on 'managerial' capacity—an efficient communication that directs the application of the skills where needed. These aspects of social insect biology have recently been discussed in great detail with respect to the social hymenoptera in the books of von Frisch (1967), Michener (1974), and Wilson (1971).

The rigours of the environment would be most acutely 'felt' in the insect colony in its initial stages before the homeostatic mechanisms resulting from the division of labour of many individuals is possible. For example, bumblebee nest temperature which affects many aspects of colony activity is imprecisely controlled when the colony contains only the queen, but later when workers are present, nest temperature homeostasis improves (Fig. 4.8). Consequently, the heaviest mortality undoubtedly occurs at or near colony founding, and selective pressures to reduce mortality in this bottleneck should be great, particularly in harsh environments. It is of interest, therefore, that in some ant species inhabiting very arid regions, given colonies are founded by several co-operating queens (B. Hölldobler personal communication). It is not a far step from such co-operative efforts for a queen to invade the nest of another that has already been started. The next possible evolutionary step is slavery. Indeed, slavery in ants appears to be most common in north-temperate regions that are probably not the most ideal for this group judging from their great preponderance in the tropics. It is also of interest that of the two *Bombus* species in the High Arctic, one is a social 'parasite' of the other (K.W. Richards 1973). It would be of interest to determine to what extent queens of the two species may co-operate in initial nest building. Even the totally parasitic bumblebees, *Psithyrus*, may at times be tolerated for a long time in the

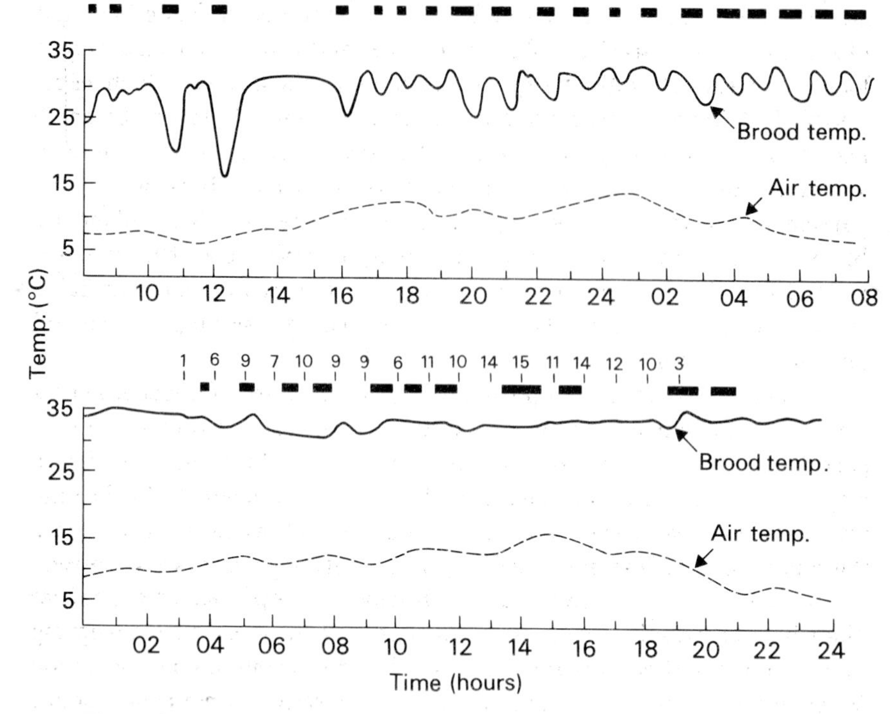

Fig. 4.8. Nest temperature regulation in *Bombus polaris* at Ellesmere Island (81°N), showing large temperature fluctuations (June 22) in a nest containing the founding queen only (top), and higher and more precisely controlled nest temperature (July 6) after 16 workers had eclosed (bottom). Duration of queen foraging trips are indicated by horizontal bars, and number of worker foraging trips each hour are given by numbers (adapted from K.W. Richards 1973).

nests of their hosts. Mixed-species colonies are commonly produced in nature when the queen of one species enters a nest and supplants the resident queen (see Plath 1934).

I suspect that a thorough examination of comparative economics of nesting behaviour in bees and other insects could shed additional light on the evolution of sociality. For example, many solitary bees place their brood at the end of tunnels dug into the ground. Ordinarily each female digs her own tunnel. However, possibly because suitable nesting sites with sufficiently soft-textured soil are sometimes few and restricted in size, numerous females may nest in aggregations. Michener (1974) describes communal nesting for 15 bee genera in which communal nesting of two to a dozen or more individually fertile females occurs.

In communal burrowing solitary bees, there should be economy of labour as only one main burrow has to be made and defended against often numerous parasites. The energy expenditure of constructing a long main tunnel could be great. Obviously, those bees which enter established burrows only need to make their nest chamber. In small bees the energy saving of digging a burrow in hard soil may be considerable. Secondly, the time involved to dig such a burrow may be particularly costly in bees specializing on the notably ephemeral blooming of many desert plants. However, as with most other matters related to colony economics, there are no quantitative data available on the energy investments and possible energy savings of several alternative strategies.

It is probably a small step from using the entrance of another bee to using its nest, which could conceivably evolve either to sociality, or to parasitism. Indeed, the distinction between sociality and parasitism may not be great. In one (sociality) the reproductive effort of other individuals of the same species is suppressed, and in the other (parasitism), the reproductive effort of individuals from another species is suppressed.

Nest initiation by swarming in honeybees, *Apis mellifera*, rather than by individual queens as in *Bombus* (and in most ants excepting army ants) could conceivably also be related to economic factors given their specialization on seasonal food resources. Honeybees cannot survive the winter without large honey stores. The larger the colony the more readily it can regulate its temperature, lay up food surpluses and produce a large worker force early in the spring that harvests the nectar and pollen during the mass flower in spring. A queen initiating a colony by herself would be severely limited in the size of the colony she can produce by fall. Food resources collected can conceivably be allocated either into producing a large population at the end of the summer (as in bumblebees and wasps), or into pollen and honey stores. The latter is a strategy that has evolved in honeybees to survive marked seasonality with an intact colony, allowing them to specialize on the mass blooming of spring flowers.

Colony founding may require large numbers of individuals for other economic factors besides the necessity to circumvent time constraints in establishing nests or laying up food reserves. For example, the specialized behaviour associated with group predation in army ants probably requires a critical mass in number of individuals to be successful. Unlike in most other ants that are not group predators, colonies of these ants are not established by lone mated queens, but in large swarms. The colonies divide, and a new queen departs with a large worker force. Two other social insects that also achieve huge colony sizes, *Atta* leaf-cutter ants and some termites, start their colonies from a single female or a pair of reproductives without the initial aid of workers. In both cases, however, the food is grown or collected from common material available in the immediate nest vicinity, and little or no co-operative foraging is required. Co-operative foraging later on may promote colony build-up, but the initial capital investment to the colony need not be great.

Co-operative foraging in insects other than the eusocial hymenoptera and the termites is rare, and/or not well documented. Frankie (1976) reports that groups of up to 300 males of the solitary anthophorid bees *Centris adani* sometimes forage in clusters, moving in 'waves' from one branchlet to another on massively-flowering trees in Central America. Flowers that have been visited are temporarily unattractive, apparently due to scent-marking. Possibly the group foraging ensures that search for flowers is in previously unexploited areas, while the pheromone marking functions in reducing the revisitation of recently emptied flowers.

Other large aggregations of unrelated insects include migrating locust swarms, pods of gyrinid beetles on the water surface, sleeping aggregations of male bees (Linsley & Cazier 1972), and over-wintering clusters of coccinellid beetles. It might be well to keep in mind that non-social insects have probably evolved to band together for defense and a variety of other reasons. The significance of these groups of unrelated individuals is in most cases unknown, but it deserves serious consideration.

4.6 Conclusions

It is probable that co-operative foraging is, in insects, not alone a sufficient cause for the evolution of sociality but, primarily, a consequence of it. If this were not so one would predict that there would be insect societies with unrelated individuals foraging co-operatively, each gaining advantages because of the association, and each increasing its fitness as a result. Although there is little conclusive evidence of co-operative foraging of unrelated individuals, there is considerable evidence that unrelated individuals of some of the 'primitive' bees and wasps nest co-operatively (Mathews 1968, Michener 1974, Evans 1977), thereby gaining advantages by mutual economy in nest building and in defence. These latter aspects of the economics of sociality are undoubtedly important in the biology of most social hymenoptera, but they are so far largely unexplored in the highly social species.

Chapter 5
Anti-predator Defence Strategies: Some Evolutionary Problems

PAUL H. HARVEY AND PAUL J. GREENWOOD

5.1 Introduction

Animals defend themselves against predators in a variety of ways, many of which are well catalogued by Edmunds (1974). The evolution of certain prey defence mechanisms poses several problems for the behavioural ecologist. For instance, many species have evolved forms of defence against predation which apparently involve signalling to the predator. It is the function and evolution of such characteristics which we shall discuss in this chapter. We have attempted to choose areas where apparent paradoxes occur, and our review of the literature has been deliberately selective because we have tried to use examples which either help towards an understanding of the relevant selective processes or which highlight problem areas requiring further investigation.

Three points should immediately be made clear when thinking about prey defence in an evolutionary perspective. First, anti-predator adaptations co-evolve with predator hunting strategies (see Gilbert & Raven 1975). The adaptive significance of the latter has received much attention recently and is reviewed by Curio (1976). Second, a genetically controlled anti-predator strategy can evolve in a population even when predation is not a density-dependent limiting factor on the population. As Haldane (1953) wrote 'if the density (of a population) is limited by food supply, a gene which makes the animals less conspicuous to a predator . . . will be favoured by natural selection, and if it occurs by mutation will spread through the population. But it will not increase the food supply, or the density'. Third, prey might not be trying to defend themselves. Ultimately, individuals have been selected to maximise their reproductive success, but they might do this most efficiently in encounters with predators by attempting to protect themselves, their relatives (who share a portion of their own genes) or even other animals, conspecifics or not (see 1.4, 1.5, 1.6).

The topics we have chosen to consider are aposematic colouration, visual colour polymorphism, rump patch signalling, alarm calling, distraction displaying and mobbing. Several other types of anti-predator response involve signalling to the predator, but these either pose no evolutionary problems (e.g. feigning death: see Edmunds 1974), or are considered in this volume (e.g. aggressive group defence: see Chapter 3) or recently reviewed elsewhere (e.g. mimicry: Turner 1977). It should not be assumed that signalling to predators always implies anti-predator behaviour. Some animals want to be eaten. Perhaps the most obvious cases where prey signal to predators for this reason are among parasites. For example, the trematode *Leuchloridium macrostomum* uses the snail *Succinea* as an intermediate host. Sporocyst sacs penetrate into the snails tentacles which become conspicuous to bird predators, the definitive hosts. The tentacles are bitten off and the sporocyst develops within the bird (Wickler 1968).

Our concluding discussion will be primarily concerned with considering the different methods available for assessing the relevance of the various hypotheses which have been advanced to explain the functional significance of these traits.

5.2 Aposematic colouration

Those insect species which have evolved distastefulness as a means of defence against predators generally obtain their noxious chemicals from food plants (Brower & Brower 1964, Ehrlich & Raven 1967), although some groups of grasshoppers and Lepidoptera may be able to synthesise their own toxins (Rettenmeyer 1970). Many such unpalatable species are brightly coloured and this is normally regarded as a signal to predators. Some predators have evolved innate avoidance reactions to particular aposematic signals (e.g. Rubinoff & Kropach 1970, Loop & Scoville 1972), but more often predators learn to avoid such prey after sampling one or a few individuals (Eibl-Eibesfeldt 1952, Brower 1958a, b, c). Two evolutionary problems are of interest here. First, how did distastefulness as a means of prey defence evolve? Second, why are distasteful prey not cryptic?

The first question was discussed by Fisher (1930, 1958). He recognised that most means of prey defence could have evolved by increasing the chances of survival of the individual in which they are found. But for those species which are distasteful to predators and in which at least one individual must be sampled (and probably killed) before the predator learns to avoid other members of the species, it is 'difficult to perceive how individual increments of the distasteful quality, beyond the average level of the species, could confer any individual advantage' (Fisher 1958, page 178). However, Fisher realised that distastefulness among insects is often associated with a gregarious habit, and further that the individuals involved would probably be full sibs. This led to an explicit formulation (probably the first) of a kin selection model. If only one or a few siblings are sampled from a large brood and the predator learns to avoid other members of the group, then a gene for distastefulness can increase in frequency through kin selection (see 1.5).

A simple answer to the second question posed above is that distasteful species sometimes are cryptic (Marsh & Rothschild 1974), but nevertheless distasteful species are very often brightly coloured. Two hypotheses have been put forward to account for this. Although they are not absolutely mutually exclusive it does seem worthwhile to present them separately. One emphasises the contrast of an aposematic prey against its background, while the other considers the adaptive value of bright colouration to be a consequence of novelty in the predator's diet. The first assumes that a predator can learn to avoid contrastingly coloured prey more readily or for longer periods than cryptically coloured prey and that a predator's association of unpalatability will, therefore, be reinforced by conspicuous prey colouration (e.g. Rettenmeyer 1970, Turner 1975, Matthews 1977). In addition, the image of the prey might be reinforced more frequently in a conspicuous prey species which will be perceived more often than a cryptic one. However, Matthews (1977) incorrectly claimed that Gibson (1974) had demonstrated the effectiveness of contrasting colour, as opposed to a background or matching one, in 'avoidance-image formation'. Gibson presented blue, red and green dyed millet seed to the star finch *Bathilda ruficauda* on a hinged platform coloured with blue and green dots. When the birds attempted to feed on either the blue or red seed, the platform was dropped from sight and the feeding attempt was unsuccessful. The birds stopped attempting to feed on the red ('aposematic') seeds before the blue ('cryptic') seeds and continued to feed on the green seeds. However, no evidence was presented that the 'cryptic' seeds were cryptic (the results for a control group showed that there was no colour preference measured by mean feeding latency). In addition, the results could be interpreted in terms of innate differences in the capacity to learn different colours. The crucial test in experiments of this sort is to reverse the roles of the cryptic and contrastingly coloured forms by changing the background. The speed at which different colours are learned might then be shown to depend on contrast with the background.

A second theory to account for the advantage of aposematic colouration has been suggested by Turner (1975) who argues that there has been 'selection in favour of looking as different as possible from those camouflaged forms for which the predators are constantly hunting'. The emphasis here is on novelty rather than contrast. If we assume that the explanation applies to naive predators searching for prey then Shettleworth (1972) had, in fact, carried out a series of experiments in support of this idea. Young domestic chicks learned to avoid drinking either quinine flavoured or electrically shocked water of a distinctive colour. They learned to do so much faster if the unpalatable water was an unfamiliar colour and the palatable water was the colour they had been raised on than if the opposite was true. The 'familiar' water colour was reversed in two groups of chicks and the above conclusion held for both groups. Clearly, further work is necessary in order to distinguish between the two hypotheses and to test the extent to which either has been important in the evolution of aposematic colouration. One series of experiments even indicates that distasteful prey which are cryptic have a selective advantage over those that are brightly coloured (Papageorgis 1975).

As with distastefulness, warning patterns must, presumably, evolve through kin selection (Turner 1971, Matthews 1977) because rare aposematic forms will not be at an individual advantage. To some extent the available data bear out this conclusion. For example, distasteful brightly coloured butterflies and their larvae tend to be gregarious while closely related cryptic species do not (see e.g. Ford 1945). Post-reproductive longevity might also be selected for since a post-reproductive animal can still form the prey of a naive predator and reinforce the prey colour pattern to other predators. In cryptic species the opposite would be expected: if the effective density is increased then the predator is likely to begin hunting for that prey species (Tinbergen 1960). Blest (1963) has produced evidence in support of this hypothesis by comparing post-reproductive longevity in species from two related genera, some of which are aposematically coloured while others are cryptic. Turner (1971) realised that, for similar reasons, within the heliconids selection might be expected to have produced a correlation between distastefulness, restricted home range and range of courtship, communal roosting, longevity, and delayed sexual maturity. The expected correlations were found to exist. Clearly, functionally related constellations of characteristics of this type are selected for simultaneously during evolution. Nevertheless, the approach is predictive and gives important insights into the selective forces which mould life history strategies and certain behavioural characteristics.

5·3 Visual colour polymorphisms

If we accept Ford's (1940) definition of polymorphism as the occurrence together in the same locality of two or more discontinuous forms of a species in such proportions that the rarest of them cannot be maintained by recurrent mutation, then many invertebrate species (particularly among the Lepidoptera and Mollusca) show distinct genetically-based colour and pattern polymorphisms (see Edmunds 1974). Furthermore, the polymorphisms in many species can be shown to be 'balanced' (i.e. actively maintained in a reasonably constant environment) rather than 'transient' (Ford 1964).

5·3·1 *Selection and crypsis*

The question arises: can such polymorphisms evolve as an anti-predator mechanism? The first point to make is that selective predation of the more conspicuous forms (or morphs) would lead to the elimination of these polymorphisms: if the best camouflaged morph differed between habitats, then the morph at fixation would similarly vary. It is certainly true that selective predation by various predators on a variety of visually polymorphic invertebrates occurs, and that the more conspicuous forms on a given background are selected (e.g. Kettlewell 1956, Sheppard 1951). It is also common for the frequencies of various colour morphs in natural populations to vary between habitats in a way that suggests that morph frequencies are influenced by selective predation (e.g. Cain & Sheppard 1954, Giesel 1970, Clarke *et al.* 1963, Kettlewell 1968).

5·3·2 *Apostatic selection*

However, the question remains: can predation be maintaining such polymorphisms? It was suggested by Clarke (1962a) that vertebrate predators concentrating on the more common morphs of a polymorphic prey population could maintain such polymorphisms. He called this form of selection 'apostatic'—rare morphs are favoured because they have a selective advantage. (Moment (1962) called the same form of selection 'reflexive'). Clarke concluded from the evidence then available that apostatic selection is likely to be a potent force in both maintaining visual polymorphism and favouring the evolution of new morphs. The constraints on the number of morphs that might be maintained in a population by apostatic selection deserve attention. One important characteristic of the predators behaviour is 'switching' (Murdoch 1969) which is defined as the absence of a linear relationship between the proportions of different prey present in a population and the proportions consumed by the predator. Clearly, in the case of apostatic selection the form of the non-linear relationship is also relevant and a stable equilibrium point must exist where the proportions present in the population and the predators diet are the same (see Fig. 5.1).

Allen and Clarke (1968) reported a series of experiments to test whether ground feeding passerines do select prey with which they are

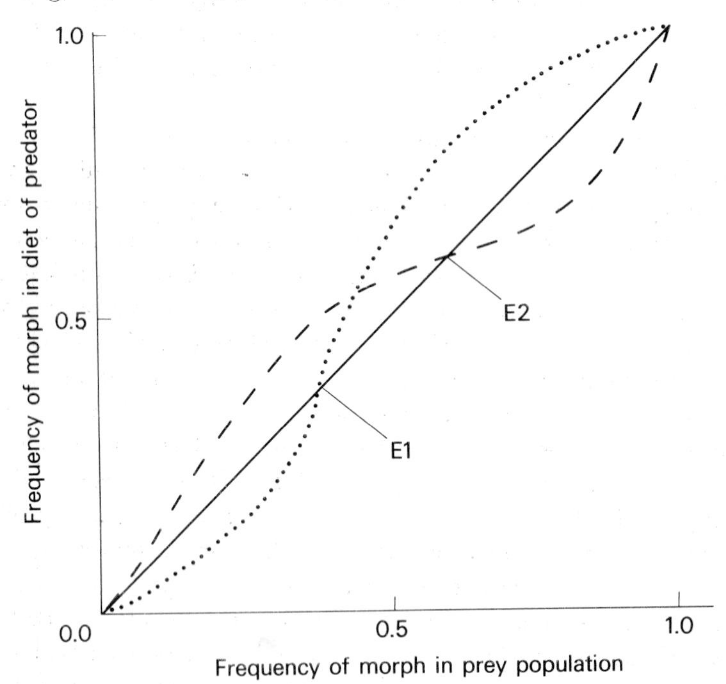

Fig. 5.1. The straight line represents a situation where the proportion of a given morph is the same in both the prey population and the diet of the predator. The line .... shows switching (see text) which would maintain the polymorphism in the prey population by apostatic selection at the stable equilibrium point E1. A predator switching to give the response - - - - - - would lead to monomorphism in the prey population. The equilibrium point E2 is unstable.

familiar, and further, whether the birds can change their preferences as a result of frequent encounters with a new prey. Their results were in accord with these predictions. One interesting point about their experiments is that the 'prey' types used (which were small cylindrical baits made from flour and lard) differed only in colour (green or brown) and were designed to mimic a polymorphic prey species. They concluded that 'ground feeding passerines, which are important predators of many polymorphic species of insects and molluscs, appear to hunt in a manner that would tend actively to maintain colour polymorphisms in their prey'. In some later experiments using painted mussel shells (*Mytilus edulis*) baited with meat, Croze (1970) was able to show that monomorphic 'populations' suffered a higher rate of predation from carrion crows (*Corvus corone*) than trimorphic ones. This raises the interesting possibility that while a rare morph is increasing in frequency in a population (because the predator is concentrating on common morphs), a predator may cease feeding on the prey species altogether because the commonest morph reaches a sufficiently low absolute density for the predator to turn to alternative prey sources.

If polymorphism does reduce the efficiency of predation to a level at which the predator switches to another species, it could be that if the density of the prey species increased then morph frequencies would also change. The relationship between apostatic selection and prey density is not yet clear (Allen 1972, Harvey *et al.* 1974, Cook & Miller 1977). The extent to which individual variation in predator 'preferences' (Allen & Clarke 1968, Cook 1965, Elton & Greenwood 1970, Järvinen 1976) influences apostatic selection is also uncertain. Although these latter problems deserve attention, it is unlikely that they will be resolved without detailed consideration of the behavioural mechanisms which lead to prey selection by predators. Krebs (1973a) and more recently Curio (1976) have distinguished between various profitable hunting strategies which predators might adopt, several of which could lead to apostatic selection. Of particular importance here is the concept of the 'specific search image' which was defined by Tinbergen (1960). He suggested that as the absolute density of a particular prey type increases, birds begin to concentrate their feeding upon it. The concept of the specific search image has been used in most discussions of apostatic selection and is normally assumed to be the underlying relevant behavioural mechanism. Murdoch and Oaten (1975) point out that this may not be the case. If animals develop searching images then as H_1 increases so should a_1 in the following equation

$$\frac{N_1}{N_0} = \frac{a_1}{a_0} \times \frac{H_1}{H_0}$$

where H_1 and H_0 are the densities of alternative prey types in a population, N_1 and N_0 are the numbers eaten by a predator in a given time interval, while a_1 and a_0 are the probabilities that one individual of the relevant prey type will be attacked per unit of time. Searching images develop, therefore, with increases in the density of particular types (or morphs in this case). However, apostatic selection occurs when c increases with the ratio of H_1 to H_0 in the following formula:

$$\frac{N_1}{N_0} = \frac{c \times H_1}{H_0}$$

and when a stable equilibrium exists (see above). Clearly, if we accept this formulation, apostatic selection is a frequency dependent model while the searching image hypothesis is density dependent. For further discussion of the consequences of search image behaviour, and its relevance to apostatic selection the reader is referred to Murdoch and Oaten (1975).

5·3·3 *Polymorphism in the snail* Cepaea

The species which Clarke had in mind when he suggested the term 'apostatic selection' and the species in which visual polymorphisms have been most intensively studied is the garden snail (*Cepaea nemoralis*). The shells are yellow, pink or brown and have up to five dark brown bands running along the whorls. The genetics of the polymorphism is known (Cain *et al.* 1968) and both banding and colour alleles at their different loci show complete or near complete dominance. Nearly all populations studied so far are polymorphic at either the colour or banding loci (or, normally, both). According to the locality, morph frequencies may be influenced by song thrush (*Turdus philomelos*) predation and possibly climatic selection (see Harvey 1976, Jones *et al.* 1977). For instance, light coloured shells have a higher albedo (reflectance of solar radiation) than darker shells (Richardson 1974) and will therefore be favoured when there is a heat load, while darker shells will be more successful when efficient solar absorption is necessary to reach temperatures required for metabolic activity. Jones (1973a, b) interprets a cline in yellow morph frequency in terms of warmer mean summer temperatures favouring the yellow morph.

Nevertheless, such selection is not likely to maintain the polymorphism (although it could in certain areas if there was differential habitat selection by the different morphs within a population), nor is it likely to keep the morphs distinct (there is adequate genetic variation in natural populations to permit selection to produce intermediate morphs). Since there appears to be random mating between morphs within populations (Lamotte 1951, Schnetter 1950), either some mechanism not related to the visible characters of the genotypes (e.g. heterozygote advantage) or a system involving predation are the most likely candidates for the maintenance of the polymorphism (Clarke 1962a). There is no evidence for the former type of mechanism. However, circumstantial evidence from patterns of morph frequency distribution in natural populations led Clarke to argue that apostatic selection might be an important factor (Clarke 1962b, 1969, but see Carter 1967). In order to test this theory, two series of experiments were performed using natural song thrush populations feeding on artificial, confined colonies of *Cepaea* (Bantock & Bayley 1973, Bantock *et al.* 1976). The results demonstrate that thrush predation was both selective and morph-frequency dependent. However, the results can be interpreted in terms of differential behaviour of the various snail morphs, rather than changes in thrush selection behaviour. In order to eliminate the possibility of variations in snail behaviour, Harvey *et al.* (1975) carried out a series of experiments using bread-stuffed *Cepaea* shells as the prey for a natural song thrush population. The experiments show that experience of a single colour morph (either brown or yellow) can produce changes in the feeding behaviour of the birds so that they select that morph when subsequently presented with a dimorphic population. The evidence, therefore, is good that song thrushes exert frequency-dependent selection in a way that would maintain the polymorphism in natural population of *C. nemoralis*. However, this cannot be the whole story (Bantock & Harvey 1976, Jones *et al.* 1977): *C. nemoralis* populations are polymorphic in many areas where the song thrush is not found. There are many other predators of *C. nemoralis* including birds, mammals (e.g. Goodhart 1962, 1963, Oldham 1929), glow worms (*Lampyris noctiluca*—O'Donald 1968) and probably beetles. Both glow worms (O'Donald 1968) and some small mammals (Cain 1953) are thought to select by the visual characteristics of the shell, and presumably avian predators other than the song thrush do as well. The incidence of predation by these other groups is not known and whether they are actively maintaining any shell pattern polymorphisms is a matter for speculation.

We have dealt with the case of *C. nemoralis* at length for two reasons. First, it is the species in which apostatic selection has been most studied. Second, the polymorphism demonstrates well the interplay of various selective forces (some of which may have nothing to do with predators). Clearly, many similar polymorphisms in a variety of groups might be maintained by frequency-dependent predation (see Clarke (1962a) for some suggestions); it remains to be demonstrated that any are.

5.4 Rump patch signalling

Many cursorial mammals possess white rump or tail patches which are used as a signal in the presence of potential predators (Fig. 5.2). Moreover, in several distantly related species this signal is accompanied by a stance or movement which enhances its perceptability (e.g.

Fig. 5.2. Rump patch signalling by white-tailed deer in the presence of a potential predator.

in members of the Antilocapridae (Seton 1953), Bovidae (Walther 1969) and Caviidae (Smythe 1970)). This behaviour makes the species involved more conspicuous to predators and, in many cases, it allows the predator to close in on the prey (Wilson 1975).

However, Guthrie (1971) has reviewed the literature on both the distribution and possible functional significance of white rump patches in a variety of mammalian groups (particularly artiodactyls, lagomorphs and primates) and it is clear that they do not function uniquely in anti-predator contexts. He argues that, as a consequence of the role of the rump patch in courtship by sexually submissive oestrous females (see Clutton-Brock & Harvey 1976), it is mimicked by males and functions as a generalized intraspecific appeasement signal. Guthrie goes on to interpret rump patch display by, for instance, the white-tailed deer (*Odocoileus virginianus*) or cotton-tailed rabbit (*Sylvilagus floridanus*) when surprised by a predator as non-functional. These species are, he believes, responding to the predator as if it were a dominant conspecific, even though this is to the immediate disadvantage of the prey because it appears more conspicuous to the predator. Hirth and McCullough (1977) discount Guthrie's hypothesis with reference to the white-tailed deer, at least, by pointing out that in this species the tail is tucked down 'between the legs' during submission behaviour: exactly the opposite of the behaviour assumed by Guthrie. They present additional evidence to suggest that Guthrie's unitary hypothesis for the evolution of white rump patch displays does not accord with the known behavioural data in other mammalian groups.

Assuming that rump patch display does have some functional significance as an anti-predator device among many cursorial mammals, perhaps the first question to ask is: who is the signal aimed at—the predator or conspecifics? Depending on the recipient of the signal, various hypotheses have been put forward which suggest a selective advantage for the behavioural patterns involved. If the display is towards conspecifics, then it presumably acts as a warning that the predator is present. Such a warning could range in function from selfish manipulation of other group members to altruism. This type of behaviour would not occur in solitary species and generally does not (Hirth & McCullough 1977), but if it is to be interpreted in this way then it is surprising that the signal has 'evolved on the part . . . that is most visible to a pursuer rather than on the flanks where it would be most visible to other group members' (Smythe 1977). Nevertheless, the warning behaviour might have several consequences: members of the group might flee, hide or aggregate. Hirth and McCullough (1977) believe that the latter is the case and that tail flashing is a signal to promote social cohesion. The functions of grouping have been discussed elsewhere (Bertram this volume, Clutton-Brock & Harvey 1977). In this particular context aggressive group defence does not seem relevant as the species involved flee from predators (although they may aggregate to flee). Another disadvantage of grouping, that of the selfish herd, was described by Hamilton (1971): individuals can reduce their 'domain of danger' by placing conspecifics between themselves and the predator. Hirth and McCullough argue that this concept is central to the evolution of the white rump patch as an intraspecific signal in, among other species, the white tailed deer. If animals respond to the presence of a predator by flight, then it may well be worth warning other group members of the presence of the predator *after you have passed them.* The predators attention might then be diverted to those other individuals. Surely, the sound or sight of your movement will be sufficient warning in that case. In addition, in the study quoted by Hirth and McCullough, average group size varied seasonally from 1·2 to 2·9 in doe groups and from 1·2 to 2·0 in buck groups. Therefore, many, if not the majority of 'groups' consisted of one individual. The selfish herd explanation cannot be invoked in those instances. Because doe groups often consist of related individuals, and such behaviour would put relatives at risk, we might expect tail flashing to be less frequent there than in buck groups. No such difference was recorded (whereas an accompanying warning snort was given more frequently in the former type of group—see below).

Smythe (1970) has suggested that the rump patches in several cursorial mammals are used as a signal to the predator, and further that they function to elicit pursuit. If the potential prey recognises a hunting predator while the flight distance is sufficient to allow safe escape, rump patch display as a signal to the predator may shorten the potentially protracted interaction. The predator may go off looking for less alert prey, or it may pursue the displaying individual and fail. Either way, the disruption of the time energy budget of the prey will have been minimised. Smythe and Zahavi (see Dawkins 1976) also

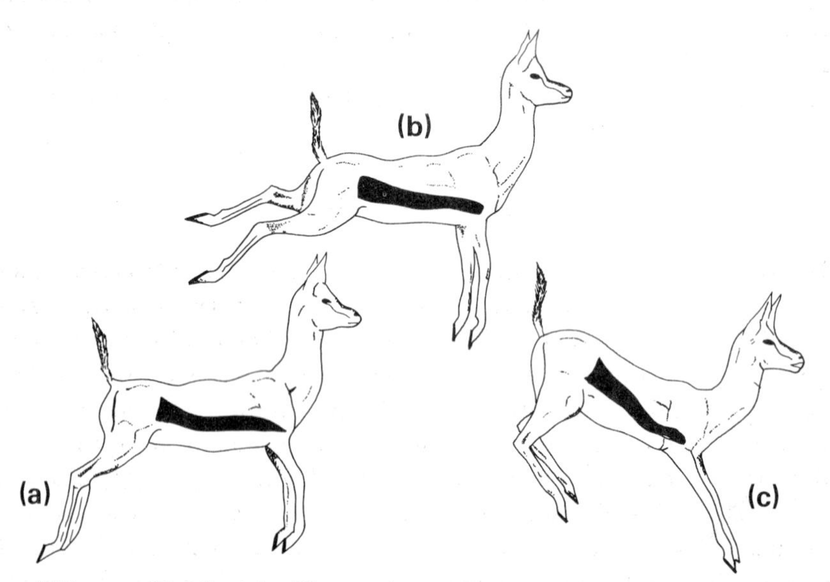

Fig. 5.3. 'Stotting' in Thomson's gazelle. (a) Normal stotting gait. (b) High stotting. (c) Landing from high stotting. (From Walther 1969.)

interpret the stotting (Fig. 5.3) or prancing movements (Walther 1969, Estes & Goddard 1967), which may be used in parallel with tail flashing, in a similar vein. Hirth and McCullough (1977) have criticised Smythe's theory, but as Smythe (1977) points out nearly all their arguments can actually be used in support of pursuit invitation. Zahavi (see Dawkins 1976) carries Smythe's theory further by arguing that these movements inform the predator that the prey is a particularly healthy individual ('look how high I can jump') and could easily escape if pursued. By implication, the predator should concentrate on either non-stotting individuals or those which are stotting ineffectually (i.e. not jumping as high). The idea has also been put forward that the predator might find it difficult to pick out a potential prey individual from a group in which many white rumps are continually moving in relation to each other (Walther 1969, see also 3.2.4). A similar explanation could explain the occurrence of white rump patches in some birds, particularly winter flocking wading birds such as the redshank (*Tringa totanus*).

Experimental methods to distinguish between the theories outlined above have not been devised (Wilson 1975). The comparative approach which relates the attributes (e.g. behaviour, presence of rump patch) of different species in a taxon to ecology and the selective pressures acting may well indicate the relevance of the different theories. Both Guthrie (1971) and Hirth and McCullough (1977) have used this approach in qualitative discussions, but no quantitative analyses appear to have been performed. It is likely, of course, that no single explanation is sufficient for any particular taxon and that the functional significance of rump patch display to predators will be different in different taxa.

5.5 Alarm calls

No single problem in anti-predator behaviour has attracted more theories and produced fewer facts than that of the evolution and function of alarm calls in birds and mammals. These calls are typically given at moments of potential danger, usually after a predator has been sighted. In some bird species the structure of the call given in response to an aerial predator shows a high degree of convergence to a form which may be difficult to locate (Marler 1955, 1957, Konishi 1973). The call is high pitched and tuned to a narrow frequency band. Similar calls are also found in mammals (e.g. round-tailed ground squirrel *Spermophilus tereticaudus* Dunford, 1977). Shalter and Schleidt (1977), however, have shown experimentally that barn owls (*Tyto alba*), which responded to the 'seet' alarm call of the clay-coloured robin (*Turdus grayi*) could accurately orientate towards the source. It may seem inappropriate to test the locatability of bird alarm calls, normally given in response to diurnal predators hunting by visual cues, on a nocturnal predator which may initially locate its prey (small rodents) by sound. In addition, a number of owls, including the barn owl, are unusual in having bilateral external ear asymmetry, which probably allows accurate discrimination of both directional and distance components of sound (Norberg 1977a). It appears, however, that a diurnal predator (goshawk *Accipiter gentilis*) can also localise alarm calls (Shalter 1978). Perhaps we should not be surprised that some predators have evolved the means of localising such alarm calls. When species (e.g. Kloss's gibbon *Hylobates klossii*, Tenaza & Tilson 1977) emit loud, repeated wide-frequency alarm calls in response to predators there are presumably no problems in localising the sound.

Since Hamilton (1963) first offered an explanation for what appeared to be altruistic behaviour on the part of the alarm caller (see below), there has been a spate of different hypotheses. Admittedly, the problem is a complex one and explanations differ partly because the initial assumptions that can be made vary so widely. For example (as with rump patch signalling), is the alarm call directed at the predator or conspecific prey? Is the caller being selfish or altruistic? Does alarm calling enhance or decrease the individual's chances of being captured? Any attempt at a unifying theory may be undermined by differences between species in the nature, context and function of alarm calls. We do not intend to add to the confusion by offering further unexplored hypotheses. Instead, a brief restatement of the theories will suffice followed by an assessment of the field data which has recently been published. We are concerned with explanations for warning calls, such as the 'seet' call of birds rather than shock or distress calls given by startled or trapped individuals.

5.5.1 *The evolution of alarm calls*

Hypotheses for the evolution and function of alarm calls fall into four categories.

(1) Alarm calling is an altruistic trait which benefits other unrelated group members, by alerting them, at some cost to the caller. Alarm calling has evolved through inter-group selection (Wynne-Edwards 1962). This proposal is largely untenable since there is no evidence to suggest that natural populations of vertebrates possess the type of

group structure and dynamics which would allow group selection to be a major force (Maynard Smith 1976a) (see 1.4).

(2) Individuals giving alarm calls are acting to reduce their own chances of being preyed upon, either immediately or on future occasions (Trivers 1971). The suggested routes by which this may be achieved differ substantially. For instance, in birds a call may not only be hard to locate but might actually be ventriloquial and thus divert the predator away from the caller (Perrins 1968). An animal which gives an unpredictable alarm call could startle the predator and escape, a vertebrate counterpart of the visual flash displays of some Lepidoptera (Andrew 1961d, Driver & Humphries 1969). The call might warn nearby conspecifics of an approaching threat and lower the chance that a predator will hunt that species or in that area in future. The benefit to an individual of preventing an immediate or subsequent hunting success may be greater than any initial cost in alarm calling (Trivers 1971). The call could be an auditory counterpart of pursuit invitation signalling (see above). Calling might well promote prey aggregation which could reduce an individual's risk *per se* (Hamilton 1971), increase prey crypticity (Dawkins 1976) or result in co-operative prey defence or group mobbing (Rohwer *et al.* 1976). Alarm calls could serve to warn mates (Williams 1966) when the cost of replacing a mate multiplied by the probability that the mate will be taken in the absence of a call are greater than the costs of calling. Finally, Charnov and Krebs (1975) have suggested that alarm callers could be selfishly manipulating other group members by warning them of the presence of danger, but not its location. If this increased the susceptibility of other group members to predation, then it may reduce the caller's own chances of being killed.

(3) Alarm calls function to warn relatives that share a proportion of the caller's genes (Hamilton 1963, 1964, Maynard Smith 1965). In such cases the caller may endanger its own life by attracting a predator to itself, although an assumption of risk is not a prerequisite for the evolution of warning behaviour through kin selection. The possibility that calling for the benefit of relatives during the breeding season is non-functional at other times (Williams 1966) is implausible (Trivers 1971).

(4) The call may assist unrelated neighbours at some cost to the caller if such behaviour is likely to be reciprocated in the future (Trivers 1971). It is difficult to envisage the evolutionary origin of such a trait in non-calling populations and, as Trivers himself points out, if established there would be a high probability of individuals cheating by not reciprocating (see 1.6).

5.5.2 *Alarm calls in Belding's ground squirrel*

The only detailed test so far of the predictions that result from the different evolutionary hypotheses is that by Sherman (1977) on Belding's ground squirrel (*Spermophilus beldingi*). The species has a matrilineal social structure in which related females live in close proximity and are extremely sedentary from one breeding season to the next. Males, on the other hand, move between breeding seasons and are highly polygynous. There is no tendency for related males to disperse to the same area or to be associated with female relatives. In the presence of a predatory mammal, females give markedly more alarm calls than males (Fig. 5.4). There are also differences between

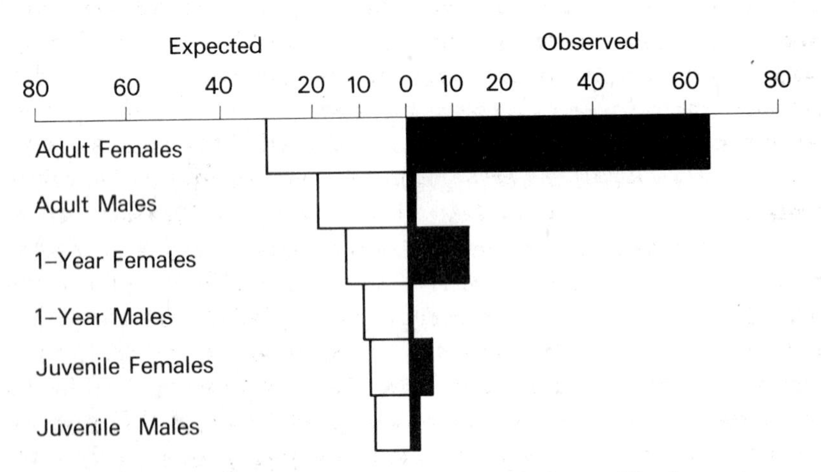

Fig. 5.4. Observed and expected frequencies of alarm calling by various age and sex classes of Belding's ground squirrels. (From Sherman 1977.)

categories of female in their propensity to call. For instance, reproductive females lacking relatives other than their own young of the year give more alarm calls than non-reproductive females which lack relatives, when the two types of female are present in the same group. However, the same category of reproductive females gives fewer alarm calls when associated with reproductive females which do have female relatives alive in the area (e.g. grand-daughter, mother, sister). The latter group call more than the former. The question arises, why does the same type of reproductive female apparently respond to the presence of a predator in different ways when associated with two different categories of non-relatives? The answer is probably due to the fact that the latency of response to an observed predator is greater in those females without relatives compared with those which do have living relatives in the area. The latency of response is about 40 seconds in the former group and 15–25 seconds in the latter (Sherman personal communication). It would seem then that in the presence of females which respond quickly other females refrain from calling. In other words, reproductive females with living relatives (in addition to their own yearling offspring) respond more quickly to danger and give more alarm calls than reproductive females without relatives, which in turn give more alarm calls and possibly respond more quickly than non-reproductive females without relatives.

The intra-sexual and inter-sexual differences in calling frequencies and response times in Belding's ground squirrel support the hypothesis that alarm calls function to warn offspring and other relatives, and consequently that the trait has evolved through kin selection. Further evidence adds weight to that proposal since predictions from other theories for alarm calling in sedentary individuals (in this case females) were not supported. For example, alarm calls did not appear to extend the return times of the predator species, and the callers put themselves at risk because the predators stalked them more often than non-callers. What was surprising though was that those males which had copulated the most did not give more alarm calls than relatively unsuccessful males. However, it may be wrong to assume that mating frequency would be correlated with successful inseminations or any measure of paternity certainty. Overall, the salient features which resulted in the distribution of alarm calling were probably site fidelity and familiarity in conjunction with the close proximity of female relatives. The comparative importance of site familiarity and the presence of relatives on alarm calling behaviour could not be determined.

5.5.3 *Other field studies*

Further, less detailed studies on other species have also implicated, on the basis of sex differences in alarm call frequency, a major role for kin selection in the functional interpretation of the behaviour. In the round-tailed ground squirrel, females usually give more alarm calls than males although, interestingly, juvenile males call as frequently as juvenile females prior to dispersal (Dunford 1977). In the white-tailed deer, does which live in matrilineal groups emit an alarm snort when disturbed that is not found in comparable groups of unrelated bucks (Hirth & McCullough 1977). Such differences in calling frequency do not in themselves exclude all alternative hypotheses particularly those that would predict the evolution of calling in sedentary individuals whether relatives were present or not. (Similar problems are inherent in the findings that sedentary species of birds are more likely to emit distress screams than migratory ones (Rohwer *et al.* 1976) and that Kloss's gibbon has a loud call which supposedly warns kin in adjacent territories (Tenaza & Tilson 1977).) In addition, one further problem with inter-sexual comparisons is that it appears to be an implicit assumption that similar selective forces act on both sexes. This would not be the case if one sex was more vulnerable to predation than the other, if one sex was more vigilant, or if the modes of predator escape or defence were different. Such factors could produce sex differences in calling propensity without invoking the presence of kin as the underlying reason.

Most of the evidence so far has suggested kin selection as the possible mechanism whereby alarm calling has evolved. The number of species studied though is small, and the actual behavioural contexts and functions of those calls are even more restricted. Less attention has been given to examining alarm calling in other circumstances, particularly where alternative hypotheses may be more pertinent. For example, Owens and Goss-Custard (1976) consider that one function of alarm calls in wintering shorebirds (Charadrii) is to promote flocking, thereby reducing an individual's risk of predation. They noted that the warning calls of species that are widely dispersed when feeding are louder than those species that habitually forage in tight-knit flocks.

We suggest two approaches to tackling the evolutionary and functional questions raised by alarm call behaviour. First a comparative one in which such behaviour is examined in related species. There is some evidence already that this may be a fruitful area in the Sciuridae. Male biased calling occurs in the Alpine marmot, which has a harem social system (Barash 1976a) whereas female biased calling is prevalent in those ground squirrels in which male investment is relatively small and paternity probably uncertain (Sherman 1977, Dunford 1977). Second, there is a need for experimental data. Natural predation is rarely observed in the field, but there would seem ample scope for using trained predators (e.g. hawks on bird flocks) to elicit alarm calling as a preliminary step towards testing the various explanations in an area that has a surfeit of conflicting hypotheses.

5.6 Distraction displays

A distraction display is broadly defined by Wilson (1975) as 'any distinctive behaviour used to attract the attention of an enemy and to draw it away from an object that the animal is trying to protect'. There is little doubt that distraction displays employed most obviously by birds have evolved through kin selection; the objects of distraction are normally offspring, the individuals distracting are parents and the enemy is a predator. This type of behaviour has been reported in many species and there is often marked temporal variation in its intensity (Armstrong 1947). In those bird species with precocial young, parental distraction is most vehement at the time of hatching, whilst in the case of altricial young it is at the time of fledging (Armstrong 1956). Early interpretations explained the temporal differences in the intensity of distraction displays in terms of the risk of predation to the offspring. Recently, however, Barash (1975) has argued that the intensity of distraction should be examined in terms of the cost to the parent of losing a brood. This should increase, he proposes, with the amount of parental investment (*sensu* Trivers 1972). Although this interpretation should be rephrased to consider the subsequent investment required to reach the same stage of breeding if the parents had to recommence the breeding cycle, rather than considering the amount already invested (Dawkins & Carlisle 1976), the prediction that follows is essentially the same (see Maynard Smith 1977). As parental investment increases so should the intensity of the distraction display. This is apparently the case in the alpine accentor (*Prunella collaris*). The incubating or brooding female takes more risks in distracting a 'human predator' from the nest during the course of the breeding cycle (Barash 1975). Unfortunately, this finding does not negate the original hypothesis. A similar result would occur if there was a gradually increasing probability of predation during the nesting cycle. A nest site may become easier to find with time and as the fledglings grow they are both more conspicuous and more valuable to a predator. Account would have to be taken of these factors before the importance of either hypothesis could be assessed.

5.7 Mobbing

Mobbing may be defined as the hustling of predators by potential prey. It has been observed in both vertebrates and invertebrates, but is most visibly manifested in birds and mammals (see Wilson 1975 pp 46–47). Mobbing may involve direct attacks upon the predator and generally tends to make nearby individuals of the prey species aware of the presence of the predator. For instance, birds frequently emit an easily located call that attracts others to the supposed site of danger (Marler 1955). Calls may elicit interspecific as well as intra-specific-mobbing responses. So far the majority of studies have concentrated on the motivation, causation and ontogeny of the mobbing response (Hinde 1954a, b, Andrew 1961a, b, c, d, Curio 1959). Further work has shown that mobbing may be well developed in some species but absent in others (Altmann 1956), although the reasons for the differences are unknown. Intraspecifically, an innate mobbing response to a specific predator may be absent in populations lacking the predator (Curio 1975—the pied flycatcher *Fidecula hypoleuca*). Conversely, ground squirrels from populations exposed to predatory venomous snakes showed an attenuated mobbing response, compared with squirrels from areas where such snakes were not present and all mobbed snakes were non-poisonous (Owings & Coss 1977). In other words there was a reduced reaction in situations where the risks to the squirrels appeared to be greater.

The evolutionary origins and functions of mobbing are obscure. Several of the supposed functions of alarm calling (see above) could equally well be employed for mobbing. For instance, it could announce to the predator that observant prey are not worth pursuing (Alcock 1975). Alternatively, a group of mobbing animals could reduce the risk of predation to themselves or their offspring by successfully driving away or deterring the predator (Hartley 1950). There is some evidence to support this hypothesis in colonially nesting birds (Kruuk 1964, Hoogland & Sherman 1976). If, however, the function is to draw the predator's attention away from offspring or relatives, then mobbing could be considered a distraction display.

Some indication of the adaptive significance of mobbing in a territorial passerine comes from Curio's (1975) detailed study of anti-predator behaviour of the pied flycatcher. He distinguished two distinct responses to predators. First, 'snarling' which consisted of close attacks on predators which were a threat to the eggs and young, but not to adult birds (e.g. the great spotted woodpecker, *Dendrocopus major*). Second, mobbing which was performed at a distance from potentially dangerous predators (e.g. the sparrowhawk, *Accipiter nisus*) often for prolonged periods. Breeding pairs increased the intensity and frequency of both snarling and mobbing during the course of nesting period, peaking just before fledging. Parental responses continued after fledging, but at unknown level, during a period in which the young were probably extremely vulnerable to aerial predators. Further evidence also suggested that the anti-predator behaviour evolved to protect offspring, possibly at some risk to the parents. For instance, prior to the breeding season unmated birds had low level responses which continued through the summer if the birds remained unpaired. Mobbing was also reduced in birds that had lost a brood or were present in a territory other than their own.

Although it seems reasonable to propose that the protection of offspring is the major reason for mobbing in the pied flycatcher, the means by which this is achieved is still not clear. Curio considered that the main function was possibly to induce silence and lack of movement in young birds, thereby reducing their chances of being preyed upon. This explanation does not account for the presence of mobbing during the egg stage. The possibility that an increase in the mobbing response at such times is non-functional (Curio 1975), but maintained in the population because it is not detrimental seems unlikely. Clearly there is a need to test the different hypotheses experimentally, not only in the pied flycatcher but in other species where the functions and contexts of mobbing may be different.

5.8 Concluding discussion

Any genetically determined anti-predator device will only be selectively favoured if individuals possessing it have a higher inclusive fitness (Hamilton 1964) than those who do not. Consequently, any measure of the adaptiveness of such a trait should be based on that difference. In field studies, inclusive fitness is normally impossible to measure and even reproductive success can only rarely be estimated, particularly in males because of difficulties in assessing the likelihood of paternity. However, because kin selection is so often implicated in the hypotheses outlined in this paper, inclusive rather than individual fitness is the necessary measure. This problem is compounded in field work on anti-predator behaviour because predation is so rarely observed. Therefore, we need some other ways either to assess the costs and benefits of anti-predator tactics, or to test between the various hypotheses.

Clearly, such approaches need to be predictive. The design of relevant experiments is discussed in this paper and elsewhere (Bantock

& Harvey 1976). Good observational work can, however, discriminate between some hypotheses before experiments become necessary, particularly if the response of individuals to the presence of predators provides sufficient data (e.g. Sherman's study on alarm calling in Belding's ground squirrel discussed above). Although such studies may clarify the ultimate functions of a behaviour, they less often give insight into the proximate functions. Curio's (1975) work on mobbing is a case in point: the temporal pattern of the mobbing response in the pied flycatcher indicates that one ultimate function of mobbing is to protect the young, but how the behaviour achieves this is not known.

A further approach is the comparative one applied between species (see 1.3). This, we believe, will be important in distinguishing between many of the hypotheses described in this paper. If anti-predator strategies incur appreciable costs or benefits then they may markedly influence ecological, life history or behavioural characteristics of the species possessing them. For instance, mid-European passerines which nest in holes and are subject to lower rates of predation than open-nesting species, have larger clutches and longer incubation and fledging periods (Lack 1968). It is presumed that selection has acted to produce rapid growth and development in species with accessible nests in order to minimise the risks of predation.

Similarly, life history variables might be expected to determine whether a particular anti-predator device could be selected for or not. For instance, by comparing variables related to individual dispersal in the heliconids, Benson (1971) found correlations with unpalatability across species which were expected if kin selection had been important in the evolution of unpalatability, but not if individual selection had been of overriding importance. Turner's (1971) study on the same group demonstrated life history modifications which are expected as a consequence of the evolution of unpalatability and aposematic colouration (see 5.2). The assumption made here is that unpalability *is* an anti-predator device.

Finally, it would be wrong to leave the impression that for each and every predator there is a prey defence mechanism. The problems of evolving an effective anti-predator device may be too costly or of little worth (and therefore not increase the inclusive fitness of the prey), or beyond the biological capabilities of certain species. The vulnerability of birds eggs to predation has not resulted in the evolution of their unpalatability (Orians & Janzen 1974). Instead, selection has tended to produce behaviours which minimise the detection of nests (e.g. distraction displays, concealment, short incubation periods) or post-predation strategies which minimise the impact of predation (e.g. rapid replacement of lost clutches). The integration of anti-predator strategies and concomitant life history variables into an evolutionary framework is one of the most challenging areas for the behavioural ecologist.

Part 2
Sex, Mating and Signals

Introduction

As Michael Innes once wrote, the male plays a 'brief but decidedly seminal part in the creation of his children'. In the vast majority of animal species, sexual reproduction is the normal method of gene propagation, and males nearly always put far less than females into each offspring. Females (by definition) make larger gametes and hence put more reserves than the male into each zygote, and in animals with parental care females usually do most of the work. This fundamental division of labour poses a great problem for the theory of natural selection. To put it simply, sexual reproduction involves parasitism of females by males: parasitism of large gametes by small ones, and parasitism of parental care in many higher animals. Why do females accept male parasitism, and indeed expend half their resources creating sons instead of daughters? In Chapter 6, Maynard Smith shows how an hypothetical asexual female in a population of sexuals has approximately two-fold advantage in terms of gene propagation. While only half of a sexual female's offspring invest in their children, every one of the asexual female's offspring does so. Of course if males play enough part in parental care to double the output of offspring by a sexual female compared with an asexual, then sex may cost a female virtually nothing. But in animals without male parental care, sex has a two-fold cost: the cost of male laziness.

Maynard Smith discusses two types of advantage of sexual reproduction which might outweigh the two-fold cost: long term group-selected, and short term individual-selected benefits. The long term advantages of sexual reproduction are in maintaining genetic variability and allowing recombination of genes, both of which mean that in the long term, sexual species are less likely to go extinct than are asexuals. This explanation for the predominance of sexual reproduction in animals is the historical one, that asexual species simply do not persist. The other possible advantage of sex is short term: if there is intense selection and the optimal genotype changes between each generation, then the greater variability of sexual offspring could conceivably more than double the chances of one of them surviving, compared to an asexual offspring. Maynard Smith also discusses other fundamental questions about the evolution of sex; the sex ratio, when it pays to be a hermaphrodite, and why males of some species but not others perform parental care.

Whatever the advantage of sex, the unequal division of investment between male and female means that females are in effect a limiting resource for males. The ensuing intense competition between males for females has in many species led to the evolution of elaborate sexual displays through sexual selection. In Chapter 7 Halliday clearly reviews the current ideas about sexual selection, particularly those seeking to explain why such extreme secondary sexual characters are often produced.

Sexual selection in higher animals is more intense in polygamous and promiscuous species than when there is monogamy. For example Kroodsma (1977) has shown that the song of male wrens, which is used in mate attraction, is more elaborate in polygynous than in monogamous species. Halliday discusses the relationship between mating systems and sexual selection. He then goes on to review the evidence that females do actually exercise choice in mating, and points out what benefits females could get from choosing.

A specific example of sexual competition is reviewed by Parker in Chapter 8. The reason for devoting a whole chapter to dungflies is that Parker's study is one of the few successful attempts to apply quantitative reasoning in behavioural ecology. It brings together many ideas from different parts of the book: the time allocation models are exactly analogous to those discussed for optimal foraging in Chapter 2; the theory of habitat choice links with Chapters 11 and 12; and ESS theory appears in many other chapters, notably 6 and 10. The problem analysed by Parker is simply this: 'How should a male dungfly allocate his searching effort in time and space in order to get as many successful matings as possible?' Female dungflies lay their eggs in cowpats but arrive by landing on the grass nearby. Males could search for a mate in either place, and the benefit from doing one or the other depends, among other things, on where other males are searching and the rate of arrival of females. Parker is able to work out in theory and test the mixed ESS at which all males are doing as well as they can, and equally well.

The temporal problem for a male is how long to stay on a dung pat.

The input of females gradually declines as the pat ages, and of course the success of an individual male depends on the degree of competition. Again there is no single optimum strategy, but a mixed ESS for stay-times: as with the spatial equilibrium, the data fit remarkably well to the predicted solution. Parker also discusses sperm competition: by taking over a female from another male, a second male can displace the first one's sperm and fertilise most of the eggs. The proportion of eggs fertilised by a particular male depends on the duration of copulation which in turn allows Parker to predict the optimal copulation time.

The classical ethological work on animal signals was based on studies of sexual displays—courtship and threat signals. Dawkins and Krebs review the traditional argument that displays are designed to communicate information to the recipient about the state of the signaller. They argue that this view does not accord with the theory of natural selection. Animals communicate because of self-interest: words such as manipulation, deception and persuasion are more appropriate than the concept of information in the analysis of animal signals.

Although reproduction is presented throughout this section of the book as a selfish rather than a co-operative affair, there are some striking examples of apparently cooperative and altruistic breeding behaviour. Social insects were discussed in 1.5 and cooperative carnivores were briefly mentioned in Chapter 3. In Chapter 9 Emlen analyses in detail cooperative breeding in birds.

The simplest form of cooperative breeding in birds (illustrated by the Florida Scrub Jay) is when one pair of breeding adults is aided by their young from a previous breeding attempt. The young are related to their younger siblings by 0·5, so at first sight kin selection seems to provide an explanation for the apparent altruism (see 1.5). But as Emlen shows, this is not the whole story. Often young birds do not increase the reproductive success of the breeders, and further, they could in theory do better (in terms of inclusive fitness) not by helping but instead by going off to breed on their own. However, a lack of empty breeding territories may prevent them from doing so. Emlen stresses that a full analysis of the benefits of cooperative breeding should take into account the differing interests of helpers and breeders. A conflict of interest is even more apparent in communal nesters, where several females lay in the same nest. In the Groove-billed Ani, one such species, females even roll each others' eggs out of the nest!

Emlen then describes a more complex system of cooperative breeding in the White-fronted Bee-eater. In this colonial species, not only do groups cooperate in rearing young, but there are also alliances between members of different breeding groups. Members of a co-operative group are often not close relatives, so kin selection may play less of a role than in Scrub Jays. The ecological circumstances favouring this cooperative system as well as the genetic benefits derived by each individual, have not yet been worked out, but when these two factors are better understood they will surely throw light on evolution of complex societies in general.

Chapter 6
The Ecology of Sex

JOHN MAYNARD SMITH

6.1 The function of sex

Although sexual reproduction is almost universal its functional significance is still a matter of controversy. In eukaryotes, its essential features are the production of gametes by meiosis (halving the chromosome number) and the production of a new individual by syngamy (the fusion of two gametes to form a zygote). In higher animals and plants, there is a division of labour between male and female gametes, egg and sperm. In many protozoa and green algae there is no morphological differentiation between the gametes ('isogamy', as opposed to the 'anisogamy' of higher forms), but there is usually differentiation into 'mating types', such that a gamete can fuse only with one of a different type. Finally, in the ciliated protozoa, no free gametes are produced, but there is a process of conjugation in which haploid nuclei are exchanged, with genetic consequences similar to the mutual cross-fertilisation of two hermaphrodites (see 6.2).

The consequence of this process is to bring together in a single cell genes from two parental cells. The process of genetic recombination which occurs during meiosis also ensures that genes from different but homologous chromosomes can be combined in a single chromosome. This has the result that a population reproducing sexually can evolve more rapidly to meet changed circumstances. This point was first made by Fisher (1930) and Muller (1932). There is some controversy about how great the effect is (Crow & Kimura 1965, Maynard Smith 1968), but it is certainly substantial, particularly in large populations. The reason is easy to see qualitatively. Suppose that in an asexual population two favourable mutations, $a \rightarrow A$ and $b \rightarrow B$, were to take place in different individuals. There would be no way in which an AB individual could arise, except by the occurrence of a second B mutation in a descendent of the original A, or of an A mutation in a descendent of B.

In a sexual population an AB individual could arise by recombination. The reason why sex is unimportant in a small population is that each favourable mutation will be fixed by selection before the next occurs.

Muller (1964) pointed out a second advantage of recombination, which has been called 'Muller's ratchet' by Felsenstein (1974). Consider an asexual population. It is inevitable that slightly deleterious mutations will occur. We can classify the individuals according to whether they have 0, 1, 2 . . . deleterious mutations. Suppose the number of individuals with no deleterious mutations is small. Then, although these individuals are fitter than average, there is a chance that in one generation they will fail to reproduce. In the absence of sex, there is no way in which an individual free of deleterious mutations can arise again (save by back mutation, which can be shown to be unimportant). The 'ratchet' has clicked round one notch. In this way, deleterious mutations will accumulate in the population. But if sexual reproduction occurs, an individual with no deleterious mutations can be produced by recombination between two individuals with different mutations; the ratchet would cease to turn.

A quantitative analysis of Muller's ratchet (Maynard Smith 1978) suggests that it could have been important in the origin of recombination very early in the evolution of life. Two individuals, each carrying a deleterious mutant, could produce one perfect individual (and, of course, one doubly damaged one) by recombination. Recombination is then seen as simply a form of DNA repair. Muller's ratchet may also be important in causing deterioration of parthenogenetic strains of higher plants and animals.

So far, the advantages I have suggested for sex (acceleration of evolution, preventing the accumulation of deleterious mutations) are advantages to the population as a whole. This has lead to the 'long-term' or 'group selection' explanation of sex. Species are, for the most part, sexual because those which abandon sex fall behind in the evolutionary race, and go extinct.

There are good grounds for being cautious about group selection explanations in biology (Williams 1966, 1975, Maynard Smith 1964, 1976a). The crucial difficulty is that a gene increasing individual fitness will spread through a population, even if it reduces the fitness of the population; if populations are not completely isolated, a gene which has spread in one population can 'infect' others (see 1.4). There are, however, two reasons why group selection cannot be ruled out as an explanation of sex, even though it is probably unimportant in most other contexts. First, parthenogenetic mutants (which would increase individual fitness) are rare; second, species really are reproductively isolated (see also 6.1.1).

I will discuss the maintenance of sex rather than its origin, because we only have evidence concerning the former. One crucial point is that in a sexual population a parthenogenetic mutant would, other things being equal, have an immediate twofold advantage. Thus if, on average, every female can produce two surviving offspring, a sexual female produces one female offspring like herself, and a parthenogenetic female produces two female offspring like herself. The point is illustrated in Fig. 6.1. Sexual reproduction entails the twofold disadvantage of producing males.

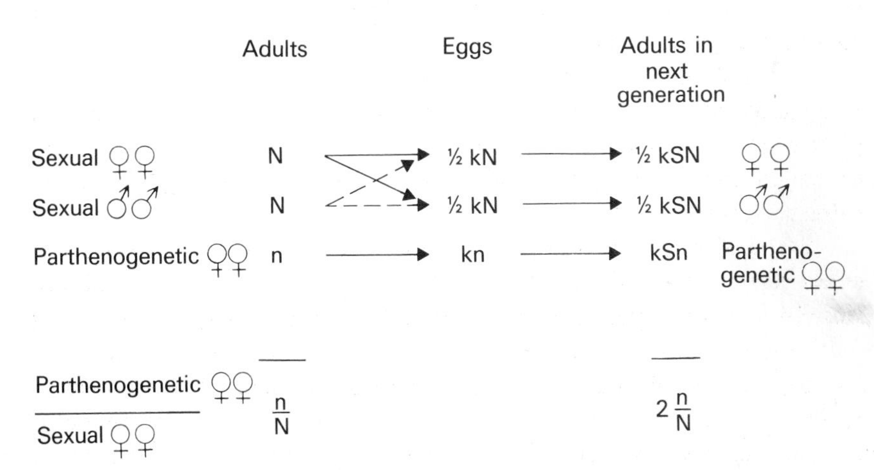

Fig. 6.1. The twofold disadvantage of producing males. k is the number of eggs laid by a female, and S the probability that an egg will survive to become an adult.

Does the twofold disadvantage arise in isogamous species? The problem is illustrated in Fig. 6.2. A gene A which suppressed meiosis would gain some advantage through not wasting time; this may be one reason why meiosis in protists often occurs at times of food shortage when growth is in any case impossible. However, the gene A does not obtain a two-fold advantage, which therefore need not be taken into account when discussing the origin of eukaryotic sex (it is fairly safe to assume that isogamous sex preceded anisogamy). In higher organisms, if there is paternal care a sexual female which pairs with a male may produce twice as many offspring as a lone parthenogenetic female; if so, sex does not carry a two-fold disadvantage. (The problem remains why, in a typical sexual species with paternal care, does not a partheno-

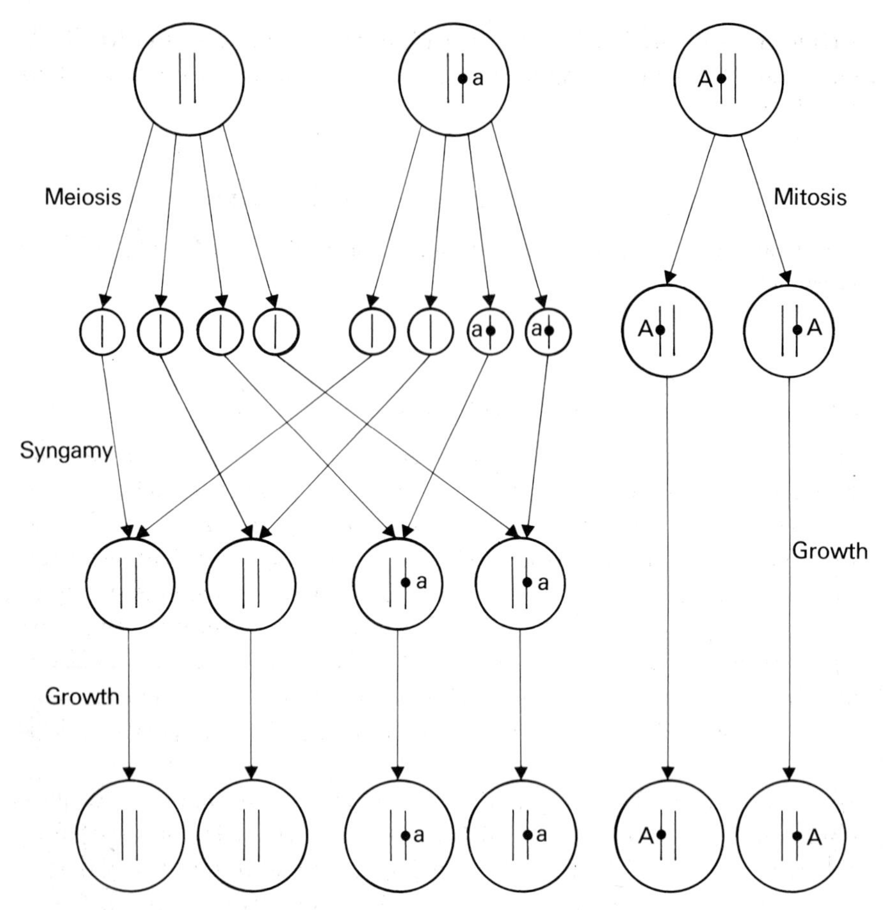

Fig. 6.2. Does meiosis confer a twofold disadvantage in an isogamous population? The gene A suppresses meiosis. After mitosis, the number of copies of A is the same as the number of copies of an allele a after meiosis and syngamy. Therefore, there is no necessary twofold disadvantage, but in a growing population there would be some advantage to A through time-saving.

genetic mutant arise which accepts the male's contribution to parental care but not his genetic contribution.) But, despite these exceptions, the selective forces which maintain sex must in general overcome this two-fold short-term disadvantage.

The difficulty of the group selection explanation should now be clear. In its simplest form (and we shall see later that this simple form may be misleading), it implies that each time a parthenogenetic variety arises, it will, because of its twofold advantage, replace the sexual species which gave rise to it. Later, the parthenogenetic species will go extinct in competition with more rapidly evolving sexual species. Thus each parthenogenetic origin is balanced by a species extinction. Can this be true?

6.1.1 *The long-term advantage of sex*

Three arguments which support the group selection theory will be considered first.

The origin of parthenogenetic strains may be a rare event

In animals, parthenogenesis can occur in one of two ways, automixis and apomixis. In automixis, the egg undergoes meiotic reduction in the normal way, and diploidy is restored either by fusion of two of the four pronuclei, or by fusion of two genetically identical cleavage nuclei. Most cases of occasional spontaneous parthenogenesis in typically sexual species are of this type. Since the process leads to wholly or partially homozygous offspring, usually of low viability, it is unlikely to give rise to a successful parthenogenetic strain, although some naturally occurring strains are automictic (e.g. in enchytraeid oligocheates).

In apomixis offspring genetically identical to the parents are produced without meiosis. The origin of such strains of animals is probably a very rare event, although it has occurred, for example, in the ancestry of the cyclically parthenogenetic aphids and cladocerans, and the obligately parthenogenetic weevils.

Among the vertebrates, parthenogenesis occurs in nature in several genera of fish, amphibia and lizards. In the whiptail lizards (*Cnemidophorus*), it leads to the production of offspring genetically identical to the mother, but by the elaborate procedure of first doubling the chromosome number, then pairing sister chromosomes and undergoing a 'normal' meiosis to produce a diploid egg nucleus.

Unhappily, the botanical terminology is totally different to the zoological. However, the essential fact is that when higher plants produce seeds which develop without fertilisation, those seeds are genetically identical to the parent. It may be that the origin of such parthenogenetic varieties is much commoner in plants than animals. One reason may be that in higher plants a single somatic cell can develop into a whole plant, whereas in higher animals it seems that it may be very difficult to initiate development without meiosis.

The taxonomic distribution of parthenogenetic varieties suggests that they are short-lived in evolutionary time

If the abandonment of sex condemns a species to early extinction, we would expect to find that most existing parthenogenetic varieties are similar to sexual species. We would not expect to find large taxa (e.g. families or orders) consisting wholly of parthenogenetic varieties; nor would we expect to find isolated parthenogenetic varieties unrelated to any sexual form. By and large, this is what we do find. In animals, existing parthenogenetic forms must represent hundreds, and in all probability thousands of separate origins. Yet with one important exception (the bdelloid rotifiers) no major taxon (subfamily or above) consists predominantly of parthenogenetic forms, and there is no taxonomically isolated parthenogenetic species. To give a particular example, Mockford (1971) states that there are 28 known parthenogenetic strains of psocids; they belong to 13 different families, and in 12 of the 28 cases there are sexual and parthenogenetic forms of the same nominal species. A fundamentally similar picture is found in plants.

In case it be thought that this taxonomic distribution is an accident of sampling, it is worth comparing it with the distribution of another sexual system, male haploidy ('arrhenotoky'). Existing arrhenotokous species may be descended from as few as eight ancestral lineages; of them, five have given rise to families or larger taxonomic groups (the Hymenoptera, Thysanoptera, Monogonont rotifiers, a group of mites and a group of homopteran bugs) and one to a taxonomically isolated beetle species, *Micromalthus debilis*. Thus, despite the problem raised by the bdelloid rotifiers, the taxonomic distribution of parthenogenesis strongly supports the view that it leads to early extinction.

An established parthenogenetic variety may not replace its sexual ancestral species

A parthenogenetic variety may arise as a unique event, and contain, at least initially, only a single genotype. Despite its twofold advantage, it is therefore unlikely to replace its sexual ancestor over the whole ecological range to which the latter is adopted. For example, the lizard *Cnemidophorus uniparens* has been shown by skin-grafting to consist of a single clone (Cuellar 1976).

C. uniparens may be of recent origin. Parker and Selander (1976) have found, using electrophoretic techniques, that another parthenogenetic 'species', *C. tesselatus*, is more variable. This species consists of diploid hybrids between two sexual species, and of triploid hybrids between diploid *C. tesselatus* and a third sexual species. All the triploids examined belonged to a single clone, suggesting a single origin, but 12 distinct diploid biotypes were found. These probably represent 5 distinct hybridisation events, with some further variability being generated within clones by mutation and recombination. The interesting point is that despite a minimum of 6 separate original clones, with their twofold advantage, the parthenogenetic forms have not wholly replaced their sexual ancestors. These lizard clones do illustrate the fact that a single established variety has too narrow a range of genetic variability to eliminate its sexual ancestors.

The fish *Poeciliopsis monacha-occidentalis* is a parthenogenetic hybrid between the sexual species *P. monacha* and *P. occidentalis*. Moore (1976) has attempted to measure three components of fitness of the hybrids: the advantage from not producing males, the disadvantage arising because the hybrid must mate with male *P. occidentalis* before its eggs will develop (although the chromosomes provided by the male are later eliminated), and the 'primary' fitness when these two factors have been allowed for. He finds that the primary fitness is high in the region where the ranges of the parental species meet, but falls to half that of *P. occidentalis* further north. Although the genetic system of *Poeciliopsis* is bizarre, Moore's experiments do indicate that a parthenogenetic variety with a narrow range of genotypes could not replace the sexual parent over its whole range.

This point is demonstrated on a larger scale by several plant complexes, of which *Taraxacum* (dandelions) is an example (A.J. Richards 1973). There are some 1000 asexual 'species' of *Taraxacum*, and some 50 sexual ones. Parthenogenetic varieties probably first arose in the Cretaceous. New ones continually arise either from sexual ancestors, or, more probably, by hybridisation between a sexual species and an asexual variety acting as pollen parent and carrying the gene for apomixis. It is a striking fact that the sexual species still survive, albeit restricted in range, despite continuing competition from their asexual descendents. Part of the reason may be that the asexual varieties do not reap the full two-fold advantages of parthenogenesis, since they produce large flowers and, in most cases, pollen. The presence of large flowers and pollen is itself a puzzle, since they serve no obvious function in a parthenogen. Even though the pollen may occasionally contribute to a new clone, this is a selective advantage which arises very rarely. A possible explanation is that these clones, which evolve slowly and which are often of relatively recent origin, have not yet had time to adapt to parthenogenesis; but I do not feel much confidence in this explanation.

To summarise, group selection cannot be ruled out as an important force maintaining sex. Parthenogenetic varieties not suffering from homozygosis may arise rather seldom, particularly in animals; their taxonomic distribution suggests that they may have a short future; a single clone is unlikely to replace the whole of a sexual species.

6.1.2 *The short term advantage of sex*

There are, however, strong arguments on the other side, which have been particularly stressed by Williams (1975).

The 'balance' argument

Some species reproduce both parthenogenetically and sexually. There must be some genetic variance in the relative frequency of the two modes, and selection would therefore eliminate the sexual mode if it did not confer some short-term advantage.

The argument applies in two contexts, cyclical and facultative parthenogenesis. *Daphnia* and other cladocerans reproduce mainly by apomixis, producing eggs which develop immediately into females. Occasionally, perhaps in response to crowding, males and females are produced apomictically. Fertilised 'winter eggs' are then produced, which sink to the bottm, and hatch only after a prolonged period. One might therefore be tempted to argue that sex is maintained in the short run not because of any genetic consequences, but because winter eggs are the main means of dispersal, and of surviving the drying up of ponds. But this will not do, because some strains of *Daphnia* produce winter eggs apomictically. The balance argument for a short-term advantage for sex seems to hold. Similar arguments apply to other cyclically parthenogenetic groups—Aphids, cynippid wasps and monogonont rotifiers.

An individual plant of the grass *Dichanthium aristatum* can produce both apomictic and sexual offspring. The relative frequency can be altered environmentally (Knox 1967), and one must suppose that it could be altered genetically. Such facultative parthenogenesis is a powerful argument for a short-term advantage for sex. The case is not unique, but it is difficult to decide how common it is. There are many cases of populations consisting of an excess of females, some of which reproduce parthenogenetically. Unfortunately, it is often not known whether these populations are a mixture of sexual and obligate parthenogenetic females (in which case the parthenogenetic variety may contain only a narrow range of genotypes), or whether they are facultatively parthenogenetic, in which case the balance argument applies.

The timing of sexual reproduction

Williams (1975) points out that if a species can reproduce both sexually and vegetatively, the sexual phase is always timed to coincide with dispersal into an unpredictable environment, when recombination is most likely to be advantageous. The winter eggs of *Daphnia* are an example. In plants which reproduce both vegetatively (by stolons, rhizomes, bulbils etc.) and sexually (by seeds), it is the sexual propogule which is provided with means of dispersal.

It is not easy to evaluate these opposing arguments. Clonal extinction, in competition with more rapidly evolving sexual competitors, plays some role in maintaining sexual reproduction. But the 'balance' argument suggests that there must also be short-term advantages to sexual reproduction. The problem is discussed at length in Williams (1975) and Maynard Smith (1978).

There is space here to describe only one possible short-term mechanism—that of 'sib competition'. Figure 6.3 illustrates sib competition in a plant. It is supposed that the environment is divided into 'patches' (which can be contiguous), each large enough to support a single adult plant. A number of seeds fall in the patch, but only one can survive; in general the seed which is genetically best adapted to the patch will be the survivor. The essential feature (first proposed by Williams 1975) is that a single parent plant contributes a number of seeds to the same patch. Given that the environment in each patch is sufficiently unpredictable, sexual plants will increase in frequency relative to the asexual ones. The reason is best seen by analogy with a raffle. Each patch can be thought of as a raffle with a single prize. An asexual parent is like a man who buys a number of tickets in the raffle and finds that they all have the same number; in contrast, the sexual parent resembles a man who buys fewer tickets, but all with different numbers.

Sib competition can provide a short-term advantage for sex, provided that (i) selection is intense and density-dependent, (ii) sibs compete with one another, and (iii) the environment is unpredictable between generations. The necessity of sib competition can be shown by modifying the model of Figure 6.3 so that each parent contributes only

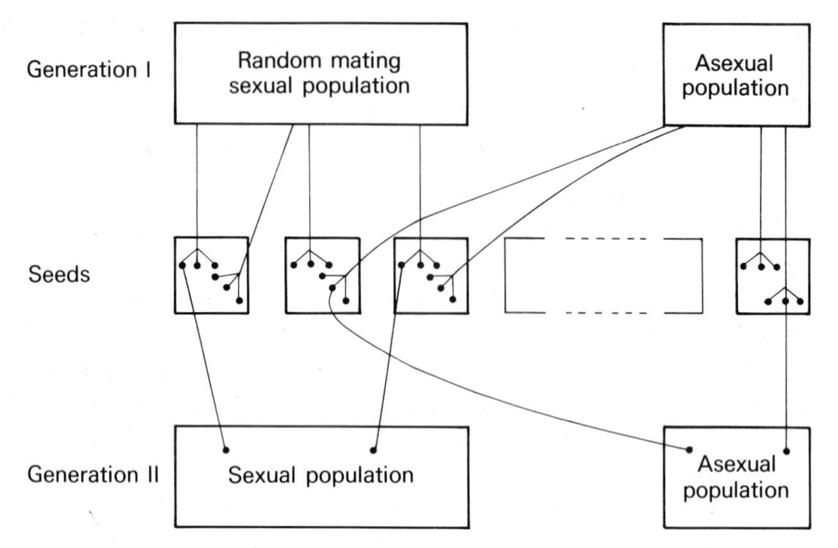

Fig. 6.3. A model of sib competition. Two parents each contribute three seeds to each patch. Only one of these seeds can survive. The sexual individuals are hermaphrodites, and produce half as many batches of seed as asexual ones. To make up for this twofold disadvantage, sexual seeds must be twice as likely to win in competition in the patches. Computer simulation suggests that for this to be the case, a parent must contribute several seeds to a patch (i.e. there must be competition between sibs). Also there must be at least 30–40 seeds in each patch (i.e. selection must be intense); for simplicity, only 6 seeds per patch are shown in the diagram. (From Maynard Smith 1976b.)

one seed to each patch; sexual reproduction then confers no advantage, and of course still suffers its two-fold disadvantage. It follows that there must be other selection pressures, long or short term, maintaining sex. The problem is related to that of how, within sexual populations, genetic recombination is maintained; this cannot be explained by group selection, because there is plenty of evidence to show that there is within-population genetic variance for recombination frequency.

Although these problems cannot be discussed in detail, a few general remarks are in order. First, selection favouring sex and recombination arises because the environment changes in space and time. Second, merely to state that the environment is unpredictable is no explanation. Third, the one well-established situation in which it has been shown that sex is advantageous is when the offspring of a single parent compete with one another; when this is so, a sexual parent has an advantage by producing a genetically variable progeny, thus increasing the chance that at least one of them will be well adapted.

Finally, and most important, the selective forces responsible for the evolution of sex and recombination are still inadequately understood.

6.2 Hermaphroditism, selfing and outcrossing

It will help to start with the meanings of some words:

'Gonochoristic': having separate male and female individuals; for example, all mammals and birds.

'Dioecious': the botanical equivalent of gonochoristic; for example, holly trees (*Ilex*) and stinging nettles (*Urtica dioica*).

'Hermaphrodite': in animals, an individual which produces both eggs and sperm. It may be 'simultaneous' as in most land and fresh water snails, or 'sequential', when an individual is first male and then female or *vice versa*.

In plants, hermaphrodite refers to the flower and not to the whole plant. A hermaphrodite flower produces both seeds and pollen. This condition is commonest in flowering plants and most certainly primitive.

'Monoecious': in plants, having separate male and female flowers on the same plant; e.g. birch trees (*Betula*).

In understanding the functional significance of hermaphroditism, a crucial question is whether an individual is self-fertile. Many hermaphrodite plants (e.g. all members of the pink family, the *Caryophyllaceae*) are self-compatible. In such cases, the likelihood that pollen will fertilise ovules from the same flower is reduced if anthers and stigma ripen at different times, but since different flowers are not synchronous this does not prevent a plant from pollinating itself. Many groups of plants, however, have evolved more effective mechanisms preventing self-pollination (for a review, see Grant 1958). A variety of genetic self-incompatability mechanisms have evolved which prevent the growth of the pollen tube down the style of the same individual. Some hermaphrodite species are divided into two (e.g. *Primula*) or even three (e.g. *Lythrum*) morphologically distinct types, such that pollination typically only takes place between members of different types; this phenomenon, known as heterostyly, was first elucidated by Darwin (1877).

Animal hermaphrodites show a similar range of ability to fertilise themselves. For example, among gastropods *Helix* and *Cepaea* are completely self sterile; *Biomphalaria* (a fresh-water planorbid) typically cross-fertilises but can fertilise itself if kept isolated; *Rumina*, a land snail, typically self-fertilises in the wild. In general, however, much less is known of animals than plants because the matter is harder to investigate; in a plant one can simply put a plastic bag round a flower and see whether it sets seed.

Three main selective forces have been responsible for the evolution of hermaphroditism: the difficulty of finding a mate or being pollinated; the genetic effects of inbreeding; the allocation of resources between male and female functions. These will be discussed in turn.

The 'low density' advantage of hermaphroditism

A single self-fertile hermaphrodite is able to colonise a new habitat, and to reproduce even if the population density is so low that it never meets another member of its species. It therefore makes sense (Baker 1955) that many annual weeds of disturbed soil are self-fertile. Families such as the Caryophyllaceae, which are primitively self-fertile, include many such colonising species, for example *Stellaria* (chickweeds) and *Arenaria* (sandworts). Many other weeds are apomicts, which have the same advantage as colonisers. Apomicts do not suffer the inbreeding depression which an initially outcrossing population undergoes when it starts selfing, but lose the possibility of genetic recombination.

Ghiselin (1969) has argued that the same selective force has been important in the evolution of hermaphroditism in animals. The case is harder to substantiate because often we do not know whether a particular species is self-fertile. However, as was pointed out by Tomlinson (1966), even a self-sterile hermaphrodite is at some advantage over a gonochorist at low density, because any two individuals are certain to be able to mate. Supporting Ghiselin's view is the fact that hermaphroditism is common among sessile animals, which cannot move around to find a mate, and among parasites (e.g. flukes, tapeworms, and many parasitic genera in predominantly gonochoristic groups, such as crustaceans, nemertines and prosobranch molluscs).

The genetic advantages and disadvantages of selfing

If offspring produced by selfing were on average as fit as those produced by outcrossing, all hermaphrodites would fertilise themselves. Thus imagine a gene A, conferring self-compatibility, arising in a population of self-sterile hermaphrodites (Fig. 6.4). An individual with gene A would always fertilise itself, and would have the same success in outcrossing as a typical member of the population. If the population size were constant, a typical individual would contribute two gametes to the next generation, one as a female and one as a male; an individual

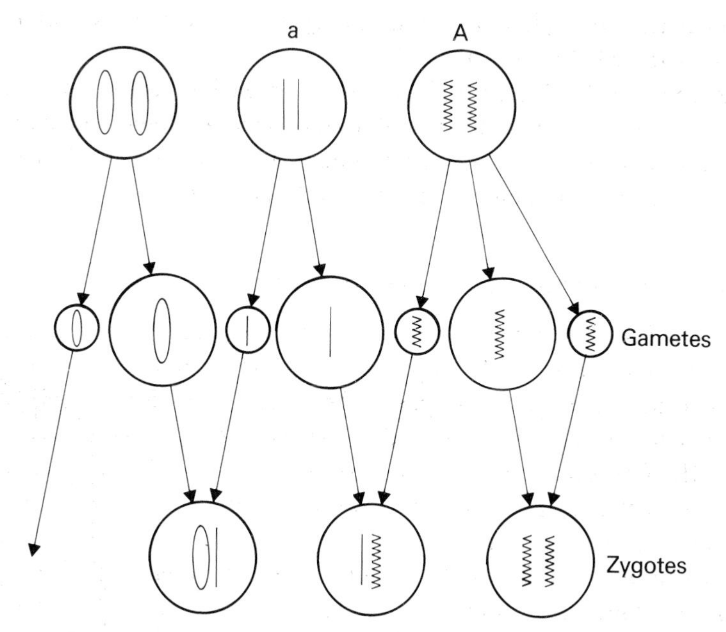

Fig. 6.4. The advantages of selfing. The individual A is a self-fertile hermaphrodite; individual a is self-incompatible. In the zygote population there are three copies of each chromosome from A for every two from a.

with gene A would contribute three gametes, one as a female, one as a male by selfing and one as a male by outcrossing. By a similar but more subtle argument (Bengtsson 1978), if inbreeding had no deleterious effects all females would mate with their brothers.

The long term consequences of selfing in hermaphrodites (or inbreeding in gonochorists) would be the loss of the evolutionary advantages of sex. Populations which took this road might ultimately be eliminated by inter-population selection. But there is a much more immediate selection pressure countering the spread of selfing and inbreeding: the offspring so produced are usually of low fitness. The exact reasons for this are not immediately relevant, and indeed are still to some extent in dispute. All that matters in the present context is that selfing does in fact lead to a dramatic loss of fitness (for a general discussion, see Lerner 1954).

In plants, it seems that dioecy has often evolved from hermaphroditism because it prevents selfing, and thus ensures that an individual produces fit outbred offspring. Monoecy may evolve for the same reason; often the male and female flowers grow on different parts of the plant, which further reduces the likelihood of selfing. This explanation makes sense only if it is in some way easier for a hermaphrodite species to evolve dioecy or monoecy than self-sterility. It is not clear why this should be so, but the taxonomic and ecological distribution of dioecy suggests that it is.

In plants, dioecy tends to occur sporadically in taxa which are otherwise predominantly self-fertile hermaphrodites (Baker 1959); for example, it has evolved at least three times among the Caryophyllaceae.

Ecologically, dioecy and monoecy are commoner in trees than in shrubs, and in shrubs than in herbs (Table 6.1). The reason is probably

Table 6.1. Percentages of forest plants with different habits.

Type of forest	Layer	Hermaphrodite	Monoecious	Dioecious	Reference
Deciduous woodland in Britain (four types)	Tree	50–57	14–30	20–33	Baker (1959)
	Shrub	60–89	0–38	0–23	
	Field	86–93	0– 9	0– 9	
Five North American temperate forests	Tree	0–27	60–83	6–17	Bawa & Opler (1975)
Seven tropical forests	Tree	41–68	10–22	22–40	Bawa & Opler (1975)

as follows. As the number of flowers on a single plant increases, the likelihood of self-pollination also increases. A small herb can rely on the movements of insects from flower to flower to ensure an adequate frequency of cross-pollination; a large tree cannot.

Resource allocation in hermaphrodites

It is possible to explain dioecy in plants as an adaptation to ensure outcrossing. But this does not explain why in higher plants hermaphroditism is the rule and dioecy a sporadically distributed derived condition, whereas in animals gonochorism is the rule. Charnov *et al.* (1976b) suggested a third selective force which may help to explain this and other features of the distribution of hermaphroditism.

Imagine a plant population consisting of males, females and self-sterile hermaphrodites. Each male produces N 'successful' pollen grains (i.e. pollen grains which contribute to a zygote in the next generation), and each female n successful seeds. Each hermaphrodite produces αN successful pollen grains and βn successful seeds. Obviously, whether hermaphrodites will replace gonochorists, or *vice versa*, depends on the values of α and β. In fact, it can be shown that if $\alpha+\beta>1$ hermaphrodites will win, and if $\alpha+\beta<1$ gonochorists will win. The reason can be glimpsed intuitively. Suppose that $\alpha=\beta=\frac{1}{2}$; that is, a hermaphrodite produces half as many successful pollen grains as a male and half as many seeds as a female. In fact, it will pass on as many genes to the next generation as a typical male or female. But if both α and β were greater than $\frac{1}{2}$ (and hence $\alpha+\beta>1$), the hermaphrodites would pass their genes on to more offspring than males or females, and so would replace them.

What factors will tend to make $\alpha+\beta>1$? For plants, three factors can be suggested. First, for any individual the period during which resources must be allocated to pollen production is earlier than the period when seeds are dispersed; consequently, a hermaphrodite may be able to allocate a greater total quantity of resources than either a male or a female. Second, in a hermaphrodite flower, there is a potential saving of resources because the same organs (petals, nectaries) serve both male and female functions. Third, and probably most important, in a sessile organism with limited pollen dispersal, a doubling of the number of pollen grains would not mean a doubling of the number of successful pollinations, because of competition between pollen grains from the same plant; similarly, a doubling of the number of seeds would not mean a doubling of the number of successful seedlings. In both cases, a law of diminishing returns would set in.

These factors do not operate with the same force in animals. The 'law of diminishing returns' may operate in sessile animals, but these often have pelagic larvae, so the law will not apply to egg production. It is unusual for the same organ to serve both male and female functions. More often, it seems likely that a doubling of investment in a particular sexual role would more than double success; this is particularly so of males which compete for females.

In gonochoristic animals, there are opposing forces favouring inbreeding and outcrossing, parallel to those favouring selfing and outcrossing in hermaphrodites. In general, the severity of inbreeding depression seems to have favoured outcrossing. There are two ways in which an animal might reduce the likelihood of mating with a close relative:

(i) by dispersing before sexual maturity in such a way that close relatives are unlikely to be encountered; (see also 3.5.2).

(ii) by recognising relatives, and refraining from mating with them. As an example of dispersal preventing inbreeding, it is common in group-living mammals for males raised in one group to transfer to another before breeding. The olive baboon, *Papio anubis*, affords a striking example. Packer (1975) found, in a 6-year study of three troops, that none of the males which reached maturity during the study remained with their natal troop. Of 41 known transfers, 39 were males. Although in most social mammals it is the males which transfer, it is sometimes the females (e.g. chimpanzees, hunting dogs).

There is no evidence that animals can recognise their relatives by their genotypes. But, as a device to avoid inbreeding, it is sufficient that an animal should recognise those individuals with which it was raised and should refrain from mating with them. Hill (1974) showed that in the deermouse *Peromyscus maniculatus*, there is a delay in reproduction if the two potential parents have been raised together, whether or not they are actual sibs. There is anecdotal evidence of this kind for other mammals and birds. Interestingly, the effect may operate in our own species. A study of Israeli kibbutz (Shepher 1971), in which children of both sexes are brought up communally, showed that of 2769 marriages between second generation kibbutz adults, not one was between members of the same peer group (see also 3.5.2).

6.3 The sex ratio

Females, by definition, produce large gametes containing substantial food reserves, whereas males produce small mobile ones. If males and females have equal capacities for acquiring resources, it follows that a male can produce many times as many gametes as a female, and can fertilise the eggs produced by many females.

For the population, therefore, the greatest production of offspring would be achieved if there were many times as many females as males. This is rarely the case, because selection acts on individuals, or more precisely on genes, and not on populations. Fisher (1930) explained why, in a random-mating population, the evolutionarily stable sex ratio is 1 : 1. Suppose, he argued, there were more females than males. Then a parent who produced only sons would have more grandchildren than typical members of the population. Therefore the gene which caused the parent to produce sons would increase in frequency. By an identical argument, if there were more males than females a gene causing parents to produce daughters would increase in frequency. Hence in a population with an excess of one sex, selection would favour the production of the other. Only the 1:1 sex ratio would be stable. More precisely, Fisher argued that a parent will expend equal resources on offspring of the two sexes. This is a difficult argument to see without a mathematical treatment, but a numerical argument may help. Suppose that females 'cost' twice as much in resources as males, and that a parent can produce 2 ♀♀ or 4 ♂♂ or 1 ♀+2 ♂♂. If the population sex ratio were 1:1, then on average males and females have equal numbers of offspring. Therefore a parent producing 4 ♂♂ would have more grandchildren than one producing 2 ♀♀, and so the population would not be evolutionarily stable. But if there were twice as many males as females in the population, males would on average have half as many children, and therefore a parent producing 4 ♂♂ would expect the same number of grandchildren as one producing 2 ♀♀. Such a population would be stable; it is one in which twice as many males as females are produced, which implies equal investment by parents in the two sexes.

Is there any evidence that the sex ratio is distorted away from 1:1 when offspring of the two sexes require different amounts of parental investment? In some birds the sexes differ in size at fledging. Howe (1977) looked at sex ratios in the common grackle, *Quiscalus quiscula*, in which males are 20% heavier at fledging. He found 52 males and 83 females at fledging; the difference appears to arise because of heavier male mortality, since he found a 1:1 ratio among embryos. These results accord with Fisher's prediction; unfortunately, they also accord with the simpler view that the primary sex ratio is 1:1 because that is the ratio produced by a regular meiosis. Thus Howe's results are consistent with Fisher's prediction, but do not prove that natural selection has modified the sex ratio away from 1:1. Equally ambiguous results emerged from the more extensive studies by Newton and Marquis (1978) on the european sparrowhawk, *Accipiter nisus*. Adult females are twice as heavy as males; they hatch from similar eggs, but a twofold weight difference exists already at fledging. The authors found 1102 males and 1061 females, in 651 broods. These data appear to contradict Fisher's prediction that the parents should equalise investment rather than numbers. However, male and female nestlings ate the same amount. Females put on weight faster, but males feather sooner and leave the nest 3–4 days earlier. So the costs of a male and a female may not be so different as their weights suggest. More data is needed on the evolutionary adjustment of the sex ratio.

Hamilton (1967) pointed out that if there is inbreeding Fisher's conclusion is modified. If a female 'knows' that her daughters will be mated by her sons, she can maximise the number of her grandchildren by producing an excess of daughters, and only enough sons to ensure that they are fertilised. He showed that in many haplo-diploid arthropods (in which a female can determine the sex of each offspring by whether or not she fertilises the egg) inbreeding is associated with a great excess of females. The extreme example is a viviparous mite, *Acarophenox*, which has a litter of one son and 10–20 daughters; the male mates with his sisters and dies before he is born.

6.4 Parental care

Species vary widely in the pattern of parental care; in particular the two sexes often contribute very unequally. Trivers (1972) interpreted these patterns in terms of 'parental investment'. In effect, he suggested that if, at some moment during the breeding period, one parent has invested more in the offspring than the other, then the parent which has invested least will be tempted to desert. He defined 'investment' as anything done by a parent to increase the chances of survival of existing offspring, at the expense of the parent's ability to invest in future offspring.

This idea was criticised by Dawkins and Carlisle (1976), who pointed out that we would expect animals to behave so as to maximise their reproductive success in the future. To argue that past investment should determine future behaviour is to commit the 'Concorde fallacy', which asserts that if one has already spent a lot of money on a project, it is wise to continue spending money. The criticism is clearly correct, yet Trivers had been able to show that past investment and future behaviour are correlated.

The real difficulty in analysing parental behaviour arises, I believe, because the optimal behaviour for one parent depends on what the other parent is doing. For example, suppose that a female has brought a particular offspring almost but not quite to the stage of independent existence. To abandon the offspring at that stage would be a bad strategy if it resulted in the death of the offspring; but if the male could be relied on to care for the offspring, the best strategy for the female would be to start on a new offspring. In other words, we are analysing a 'game', in which the interests of the parents, although similar, are not identical. The discussion which follows is based on a game theory analysis (Maynard Smith 1977). The idea is to seek for a pair of strategies, M for the male and F for the female, which together form an 'evolutionarily stable strategy' or ESS (see also 1.2 and 10.3 for discussion of the ESS concept), in the sense that it would not pay a male to depart from strategy M so long as females continue to adopt F, and it would not pay a female to depart from strategy F so long as males continue to adopt M.

The main factors determining whether there will be parental care, and if so which parent will provide it, are:

(i) the effectiveness of parental care by one or two parents,

(ii) the chance that a male deserting a female after mating will be able to mate again,

(iii) the fact that a female which exhausts her food reserves in laying eggs is less able to guard them, and

(iv) whether a male can be confident that a particular batch of eggs were fertilised by him.

The ways in which these factors interact can be most easily illustrated by reviewing some of the patterns observed in birds and fishes, which are the best studied groups from this point of view. Some idea of the frequency of different types of behaviour among fishes is shown in Table 6.2 (the data are extracted from Breder & Rosen 1966); more

Table 6.2. Number of families of bony fishes in which (in some or all species) one or other parent cares for the young (data from Breder & Rosen 1966).

Fertilisation external			Fertilisation internal			No parental care
♂	♀	both	♂	♀	both	
28	6	8	2	10	0	191

extensive data are now available in the literature). Several points are worth making:

(i) The commonest pattern is no parental care.

(ii) If there is parental care, it is usually by one parent only. Fish do not bring food to their young. Care consists of fanning the eggs, cleaning them of parasites, and protection against predators. These are tasks at which one parent may be almost as effective as two.

(iii) If there is internal fertilisation, care is usually by the female. There are two possible reasons. With internal fertilisation the male is usually not present when the eggs are laid; if he is present he may have no guarantee that the eggs were fertilised by his sperm.

(iv) If there is external fertilisation, either parent may care for the young, but it is most often the male. The male may construct a nest (e.g. the stickleback, *Gasterosteus*) or protect a particular egg-laying site (e.g. *Cottus*. *Cyprinodon*) or the male may brood the eggs (e.g. sea horses and pipefishes, family Syngnathidae).

Two theoretical points should be borne in mind. First, for a species with a given basic ecology, either male or female care might be evolutionarily stable, once it had arisen. Thus consider a species with male care. If, in the absence of care, eggs have a small chance of survival, a male which abandoned the eggs it had fertilised would have a low fitness, even if by abandoning the eggs it increased its chance of remating. The same argument, *mutatis mutandis*, applies to species with female care. The questions therefore arise, why did evolution originally take one path rather than the other, and why did it most often take the path of male care?

The question is most readily answered if we remember that in fishes uniparental care usually evolved from no care. In the absence of parental care, females will exhaust their food reserves in egg-laying. Since, during parental care, a parent fish feeds little or not at all, a female could only evolve this behaviour if, at the same time, she reduced her fitness by reducing the number of eggs she laid. In contrast, a male can fertilise the eggs and still have reserves left to live on during parental care. Also, in many cases (e.g. *Gasterosteus*) a male which guards the eggs laid by one female does not lose the chance of mating a second one.

Whereas in fishes the commonest pattern is no parental care, followed by male care, in birds the commonest pattern is for both parents to care for the young, followed by female care. In birds, the parents usually bring food to the young. In almost all such cases, both parents care, more or less equally, for the young. If the young are fed, two parents can raise, approximately, twice as many offspring as one. Even in species which do not feed their young, it is not uncommon for both parents to protect them from predators. In the water fowl, for example, if the parents are large enough to drive off most predators (geese, *Anser*, and swans, *Cygnus*) both parents care for the young, whereas in the smaller ducks (e.g. *Anas*, *Aythya*) only the female does so.

There seem to be two reasons why, if only one parent guards, it is almost always the female. First, fertilisation is internal. Hence, if we suppose the primitive condition was biparental care (as is almost certainly true, for example, for the water fowl), a male could desert immediately after mating and seek a second female, whereas a female could not seek a second male to care for a second clutch of eggs until she had laid the first. The remarkable condition in the American jacana (Jenni 1974) in which the female is polyandrous, and lays clutches of eggs to be cared for by several males, probably evolved from the habit of 'double-clutching' found in some waders. In the sanderling (*Calidris, alba*), for example, it is common for the female to lay one clutch of four eggs which is incubated and reared by the male, and then to pair with a second male and lay a second clutch which she cares for herself. Once males have evolved the habit of caring for a clutch of eggs after the female has laid them, it could pay females to compete in order to acquire more males; this is what has happened in the jacana.

The other feature of bird biology which has favoured female rather than male care is that the productivity of a pair is not usually limited by the number of eggs a female can lay, but by the number which can be incubated and by the number of young which can be fed and protected (Lack 1968). Since a female does not exhaust her reserves in laying eggs, as do fish, she is able to care for the young. In some species, egg production is limiting, and only the male cares for the young. The most striking example is the mallee fowl, *Leipoa ocellata* (Frith 1962). The male constructs a large mound of sand and leaves as an incubator for the eggs. The female lays eggs over a period of three to four months, laying up to 30 eggs each 10% of her own weight. During the laying period only the male constructs and tends the mound. The pair are monogamous for life. Despite the examples of the mallee fowl and the jacana, it is much rarer in birds than in fish for the female to get the male to look after her eggs. There seems to be two main reasons. First, since fertilisation is internal a male bird may not be present when the eggs are laid, and in any case cannot be certain he is the father. Second, in most birds a female could not greatly increase her fitness by abandoning parental care in favour of laying more eggs, because breeding success is limited not by the number of eggs laid but by the number of young birds which can successfully be raised.

Chapter 7
Sexual Selection and Mate Choice

TIMOTHY R. HALLIDAY

Woman wants monogamy;
Man delights in novelty,
Love is woman's moon and sun;
Man has other forms of fun.
Woman lives but in her lord;
Count to ten, and man is bored.
With this the gist and sum of it,
What earthly good can come of it?

Dorothy Parker

7.1 Introduction

7.1.1 *Mate choice*

The individual animals that make up a breeding population will not be equally fit; they will vary in their abilities to survive from day to day and to reproduce. An animal that mates with a partner of high quality may gain a long-term genetic advantage, a short-term material advantage, or both. For example, a female paired with a male in possession of a large territory may thereby pass on to her offspring male genes that confer some advantage in territory acquisition. More immediately, she will gain the advantage of the male's large territory with its large supply of resources, giving her offspring an immediate material advantage over the offspring of other pairs.

The term 'mate choice' is not intended to imply that animals make a conscious or rational choice from among a number of potential partners. It is simply a short-hand for any mechanism by which animals mate selectively with some members of the opposite sex and not with others. In an analogous way, an animal that responds to the courtship displays of members of its own species but not to those of another species may be said to be exercising 'species choice'.

7.1.2 *Female choice*

Bateman (1948) showed in a population of *Drosophila* in which individuals were marked by different genetic markers that the contribution of males to the next generation was much more variable than that of females. While virtually all females mated, some males achieved several matings, others none. Since females produce only one batch of eggs in each breeding cycle they need only mate once to achieve their full reproductive potential whereas male reproductive success is determined by the number of matings achieved. Bateman argued that this will be a general phenomenon among animals and will lead to 'undiscriminating eagerness' to mate in males and 'discriminating passivity' in females. Trivers (1972) has expressed this effect in terms of parental investment (see 6.4). Members of the sex that invests little in each offspring, usually males, will compete for members of the high-investing sex, usually females. It can thus be argued that animals are fundamentally polygynous (one male mating with several females) and that monogamy is a secondary condition that has evolved in response to certain environmental pressures (Wilson 1975). In monogamous species males often make substantial parental investment and may also exercise mate choice.

7.1.3 *Sexual selection*

When mechanisms of mate choice evolve in females they will constitute powerful evolutionary pressures acting on various aspects of the reproductive biology of males, particularly on those characters, behavioural or anatomical, that are used by females as criteria for their choice. This is the essence of the theory of Sexual Selection which is discussed in the first part of this chapter (7.2 to 7.8).

The argument that mechanisms of mate selection can be expected to have evolved is a very persuasive one. However, it is quite another matter to demonstrate that they do actually exist in nature. The second part of this chapter (7.9) reviews evidence from a number of species that animals select their mates and that the choice they make does enhance their reproductive success.

If members of one sex share a preference for a particular mate there will tend to be competition between them for sexual access to that mate.

Thus mate choice and intra-sexual competition are likely to be interrelated phenomena. The third part of this chapter (7.10) discusses some of the forms that sexual competition takes.

7.2. Theories of sexual selection

In many species males and females differ in appearance and behaviour, particularly during the breeding season. In such species as peacocks and birds of paradise these differences are very pronounced. The purpose of Sexual Selection theory is to explain the evolution of such differences.

7.2.1 *Darwin's theory*

Darwin (1871) invented the term Sexual Selection to describe the selection that arises when individuals of one sex, usually males, gain an advantage over others of the same sex in obtaining mates. He envisaged this happening in one of two ways. In the first, males fight with one another, the winner claiming the disputed females. Such a system creates a powerful selection pressure favouring an increase in male strength and the elaboration of limbs or organs used as weapons. This form of selection, called Intrasexual Selection by Huxley (1938), has not seriously been disputed as an explanation for the evolution of special male weapons. Darwin's second category involves competition between males to attract the attention of females and leads to the elaboration of structures or behaviour patterns that attract females. This form of selection, called Epigamic Selection by Huxley and Intersexual Selection by many other authors, depends on the assumption that females show a sexual preference for males ornamented or behaving in a particular way. This assumption has been the subject of much discussion and disagreement since Darwin first proposed it. Darwin simply assumed that females have preferences for certain types of males, without suggesting how such preferences might have arisen or how they might be maintained in a population by selection. For this reason this part of his theory was widely rejected.

7.2.2 *Fisher's theory*

Fisher (1930) effectively closed the loopholes in Darwin's theory of intersexual selection. He pointed out that for sexual selection based on a mating preference in one sex to lead to the evolution of a sexually dimorphic character in the other sex, some advantage must accrue to an individual exercising the preference. The advantage to a male possessing a character that is preferred by females is clear; he attracts more mates and has more progeny. The advantage accruing to a female that prefers that male is less obvious but was clarified by Fisher. Females that mate with males that are attractive to females in general will have attractive sons, provided that the attractive male character is inherited. Being more attractive, these sons will acquire more mates and will leave more progeny so that the advantage accruing to females that choose attractive males is that they have more grandchildren. There is thus a dual process in which male attractiveness and female preference evolve together. Variation in male attractiveness creates the conditions favouring selectivity in females; as female selectivity evolves selection will favour any increase in male attractiveness.

While this argument explains how a female preference can be maintained in a population it does not explain how such a preference becomes established in the first place. Fisher suggested that there must be a preliminary stage in which males possessing a character attractive to at least some females must have some other heritable advantage, however slight, over other males. In effect, the preferred character must be a marker of some other characteristic that increases male fitness. Once the character used as a basis for female mating preference has become established through its link with greater male fitness, it will then become subject to selection simply for its attractive properties as described above. This secondary process will proceed even though it may run counter to the selective advantage that provided the basis for the first stage of the process.

For example, an ancestral male peacock with a slightly larger than average tail might, because of that tail, have been better at flying and escaping from predators. Females preferring longer-tailed males would pass this advantage on to their sons. As males with longer tails, and females preferring them, became more common, the latter gained the additional advantage of having more attractive sons. Then sexual selection would proceed, favouring males with ever longer tails, even though, beyond a certain point, longer tails might make them less able to fly. This process would continue only if the advantage of the longer tail in attracting females more than compensated for the disadvantage in reduced powers of flight. Fisher regarded this second phase as a 'run-away' process which proceeds at an ever-increasing rate until it is checked by some counteracting selective pressure. This counter selection

will be especially effective if the character being selected for by female preference reduces a male's chances of surviving up to the time of breeding.

Selection based on female choice will only produce a sustained change in the form of a male epigamic character from one generation to the next if females select by a relative, rather than an absolute criterion (Trivers 1972). They must choose males from one extreme of a distribution, for example those with the longest tails or the brightest plumage. Over several generations such selection will cause the distribution of the preferred character to shift towards the chosen extreme.

7.2.3 *The handicap principle*

Zahavi (1975) dismisses Fisher's argument that male characters may evolve simply because they are attractive to females. He argues that females should be expected to select males on the basis of characters that are true indicators of their overall ability to survive. Further, he suggests that females should not base their choice on only one character, since any male, regardless of his true quality, may 'cheat' by developing the character preferred by females to the same degree as the fittest males. His theory for the sexual selection of conspicuous male characters is based on the idea that such characters represent a handicap to any male possessing them, that is, they reduce his capacity to survive. Thus, by choosing a male carrying a handicap, a female selects a mate who has demonstrated that he is fit enough to survive with that handicap. Males not carrying the handicap may be more or less fit but, since they have not undergone the same test, a female has no indication of what their relative fitness might be. Essentially Zahavi argues that characters that handicap males are selected for, precisely because they are handicaps. Thus the peacock's tail evolved because it handicaps a male, perhaps by hindering his movements, and so demonstrates his ability to survive despite it.

At no point in his argument does Zahavi make any allowance for male characters conferring an advantage on their bearers in terms of attracting more mates, the central point of Fisher's theory. Moreover, Zahavi presents no cogent arguments against Fisher's theory.

Superficially this theory may seem attractive because it appears to explain how the benefit that females derive by choosing decorated males represents something more than having attractive sons. However, the theory is logically false. As Maynard Smith (1976d) has pointed out, there is good reason to doubt the theory simply on the grounds that females choosing to mate with handicapped males will pass this handicap on to their sons, thereby actually reducing their fitness.

Any acceptable theory of sexual selection must adequately explain two things. First, how does a sexually dimorphic character become established in a population and, second, why does the magnitude of that character increase over successive generations? Can the handicap principal explain these two processes?

Figure 7.1a represents a population of males of a hypothetical animal whose quality is indicated by the length of a bar. For the purposes of the following argument male quality is defined as 'ability to

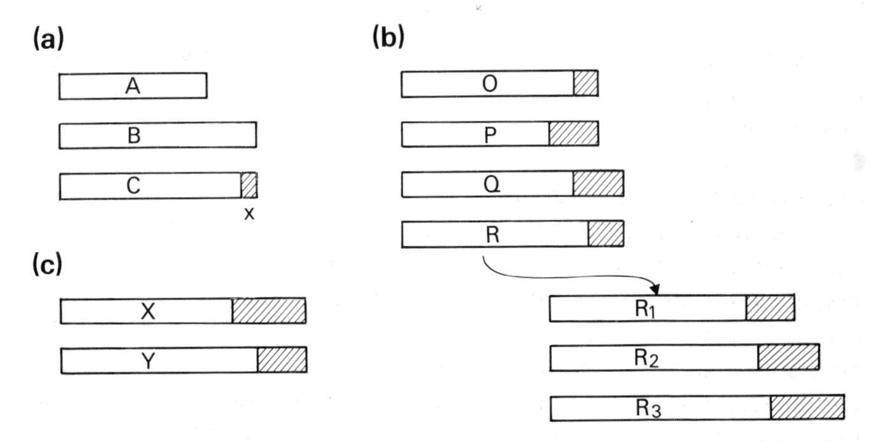

Fig. 7.1. Schematic representation of the Handicap Principle. Each bar represents an individual male, the length of the bar representing his quality (ability to survive). The shaded portion of each bar represents the decrease in quality due to a handicapping character so that the unshaded portion represents the quality that is passed on to sons. See text for explanation.

survive'; it does not include ability to attract females, since Zahavi discounts this as a factor in sexual selection. We must also assume that females cannot assess male quality directly. Males A and B differ in quality but are identical in appearance and thus cannot be differentiated by females. Male C is as good as B but carries a marker which constitutes a handicap, reducing his quality by an amount x. Females choosing C will pass on more quality to their sons than those choosing A, but less than those choosing B. However, since females choosing B are equally likely to choose A, being unable to distinguish between them, they will, on balance, pass on less to their sons, provided that the value of x is less than half the difference in the qualities of B and A. By this argument the handicap principle can explain how a handicapping character can become established in a population, provided that it is carried only by high quality males and provided its handicapping effect is small. However, the character becomes established, not because it imposes a handicap on the males carrying it, but because it is a marker of superior quality. In this respect Zahavi's theory does not differ from Fisher's.

Moving on to the second question which the theory must answer, Fig. 7.1b represents a population in which at least some males, represented by O, carry a small handicapping character. How might the magnitude of this character increase in subsequent generations? Male P, who has the same overall quality as O, increases his handicap at the expense of his overall quality. Females choosing P will pass on less quality to their sons than those choosing O so that P males will be selected out of the population. Male Q increases his handicap by increasing his overall quality and has the same residual quality as O. Females choosing Q will pass on the same quality to their sons as those choosing O and so Q males will not increase at the expense of O males. Male R also has greater overall quality than O but, while investing more of it in his handicap than O, does not invest as much as Q, so that the residual quality passed on to his sons is greater than either O's or Q's. Females choosing R will thus have an advantage over those choosing O so that R males will tend to increase in the population. Selection could thus produce ever larger handicaps through stages R_1, R_2, etc, provided that males of superior quality only invest part of their superiority in a larger handicap. This argument is unacceptable for explaining the evolution of very extravagant male characters since it is dependent on the assumption of a continual increase in the quality of males from generation to generation and a consequent continual advantage to females choosing males with the largest handicaps. As pointed out by Davis and O'Donald (1976a), male quality cannot increase indefinitely but must approach some optimum value. Figure 7.1c represents two males, X and Y, that have reached this optimum value. In this situation females choosing the male with the larger handicap, X, are at a disadvantage since they pass less residual quality on to their sons than those choosing Y. Thus, in a population in which male quality is evolving towards its optimum value, selection will favour the evolution of smaller, not larger handicaps.

Davis and O'Donald point out a further fallacy in Zahavi's argument. The handicap principle depends on the assumption that the combination of characters that give a handicapped male the high quality required to survive his handicap is passed on to his offspring. In reality the genetic recombination and reassortment that are intrinsic to sexual reproduction will tend to cause paternal combinations of characters to be broken up. Thus, even if a handicap is a marker of male quality, it is no guarantee that that quality can be passed on. A female mating with a male who has survived his handicap will have offspring that are as likely to die through carrying the handicap as they are to survive it.

Maynard Smith (1976d) has analysed the handicap principle by means of a genetic model. Even allowing for the fact that the handicap is sex-limited, so that a female choosing a handicapped male would pass on to their daughters the advantages of their father's superior quality without the handicap, the genetic model shows that handicapped males will only increase in a population if females preferring them as mates are relatively common. Once female preference for a particular type of male is established in a population, the Fisher effect must operate and the handicap principle is irrelevant.

There are situations in which the handicap principle might work. A female may well benefit by choosing a male who possesses a handicap that is not inherited, such as a broken leg (Maynard Smith 1976d). Only males of the highest quality are likely to survive with such a handicap and females mating with them will pass their superior quality, but obviously not the handicap, on to their progeny. While this might conceivably lead to the evolution of a female preference for injured males, it cannot provide an explanation for the evolution of male epigamic characters.

There is no dispute about a basic point in Zahavi's argument, that exaggerated male characters constitute a handicap to their bearers. This point was recognised by Fisher who argued that selection favouring greater development of a character resulting from female preference will be countered by the disadvantages caused by that character in day to day life. When the advantage to a male in terms of greater mating success is balanced by his decreased chances of surviving to breed, further development of the character will be prevented.

In a reply to his critics, Zahavi (1977a) has suggested that the handicap principle will operate if the manifestation of a male character is not genetically fixed but is correlated with the general quality of the individual. Thus a male who copes well with the rigours of everyday life will develop a larger handicapping character than one who does not. Males who over-develop their handicaps will not survive. Such a mechanism may exist in many species. Male newts of the genus *Triturus* develop large decorated dorsal crests in the breeding season but there is considerable variation between individuals in crest size (Halliday 1977b). When male newts are deprived of food, their crests rapidly regress. If the resources that male newts invest in crests are in excess of those required for the basic functions of life, the size of a male's crest will be an indicator of how successful he is at acquiring such resources. Even if mechanisms of this sort do exist in animals they cannot be considered to be evidence for the handicap principle as opposed to Fisher's theory. What it would mean is that each individual male arrives at his own compromise between the energetic cost of developing an epigamic character and the advantage that that character gives him in attracting females.

7.2.4 *The resource accrual theory*

Trivers (1976) proposes a theory of sexual selection which he incorporates into an argument relating to the more general problem of the evolution of sex. In species in which males make no parental investment females seem to gain little from sexual reproduction. Indeed, as explained in Chapter 6, they may suffer an immediate two-fold disadvantage compared with asexual females. Trivers suggests that sexual reproduction may still be advantageous in such species if female choice is sufficiently discriminating to ensure that the males chosen possess genes that are on average twice as good in terms of producing fit daughters as are the mother's genes. In other words, by pairing half her genes with the genes of a super-fit male, a female so enhances the chances of their survival as to offset the two-fold disadvantage of sex. It follows from this theory that the criteria used by females in mate choice must indicate qualities that are beneficial to daughters. This is quite different from Fisher's theory which explains sexual selection in terms of a benefit accruing to sons.

As an example, the antlers of male deer might indicate by their size a male's ability to acquire and process calcium in the synthesis of bone. While males use much of this ability to make antlers, females inheriting it from their fathers would use it to make larger and stronger bones.

A problem with this theory is that it cannot account for the increasing development of a character over more than a few generations. If females are to mate with males who are twice as fit as they are they must mate only with a tiny proportion of males at the upper end of the fitness distribution. It has been a common experience in the selective breeding of domestic animals that such very intense selection rapidly reduces the genetic variation in the character being selected, so that after only two or three generations of selective breeding there is virtually no variation on which further selection could act (Falconer 1960).

7.2.5 *Synthesis*

Recent theories such as the handicap principle have done nothing to refute the central point of Darwin's theory of intersexual selection, that female mating preference can lead to the evolution of male secondary sexual characters. Fisher argued that before female preference can operate as a selective force in its own right a male character that is preferred must be associated with some other adaptive male attribute. Continuous evolution of the male character requires that females exert a general preference for males who show the character to an extreme degree. What initial advantage might there be for females that prefer extreme males? One situation in which such a preference might well be advantageous is where two closely related species have overlapping ranges (Trivers 1972). Females who fail to discriminate between males of their own and of the sympatric species are likely to have inviable hybrid offspring. Selection will thus favour females who choose those males who are most obviously conspecifics, i.e. those who develop a species-specific character to an extreme degree. Such selection frequently results in the phenomenon of character displacement (Grant 1972) by which a character used by males to attract females becomes more distinctive in the area of overlap. In a number of sympatric species of frogs mating calls differ more in some parameter such as call frequency in the overlap zone than elsewhere (Blair 1974, Littlejohn 1969).

A preference for unusually large or conspicuous stimuli has been described in a number of contexts; such stimuli are described as 'supernormal'. For example, many birds are more likely to brood larger than normal dummy eggs than they are natural eggs. Why such preferences exist is not clear, but if a preference for exceptionally conspicuous stimuli is a general property of animal perceptual systems it would provide a mechanism for sexual selection; Fisher's argument that a conspicuous male character must be associated with some other adaptive attribute need not apply.

It may be wrong to assume that female preference is the only selective factor operating in the evolution of a male epigamic character.

For example, the dorsal crests developed by male European newts in the breeding season seem to provide an excellent example of the effects of intersexual selection. In possessing crests these newts differ from all other Urodeles (tailed amphibia). They are also unique in that the female is not grasped by the male during courtship and so is free to move about and, at least in theory, select which males she will respond to (Halliday 1977b). It seems probable, therefore, that the emancipation of the female newt has produced the conditions in which the sexual selection of a male epigamic character can occur. The male newt's ability to sustain courtship for a prolonged period is limited by his capacity to hold his breath; from time to time he must ascend to the pond surface to take in air. However, he augments the air supply in his lungs by skin respiration and his crest, being vascularised, probably acts as an additional respiratory surface. It has been shown that newts deprived of access to the atmosphere survive for longer if they are in breeding condition than if they are not (Foxon 1964). Thus, while the male's crest may well be important in attracting females it has probably also evolved as a respiratory structure.

7.3 Sexual selection and natural selection

Darwin made a clear distinction between Natural Selection by which individuals are selected according to their abilities to survive and to reproduce and Sexual Selection by which they are selected solely according to their abilities to obtain more mates than other individuals. Much of the development of evolutionary theory since Darwin has been based on the realisation that selection acts not on individuals but on genes (Dawkins 1976). Thus the ability to survive and the ability to attract mates become merely different aspects of the single goal of perpetuating an individual's genes and Darwin's original distinction becomes trivial. It may even be misleading and Zahavi's attempt to explain sexual selection in terms of characters solely related to individual survival may be seen as an example of the kind of conceptual difficulty to which Darwin's distinction can lead. Some recent authors have suggested that the distinction is still a useful one (e.g. Mayr 1972) but it is perhaps better to regard sexual selection simply as a sub-set of natural selection rather than as a special process. Lack (1968) has suggested that, rather than talking in terms of sexual selection, we should simply consider the adaptive value of secondary sexual characters.

7.4 Intrasexual and intersexual selection

Darwin differentiated between intrasexual selection which favours 'the power to conquer other males in battle' and intersexual selection favouring 'the power to charm the females'. In nature both processes may be acting simultaneously on the mating behaviour of a species and it may therefore be difficult to differentiate the role of each. For example, male displays that are used in the establishment and maintenance of territories may, while serving to intimidate rival males, also attract females (see Chapter 11).

In elephant seals *Mirounga angustirostris* rivalry between males for the possession of females is intense and frequently involves violent fighting. Males establish a dominance hierarchy and those of high rank monopolise females to such an extent that only 4% of the males achieve 85% of all matings (Le Boeuf & Peterson 1969, Le Boeuf 1972, Le Boeuf 1974). Male elephant seals are markedly larger and more powerful than females and provide a clear example of the effects of intrasexual selection. However, the females do not play an entirely passive role in mating. When a male attempts to mount her, a female elephant seal protests loudly, attracting the attention of other males nearby (Cox & Le Boeuf 1977). Males respond by attempting to dislodge the male who is trying to mate who will only be able to complete copulation if he is a dominant individual. The female's protests clearly increase the probability that a mounting male will be displaced (Fig. 7.2a). Her behaviour thus intensifies inter-male rivalry and makes it more likely that she will be mated only by the most high-ranking males. This effect is further enhanced by the fact that the probability that a female will protest is related to the rank of the mounting male (Fig. 7.2b). It is clear that, while the issue of which male mates with any one female is largely determined by inter-male rivalry, female choice also plays an important part.

7.5 Sexual selection and mate choice in polygamous species

Sexual selection will be most intense in those species that have polygamous mating systems, that is, species in which some members of one sex mate with several members of the other sex. In polygynous species some individual males mate with several females; in poly-

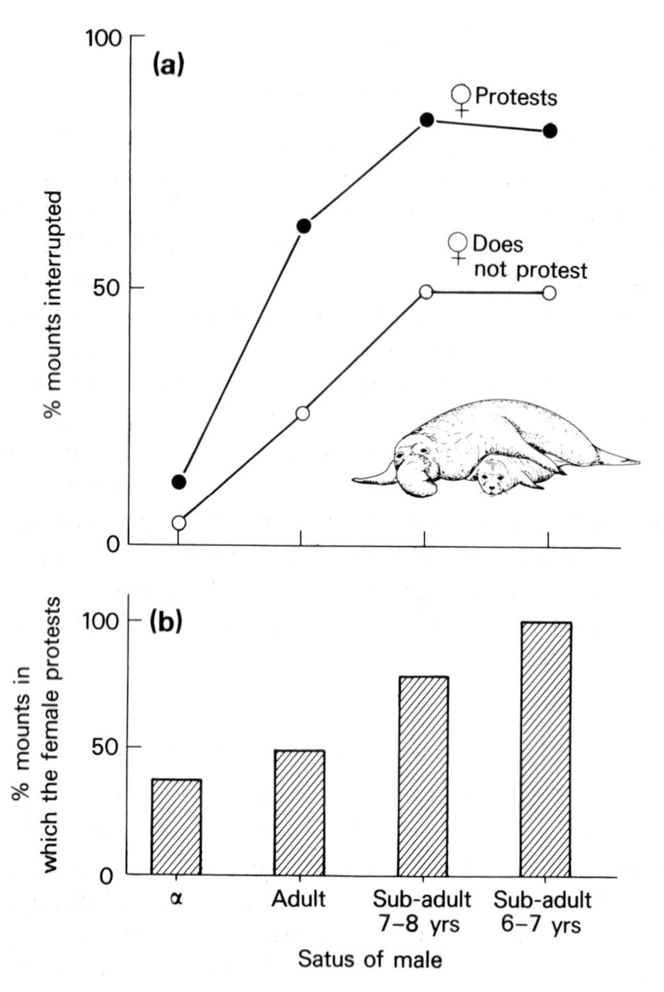

Fig. 7.2. The influence of female behaviour on the mating success of male Elephant Seals of different rank (a) and the probability that females will protest when mounted by males of different rank (b). Based on data in Cox & Le Boeuf (1977).

androus species certain females mate with several males. Sexual selection will act on members of that sex, males in polygynous species, females in polyandrous species, that are competing for access to the other sex. The nature of the mating system in any given species is determined primarily by ecological factors such as the distribution of resources and the following types of polygamous systems are taken from an ecological classification of mating systems devised by Emlen and Oring (1977).

7.5.1 *Resource defence polygyny*

In this type of mating system males gain access to females indirectly by controlling certain vital resources that females require (see 11.2.2). For example, females of the American bullfrog *Rana catesbeiana* show a preference for those territories that are defended by older and larger males (Howard in press). These territories are of high quality in that mortality among developing embryos is lower than in other territories. This is both because they are less subject to extreme temperature variations that cause developmental abnormalities and also because they are less infested by predatory leeches that eat the embryos. Female choice is directed not towards the males themselves but towards the quantity or quality of the resources that the males are defending. Mating systems of this type will thus generally not lead to the evolution of male epigamic characters through intersexual selection but will provide the conditions in which intrasexual selection will favour the enhancement of male strength and fighting ability.

7.5.2 *Female defence polygyny: harems*

In this form of polygamy males gain access to females by directly defending them against rivals. Such a system is greatly facilitated if females show an independent tendency to be gregarious, perhaps as a defence against predators or to increase feeding efficiency. Female defence may arise as an extension of resource defence, male aggressive behaviour being directed not only to the defence of an area but also to any females attracted to that area.

Harems are found in some ungulates such as the impala (Jarman 1974), a very few birds such as pheasants (Emlen & Oring 1977) and in some primates (Clutton-Brock & Harvey 1977). As described above, harems are maintained by dominant male elephant seals and this is made easier for them by the fact that females naturally tend to form dense groupings because of the scarcity of suitable sites at which they can haul themselves out of the sea. As we have seen, female choice cannot be ruled out as an important factor in the distribution of matings in such species, but in general sexual selection in harem species, as in resource defending species, will be primarily intrasexual.

7.5.3 *Male dominance polygyny: explosive breeding and leks*

In these mating systems males do not defend either resources or females. Instead they establish dominance relationships among themselves and females then choose them, largely on the basis of their status. The form that this behaviour takes varies according to the extent to which females show synchronous sexual activity. Explosive breeding occurs when all the females come into breeding condition simultaneously and arrive at the male aggregation together. Such behaviour is shown by a number of frogs and toads (Wells 1977b), including the European frog *Rana temporaria* in which mating may occur only on one night of the year (personal observation). Mating behaviour may involve some intrasexual competition but since the sex ratio is likely to be close to 1:1 throughout the mating period and since the brevity of the mating period limits the opportunity for males to monopolise several females, there is likely to be little variance in the number of females acquired by individual males and so sexual selection will be weak. Although sexual selection may be insignificant in explosive breeders, it does not follow that individuals of such species do not choose their mates (see 7.9.4).

If females are not synchronised in their periods of sexual activity only a small proportion of them will visit the male aggregation at any one time. Thus, while the sex ratio in the population as a whole may be 1:1, the ratio of males to females actively engaged in mating at any one time will be skewed, showing an excess of males. As a result intermale competition will be intense. The resulting mating system is called a lek, which is defined as 'a communal display area where males congregate for the sole purpose of attracting and courting females and to which females come for mating' (Emlen & Oring 1977).

Since males at a lek control neither a resource useful to females, nor the females themselves, females are free to choose which males they will mate with purely on criteria shown by the males. Sexual selection in lek species thus tends to be intersexual and has resulted in the evolution of some very extreme male epigamic characters and elaborate male behaviour. In a number of lek species those mating sites most preferred by females are defended by the older males. Whether this is because older males are stronger or more skilful at fighting, or whether they have simply learned by experience which positions are preferred by females, is not clear. By mating with older males, females are combining their genes with those of males who have demonstrated their ability to survive for several years. (For further discussion about leks, see 11.2.2.)

7.5.4 *Polyandry*

Polyandry, in which individual females mate with several males, is a very rare form of mating system and has been studied in most detail among certain birds. The males in polyandrous species tend to assume all parental duties, the emancipated females devoting all their reproductive efforts to producing eggs. In ecological circumstances in which egg loss is high, through either predation or climatic changes, the ability of females to rapidly produce replacement clutches is highly adaptive and such conditions seem to characterise the habitats of polyandrous birds (Emlen & Oring 1977). In polyandrous species sexual selection operates among females rather than males, since it is they whose reproductive success is the more variable, and favours female competitive ability. Female jacanas are 50% to 75% heavier than males (see 6.4).

No examples of male defence or 'harem' polyandry are known but northern phalaropes *Phalaropus lobatus* have a mating system that can be categorised as explosive female access polyandry in which females compete for males (Emlen & Oring 1977). Males and females congregate at highly ephemeral feeding areas and females, who show a great capacity to lay replacement clutches, compete with each other for males who look after the eggs. Sexual selection acting on females has resulted in a reversal in the normal pattern of plumage dimorphism; female phalaropes are brightly coloured while males are dull and cryptic.

7.6 Sexual selection in monogamous species

In monogamous species each individual mates with only one other individual during each breeding cycle. In many species males and females form stable pair bonds. Since each individual is potentially assured of an opportunity to mate there seems to be little scope for competition leading to sexual selection to occur. However, the fact that many monogamous species show marked sexual dimorphism suggests that sexual selection does operate in such species, although Selander (1972) has argued that much of the dimorphism shown by monogamous birds is due to ecological pressures.

7.6.1 *Darwin's theory*

Darwin proposed a theory for sexual selection in monogamous species based on the idea that differences in the fitness of individuals will manifest themselves as variations in the time at which those individuals are ready to breed. Males are generally ready to breed before females, perhaps because, as Darwin suggested, they incur relatively little energetic cost in preparing to breed, or because selection for early readiness is more intense among males because they have to compete for females who represent a limited resource. The fittest females will attain sexual maturity first and so either will be able to choose from among the males or will mate with those males who have established a position of dominance over the others. As the less fit females come into breeding condition they will have to pair with the less attractive or less strong males that are left unpaired. On the face of it this theory seems to predict that sexual selection will favour ever-earlier preparedness to breed. Lack (1968) has criticised the theory, saying that breeding date is determined, not by sexual selection, but by ecological factors, particularly the availability of food. Certainly, there must be a powerful selection pressure against pairs that breed too early, before ecological conditions are favourable. O'Donald (1972, 1973a) has shown by means of a genetic model that sexual selection will always operate as Darwin suggested if females breeding early in the season gain a reproductive advantage as they do in several species (Perrins 1970). However, even when early pairs do not have a higher reproductive success the model shows that there is still a selective advantage favouring males that are preferred by at least some females. In O'Donald's model, the mean breeding time is the optimum breeding time and, if female preference is only partial, preferred males are sufficiently numerous at the optimum breeding time to gain an advantage from female preference.

7.6.2 *The arctic skua*

The influence of sexual selection in a monogamous mating system has been analysed in detail in the arctic skua *Stercorarius parasiticus*, a seabird that breeds on a number of islands in sub-arctic latitudes. The arctic skua occurs in three plumage types, dark, intermediate and pale; the pale form is more common in the more northerly parts of its range, the dark form is most common in the south (Berry & Davis 1970). The colour variation is largely due to two alleles at a single locus; heterozygotes are variable in colour. Intermediate birds are mostly heterozygotes but may also be homozygous for the pale or the dark character. The mechanisms by which the three plumage types are maintained in the population have been analysed by means of a series of models by O'Donald and coworkers. The dark form is maintained by a preference shown by a proportion of females for dark coloured males. The greater attractiveness of dark males to these females may be due to their being generally more 'masculine'; they may be more aggressive and have larger territories than pale males and may be more vigorous in their courtship (Davis & O'Donald 1976b). Alternatively, some females may simply have lower response thresholds to males who are dark (O'Donald 1977). Whatever the mechanism underlying the female preference, it has the result that a female in a pair that is breeding for the first time and in which the male is dark tends to breed earlier in the season than other females and, as a result, succeeds in fledging more chicks. Why then, if dark males hold such an advantage, are not all males dark? Pale birds of both sexes have an advantage over dark birds in that they are younger when they breed for the first time. This advantage arises in three ways. First, pale birds have a greater chance of surviving to breed (O'Donald & Davis 1975), second, they have more breeding seasons in which to produce young and, third, they gain experience which will increase their breeding success in subsequent years (Davis 1976). The pale character is thus maintained in the population by the greater lifetime breeding success of pale birds. While there does not seem to be a female preference for pale males comparable to that for dark males, a proportion of intermediate females show assortative mating, that is, they mate preferentially with males of their own, intermediate type (Davis & O'Donald 1976c). It is possible by means of models to put precise figures on the proportions of females exercising the various mating preferences. Of all females nesting on the island of Fair Isle, regardless of their colour, 14% seem to prefer dark males (O'Donald & Davis 1977); of intermediate females, who make up 60% of the total population, 45% (29% of all females) prefer intermediate males (Davis & O'Donald 1976c). Those females that do not exercise a mating preference apparently pair at random with males of the three colour types.

The arctic skua provides an example of a breeding system in which different mating preferences and conflicting selection pressures are opposed. At the present time the frequency of dark birds in the arctic skua population is declining, particularly at the more northerly latitudes.

It is suggested that the female preference for dark males evolved at some time in the past when dark birds had some greater selective advantage over pale birds. Now, perhaps because of general climatic changes, this advantage has been reduced and pale birds, because of their greater lifetime breeding success, are increasing at the expense of dark birds (O'Donald & Davis 1977).

It is important to note that the evidence for female preference in the arctic skua is not based on direct observation of female behaviour. Rather, it is based on the fact that mathematical models that assume female preferences provide a very good fit when applied to field data on the breeding success of individual pairs of arctic skuas.

7.7 Sexual selection and polymorphism

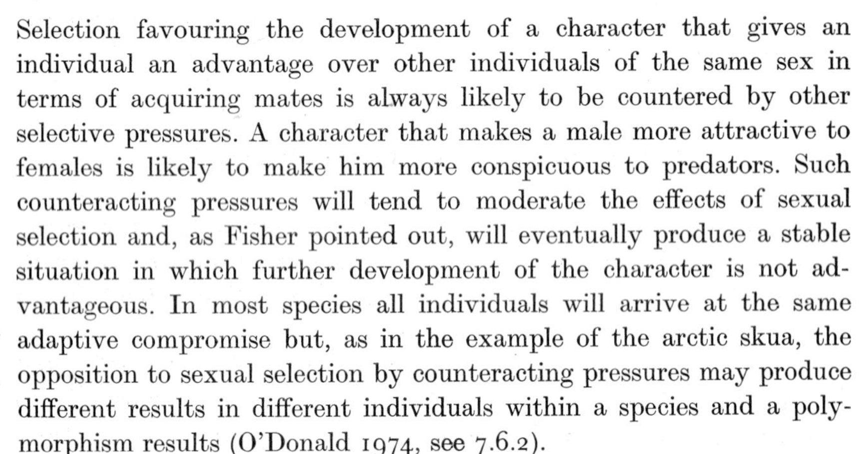

Selection favouring the development of a character that gives an individual an advantage over other individuals of the same sex in terms of acquiring mates is always likely to be countered by other selective pressures. A character that makes a male more attractive to females is likely to make him more conspicuous to predators. Such counteracting pressures will tend to moderate the effects of sexual selection and, as Fisher pointed out, will eventually produce a stable situation in which further development of the character is not advantageous. In most species all individuals will arrive at the same adaptive compromise but, as in the example of the arctic skua, the opposition to sexual selection by counteracting pressures may produce different results in different individuals within a species and a polymorphism results (O'Donald 1974, see 7.6.2).

Three-spined sticklebacks *Gasterosteus aculeatus* in Lake Wapato, U.S.A., are polymorphic with 14% of males developing a bright red colouration in the breeding season (Semler 1971). In experiments in which they were offered a simple choice between red and non-red males, females showed a significant preference for red males. From this it may be inferred that the red morph is maintained in the population by sexual selection. The advantage to females preferring red males is thought to be that red males are more effective at defending their nests against both male and female sticklebacks, who steal eggs from their conspecifics' nests. The low frequency of red males in the population is probably due to their being more visible, and therefore more susceptible to predation by fish such as trout than are the non-red males. In another locality in Canada it has been shown that a predatory minnow *Novumbra* is differentially attracted to red males and that young sticklebacks of a black form possess a behavioural adaptation against this predator (McPhail 1969).

Gadgil (1972) cites a number of insect species in which there are two types of males, differing in the extent to which they develop weapons used in inter-male competition. He suggests that such polymorphisms may be genetically based and that they represent alternative adaptive strategies. Males with weapons gain in combat but invest a lot in terms of energy; males without weapons save energy for other important activities but tend to lose in combat (see also 11.2.2).

These genetically based polymorphisms are not to be confused with changes in male strategy related to age, such as those occurring in many polygamous species such as bullfrogs (Howard in press b). Nor are they comparable to variations in the development of epigamic characters such as the crest of male newts which, while they may reflect differences in genotype, do not fall into distinct classes but take the form of a continuum.

7.8 Assortative and disassortative mating

In species in which there are a number of distinct forms or races, individuals may show assortative mating, a tendency to mate with individuals of the same type, or disassortative mating, a tendency to mate with individuals of a different type.

Assortative mating has been demonstrated among artificially bred strains of domestic fowl (Lill & Wood-Gush 1965, Lill 1968a, 1968b) and among blue and white morphs of the lesser snow goose *Anser caerulescens* in the wild (Cooke *et al.* 1976). In both examples birds show a mating preference for the type with whom they were raised. Three-spined sticklebacks *Gasterosteus aculeatus* in Canada (Hay & McPhail 1975) which live in streams running into the sea exist in two forms, an exclusively freshwater form and an anadromous form that lives primarily in the sea but breeds in fresh water. In experiments in which individuals were offered a choice of two mates, one of each type, both males and females selected a partner of their own type in about 62% of tests.

The selective advantage of assortative mating seems clear. It will generally be the case that hybrids between two types will be less viable than pure-bred individuals. In the stickleback example, hybrid offspring of freshwater and anadromous parents will be less able to survive in either habitat than will pure-bred offspring of each type. Assortative mating, if maintained for long enough, will eventually lead to complete reproductive isolation between two types which will thus become distinct species. As described above (7.2.5) it may be associated with the phenomenon of character displacement by which those characters used in mate identification become accentuated. There may thus be a very close link between assortative mating and sexual selection (O'Donald 1973b), an adaptive preference for mating with one's own type leading to the evolution of sexually dimorphic characters.

The phenomenon of disassortative mating has been studied in *Drosophila* (Petit 1958), in which it is a widespread phenomenon that manifests itself as a mating preference for individuals belonging to a rare type. In *D. melanogaster* the mating success of white-eyed mutant males is dependent on the frequency of such males in a population.

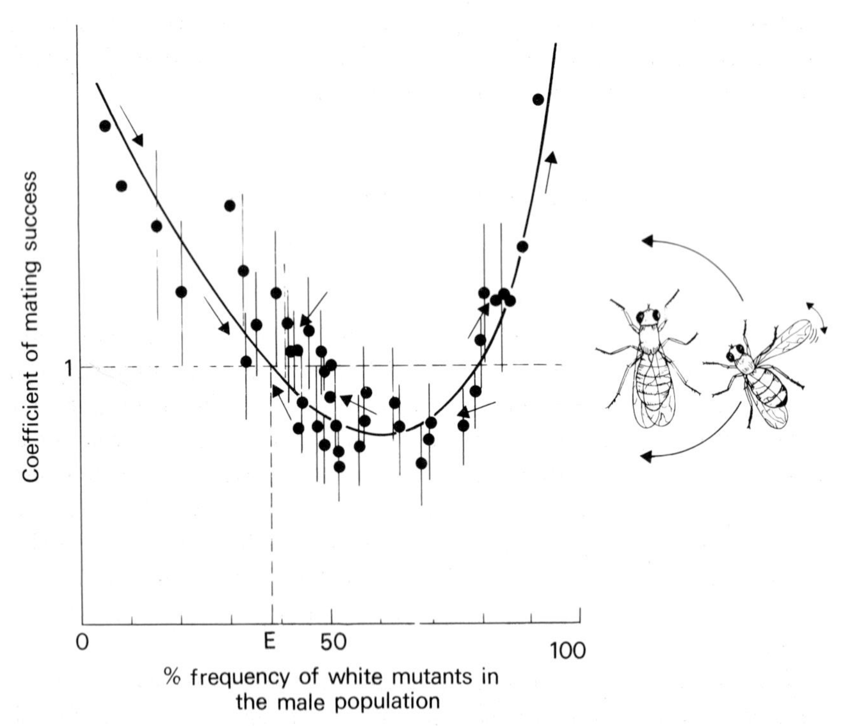

Fig. 7.3. Disassortative mating in *Drosophila melanogaster*. Mating success of white mutant males plotted as a function of their frequency in the population. A coefficient of mating success of 1 means that, on average, each white male mates once. Arrows indicate the direction in which frequency shifts towards equilibria at 40% (E) and at 100%. From Petit (1958).

As the frequency of mutants decreases, their mating success increases, up to the point where they form about 40% of the population. At higher frequencies their mating success is less than would be expected on the basis of random mating and so they tend to decline in frequency towards the equilibrium value of 40%. However, if their initial frequency is greater than 80%, white-eyed males will increase until they form 100% of the population (Fig. 7.3).

What the adaptive advantage of mating with a member of a rare type might be is not clear. It may be a mechanism to prevent inbreeding and to promote the genetic diversity of the offspring. In some polymorphic species rare morphs may be less likely to attract the attention of predators than common morphs; this is called Apostatic Selection (Allen & Clarke 1968) (see 5.3.2). In such species it may be advantageous to mate with a member of a rare morph since this will mean that some of the offspring will be of the rare, less heavily predated type. If disassortative mating is adaptive it may, like assortative mating, lead to the evolution through sexual selection of characters by which different types may be recognised by potential mates.

7.9 Female coyness and mate choice

7.9.1 *Female coyness*

Males must usually display to females to obtain them as mates and courtship can be regarded as a contest between male salesmanship and female sales resistance (Williams 1966). Given a choice of potential mates, females should choose the most fit, but often the only evidence they have of a male's fitness is his courtship behaviour. If male courtship is simple and brief it may be easy for a low quality male to perform as well as one of high quality. There will thus be powerful selection favouring female coyness. Females should be hesitant and cautious in their responses so as to elicit as much display from courting males as possible.

Female coyness takes a rather unusual form in European newts *Triturus*. For courtship to lead to successful fertilisation the female must pick up a spermatophore deposited on the pond floor by the male. In the course of a sexual encounter a male usually puts down two, three or more spermatophores. The probability with which females pick up spermatophores increases during the course of an enounter (Fig. 7.4a) so that they are more likely to pick up the third than the first (Halliday

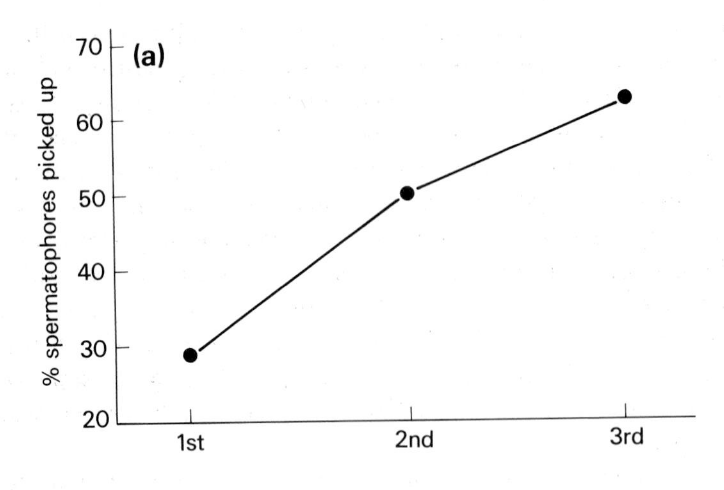

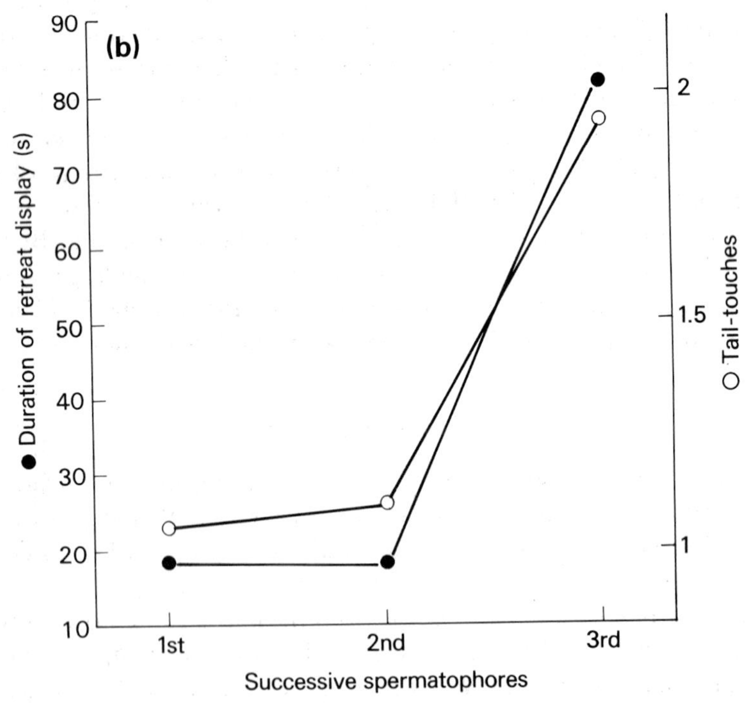

Fig. 7.4. Variations in male and female behaviour in successive sequences of a courtship encounter in the newt *Triturus vulgaris*. (a) The success with which females pick up spermatophores. (b) The mean duration of Retreat Display that precedes each spermatophore deposition and the mean number of tail-touches by the female required to elicit deposition. Data from Halliday (1974) and Halliday (1976).

1974). This can be seen as an adaptation by females that increases their chances of acquiring sperm from males who can produce several spermatophores. To do this a male must possess at least two qualities, the ability to produce spermatophores and the capacity to sustain courtship for a long time without breathing (Halliday 1977a). Exactly how these qualities might be advantageous to females is not clear. Variations in the behaviour of the male newt from one sequence to the next could be adaptive responses to female coyness. In successive sequences he prolongs his display and requires more tail touch stimuli from the female before he will deposit a spermatophore (Fig. 7.4b). Thus, as the probability of the female picking up the spermatophore increases, so he raises the criteria that must be met before he will release his sperm.

Females can facilitate their choice of a fit mate by forcing males to display in groups. This both enables them to compare males directly and also promotes inter-male competition, in effect forcing males to invest more in their display, exposing the deficiencies of weaker males. Females can do this by gathering in groups to which males must come if they are to acquire mates. In guppies *Poecilia reticulata* males display more vigorously in the presence of other males than they do alone (Farr 1976). This effect is enhanced if males in a courting group belong to different colour morphs. Males of different morphs are likely to be less closely related to each other than males of the same morph. By the argument of kin selection, a male is less fit if he loses matings to unrelated males than to related males and is therefore prepared to invest more in courtship display in the presence of unrelated males.

7.9.2 *Female choice*

The criteria on which female choice is based will depend on the nature and extent of male parental investment (Trivers 1972). Females of all species should ensure that the individuals they mate with belong to the correct species and sex and that they are sexually mature and competent. In species in which the male makes little or no parental investment, the only benefit the female derives from mating is male genes. In such species a female's choice should ensure that male genes contribute to the survival and reproductive ability of her offspring and that they are complementary to her own genes. In species in which male parental investment is high her choice should also be based on the male's willingness and ability to make his investment in her offspring and on the complementarity of his parental attributes with hers.

The recognition of species and of sex falls outside the scope of this chapter and, as one of the major functions of courtship, is discussed elsewhere (e.g. Bastock 1967).

Choice for sexual competence

Females may ensure that they mate with fully mature and sexually competent males by responding only to males who have fully developed epigamic characters and who display vigorously and persistently. The intensity of a male's display is likely to be correlated with his sperm supply and his preparedness to transfer it. The rate at which male newts perform display actions in the early phases of a courtship encounter is strongly correlated with the number of spermatophores that they subsequently produce in that encounter (Halliday 1976). In *Drosophila melanogaster* yellow mutant males show lower levels of certain courtship activities than normal males and consequently have reduced success in fertilising normal females (Bastock 1956).

Choice for good genes

Females can ensure that they mate with males who carry good genes for survival if they select older males. As discussed in section 7.5, this is effectively what happens in many lek systems. Old males have demonstrated an ability to survive for longer, and thus probably through a greater variety of environmental conditions, than younger males (Trivers 1972, Howard in press b). In species in which males compete for mates, females will tend to mate with the strongest males, though they may have little opportunity to exercise any choice. Strength in itself is an adaptive character; when passed on to sons it will increase their chances of reproductive success. Strength may also be a good indicator of other qualities because, to be strong in the mating season, a male must possess other abilities such as skill in acquiring food and avoiding predators that are adaptive throughout life.

Females of the pacific tree frog *Hyla regilla* are attracted by the calls of males who congregate at breeding ponds (Whitney & Krebs 1975a). Calling is performed in bouts alternating with periods of silence and certain males can be identified as bout leaders. These not only initiate bouts, but are usually the last to stop calling at the end of a bout. They also call at a faster rate and are more likely to call between bouts than other males. They thus expend more energy on calling than other males. In experiments in which a female was placed in the centre of four loudspeakers, all four speakers produced identical calls but one was the 'leader', initiating and terminating bouts. In 14 out of 18 experiments the female approached the leader speaker. Whether females gain anything from their preference for bout leaders other than gaining a mate who is good at calling is not known.

To show that individuals gain some advantage by mating with the partners that they choose it is necessary to obtain a comparison between their fitness with their chosen partner and the fitness they would have achieved with other potential partners. Maynard Smith (1956) found that female *Drosophila subobscura* show a preference for males of an outbred strain over those of an inbred strain. Their discrimination was based on differences in male courtship behaviour; inbred males perform a particular step in the courtship less quickly than outbred males. Females mating with inbred males left only about 25% as many viable offspring as those that mated with outbred males. Clearly, a preference for outbred males greatly increases a female's fitness (see also 3.5.2).

The argument that females might choose males with good genes may seem to be similar to Trivers' Resource Accrual theory for sexual selection (7.2.4) and thus to suffer from the same objection, that it will lead to a reduction in the heritable variance of male characters. However, the good gene argument differs from the resource accrual theory in that it requires females to choose males by a relative criterion, so that they mate with the best males available to them. Resource accrual theory says that female choice should be on the absolute criterion that a chosen male must be one of a small number of males at the extreme end of the fitness distribution.

Choice for genetic complementarity

Ensuring the complementarity of male and female genes may be the adaptive value of assortative and disassortative mating preferences. Assortative mating will tend to maintain the integrity of a parent's own characteristics in the next generation; disassortative mating will generate greater diversity in the offspring. Which kind of preference is more adaptive will vary from species to species.

A remarkable example of disassortative mating has been reported among laboratory mice (Yamazaki *et al.* 1976). Males show a preference for females who differ from them in their histocompatability antigens, probably recognising them by their pheromones. Such a preference will be adaptive in that it will tend to prevent inbreeding. Further, by producing heterozygosity in the immune system of the offspring it may somehow improve their immune response (Howard 1977).

Choice for reproductive potential

Female choice based on a male's reproductive ability can only be related to his performance in previous seasons in species in which the pair bond is maintained for more than one season. In the kittiwake *Rissa tridactyla* pairs that stay together generally show enhanced reproductive success as the result of their accumulated experience but pairs will break up and find new partners following a season of low reproductive success (Coulson 1966). A female may enhance her long-term reproductive success by mating with those males who are most attractive to females in general; by doing so she is likely to have more grandchildren (see 7.2.2). In communal mating systems like leks she could do this by going to those males that have already attracted the most females.

In species that practice external fertilisation there is a high risk that gametes will become dispersed before they can meet. The size matching shown by some species of frogs and toads appears to be an adaptive mechanism that minimises this risk (see 7.9.4).

Choice for good parental ability

In species in which males hold territories females may assess male parental ability by territory size or quality, though they may be more interested in resources defended by the male than in the male himself (see 11.2.2). In the common tern *Sterna hirundo* a male's performance in courtship feeding may be used by females to assess his future performance as a father (Nisbet 1973, see 10.3.4).

In the insect *Bittacus apicalis* the female chooses a male on the basis of his courtship feeding performance (Thornhill 1976). Though the male does not subsequently feed the young, her choice acts as a mechanism that forces him to maximise his parental investment. The male brings the female an item of arthropod prey and she accepts or rejects him on the basis of prey size, showing a preference for larger prey. This preference increases her reproductive success in three ways. Firstly, a large meal takes longer to eat than a small one and, since the male copulates while the female eats, a large meal means a prolonged copulation. Copulation stimulates oviposition and long copulations lead to accelerated deposition of the female's eggs. Secondly, a large meal will sustain a female through the refractory period between one oviposition and the next. She thus does not have to obtain food for herself during that time, making herself less vulnerable to predators, and she can devote all her energy to the production of her next batch of eggs. Thirdly, if the ability to catch large prey and the willingness to feed females have a genetic basis, she will pass these attributes on to her sons, enhancing their reproductive success.

7.9.3 *Male choice*

In the freshwater crustacean *Asellus aquaticus*, copulation is preceded by a prolonged 'passive phase' in which the male carries a female around between his legs until she is sexually receptive (Manning 1975). Males show a preference for large females when adopting the passive phase. This preference increases their reproductive success in two ways. Firstly, the eggs of large females mature more quickly than those of small females so that the male has to wait less long before copulating and moving on to a new female; by choosing large females, males achieve more copulations per unit time. Secondly, large females produce more eggs than small females.

The preference of male ring doves *Streptopelia risoria* for females who have not been sexually aroused by exposure to other males can be seen as an adaptation to prevent cuckoldry (Erickson & Zenone 1976, see 10.3.4).

7.9.4 *Conflict between the sexes*

In some species, particularly those in which both sexes make substantial parental investment, both males and females may exercise mate choice. It may be that an optimal pairing for a male will not be an optimal pairing for a female, and vice versa, so that there will be a conflict of interests between the two sexes. In the yellow-bellied marmot *Marmota flaviventris* a male will maximise the number of offspring he produces in a year by acquiring a harem of two or three females (Downhower & Armitage 1971). The yearly reproductive success of females, however, declines with increasing harem size and the optimal female mating strategy is monogamy. In nature a compromise is achieved, the commonest mating relationship being one male with two females (Fig. 7.5a).

In toads *Bufo bufo* two factors influence the reproductive success of a pair, female size and the degree to which male and female are

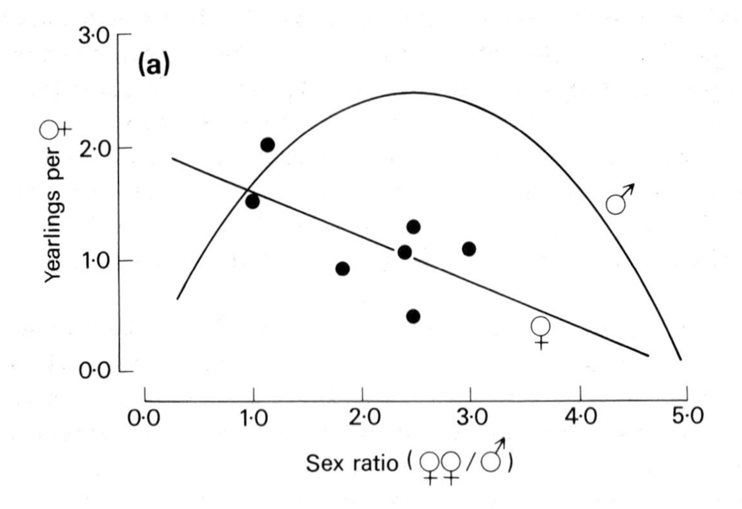

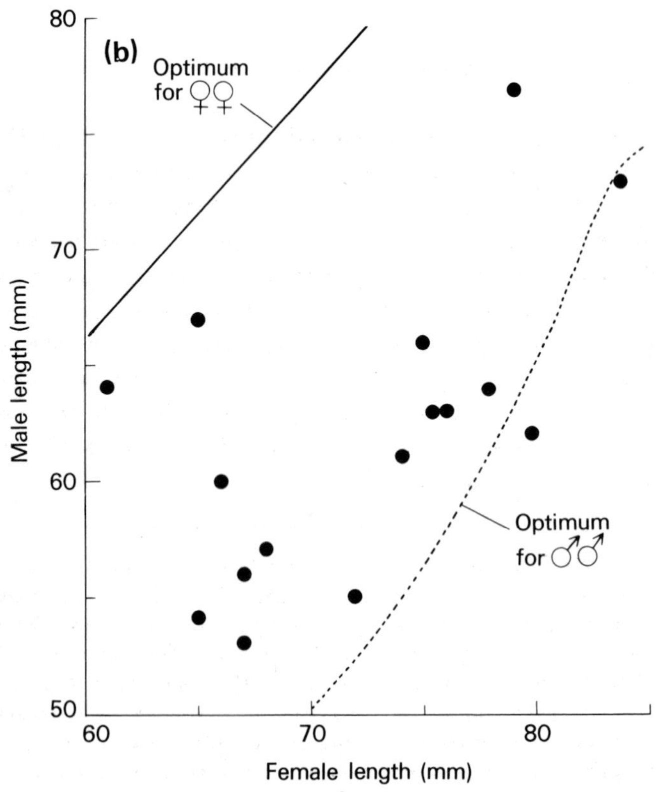

Fig. 7.5. Conflict between the sexes. (a) The relationship between harem size and reproductive success for male and female yellow-bellied

matched for size (Davies & Halliday 1977). The bigger a female the more eggs she lays; the better a pair are matched in size, the greater the proportion of eggs that are fertilised. It can be calculated that, at each mating, a male toad achieves maximum reproductive success when paired with a female 10 to 20 mm larger than himself, whereas a female should prefer the largest male she can find. There is thus a conflict between male and female since each should prefer a mate larger than itself. Observed pairings in the wild fall between the two optima (Fig. 7.5b).

Both marmots and toads are long-lived animals and the calculation of male and female optima on the assumption that animals maximise their annual reproductive output may be invalid. What is important in terms of natural selection is reproductive success over a lifetime and it may be that, by sacrificing maximum annual reproduction, animals may increase their chances of survival and thus their lifetime fitness (Elliot 1975).

7.10 Intrasexual Competition

Whenever a mating system creates conditions in which an individual can increase its fitness by obtaining more mates, or a mate of higher quality, than other individuals, selection will favour the evolution of mechanisms that enhance its ability to compete with members of its own sex. Such mechanisms may operate either before or after mating (Wilson 1975). The most widespread form of pre-mating competition is the aggressive defence by males of females or of territories that attract females. Aggressive competition leads to the evolution of greater male strength and of more effective male weapons such as horns. Among primates, the more distorted the sex ratio, the greater the degree of sexual dimorphism in size.

---

marmots *Marmota flaviventris*. Solid dots represent observed yearlings per female; female line is a regression through these points. Male curve is calculated from the data on female reproductive success. From Downhower and Armitage (1971). (b) Optimal and observed pairings in the toad *Bufo bufo*. Optima are calculated on the assumption that what is maximised is the number of fertilised eggs produced in a season. The female optimum curve represents a minimum because no experimental pairings with males larger than females could be carried out. Observed pairings are represented by the solid dots. Correlation coefficient, $r = 0.567$, $p < 0.05$. From Davies and Halliday (1977).

Intrasexual competition will also favour refinements in techniques by which prospective mates are located. The male silk moth *Bombyx mori* has evolved antennae that are so sensitive to the female sex attractant Bombykol that a receptor cell can be triggered by a single molecule of attractant (Schneider 1974). As a result males can detect females from a range of several miles.

Post-mating forms of intrasexual competition are most diverse among the insects (Parker 1970). This is because female insects generally deposit their eggs over a prolonged period, fertilising them with sperm stored in their spermatheca. It may therefore be advantageous to males to inseminate females that have already mated. In species such as the dungfly *Scatophaga stercoraria* (Parker 1970e) the last male to mate with a female fertilises the greatest proportion of her eggs, his sperm displacing that of earlier males (see Chapter 8). Sperm displacement is but one of many mechanisms by which males can compete for fertilisations with other males.

If a pregnant female mouse encounters a strange male, his odour will cause her to abort the embryos fathered by a previous mate, so that she quickly becomes sexually receptive to the new male. While this 'Bruce effect' (Bruce 1966) is clearly advantageous to the new male, it is not clear how it could benefit the female and thus how it could have evolved. Male langurs *Presbytis entellus* (Hrdy 1974) and lions *Panthera leo* (Bertram 1976) practice infanticide when they take possession of females, killing their existing offspring under a certain age (see 3.5.4). This results in the females coming into oestrous more quickly.

In many insects and mammals males leave mating plugs in the female's genital tract after copulation. This has usually been assumed to be a mechanism preventing sperm leakage but a more important function may be to prevent copulation by other males.

The Uganda kob *Adenota kob thomasi* is a lek species in which oestrous females move from one male territory to another, usually mating with several males (Buechner & Schloeth 1965). Males make no attempt to prevent females mating with other males. Instead they perform an elaborate post-coital display in which they nuzzle the female's genitalia. It is thought that this stimulates the secretion of the hormone oxytocin in the female, inducing uterine contractions that help the passage of sperm up her genital tract. It may be that the probability that a male will fertilise a female is related to the effectiveness of his post-coital display.

Male mosquitoes of the genus *Aedes* secrete a hormone called matrone during copulation that reduces the female's receptivity to other males (Craig 1967). A similar mechanism is shown by male house flies *Musca domestica* (Riemann *et al.* 1967).

Many insects prevent other males from copulating with a female they have mated by prolonging copulation after insemination is completed (Wilson 1975). In many species the male genitalia are equipped with hooks and other devices that prevent the male being dislodged by rivals. By prolonging copulation in this way a male incurs some reproductive cost because it reduces his opportunities to copulate with other females.

Another mechanism used by male insects to monopolise a female is a prolonged 'passive phase' which may occur before or after copulation (Parker 1970e) (see Chapter 8). The male attaches himself to the female until she is ready to mate or until she is no longer receptive to other males. In some dragonflies the male actively defends the female against other males while she is depositing her eggs.

Finally, intrasexual competition can be reduced by a mating pair moving away from potential rivals. In the salamander *Ambystoma*

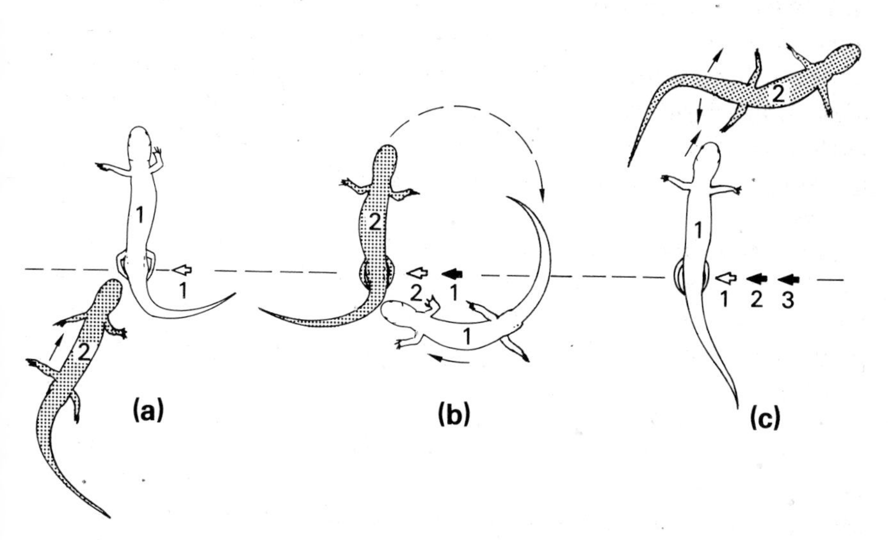

Fig. 7.6. Sexual interference in *Ambystoma maculatum*. (a) As male 1 deposits a spermatophore (indicated by an open arrow), male 2 approaches. (b) Male 2 deposits a spermatophore on top of male 1's (solid arrow). Male 1 turns back. (c) Male 1 deposits a spermatophore on top of male 2's. There are now three spermatophores on top of each other; only the top one, deposited by male 1, is accessible to the female. From Arnold (1976).

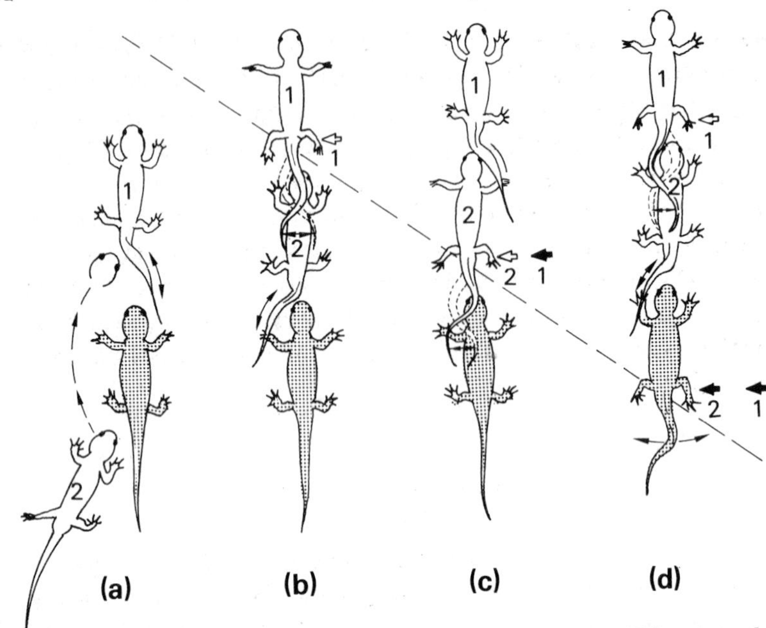

Fig. 7.7. Sexual interference in *Ambystoma tigrinum*. (a) Male 2 intrudes as male 1 leads the female (stippled) in a 'tail-nudging walk'. (b) Male 2 mimics the female and nudges male 1's tail. Male 1 deposits a spermatophore (arrowed) in response. (c) Male 2 deposits his spermatophore on top of male 1's. (d) The female picks up male 2's sperm, that of male 1 being inaccessible. From Arnold (1976).

tigrinum males physically push females away from the vicinity of other males (Arnold 1976).

7.10.1 *Sexual interference and sexual defence*

The mechanisms described above enable individuals to increase their reproductive success and thus their fitness. What is important in natural selection is not an individual's fitness measured in absolute terms but his or her fitness relative to that of other individuals. In some species individuals increase their relative fitness by detracting from the reproductive success of others, while not necessarily increasing their own absolute fitness. Such behaviour has been called Sexual Interference (Arnold 1976).

Some remarkable forms of sexual interference are shown by certain North American salamanders (Arnold 1976). In *Ambystoma maculatum* and *A. tigrinum* males will deposit spermatophores on top of those of other males, making it impossible for the other male's sperm to be picked up by a female (Fig. 7.6). In *A. tigrinum* and *Plethodon jordani* males sometimes move in between a male and a female engaged in mating. They then mimic the behaviour of the female, touching the rival male's tail to elicit the deposition of a spermatophore which they then cover with a spermatophore of their own (Fig. 7.7).

Sexual interference leads to the evolution of counter-measures or Sexual Defence. Male *A. triginum* push females away from rival males and male *P. jordani* actively defend females against other males. In *A. maculatum* males do not show sexual defence but, apparently as an adaptation against sexual interference, produce enormous numbers of spermatophores.

7.13 Conclusion

Darwin's theory of sexual selection sought simply to explain the evolution of sexually dimorphic characters. This chapter has shown that sexual selection has evolutionary effects much more diverse and complex than differences in the appearance of males and females. Ecological factors such as food availability and basic differences in sexual strategy resulting from the differences between male and female gametes combine to determine the form of mating systems and the distribution of parental investment between the sexes. Together these two factors determine the extent to which individuals can achieve greater reproductive success than fellow members of their sex and thus the direction and intensity of sexual selection.

A recurrent theme throughout this chapter is that sexual selection is but one of many selective forces acting on animals. As a result, while the theory of sexual selection is relatively simple, the study of its effects in the field is very complex. It is only in those studies in which data has been collected over many years, such as that on the arctic skua, that the intensity with which sexual selection acts can be assessed with any accuracy. As a result of the recent burgeoning of interest in the relationship between ecology and social behaviour, we may look forward to further studies that will make our understanding of sexual selection more precise.

Chapter 8
Searching for mates

GEOFFREY A. PARKER

8.1 Introduction

An inevitable consequence of anisogamy and separate sexes is that males and females are likely to diverge in their characteristic patterns of life. In particular, sexual selection will generally operate at least initially to favour enhanced mate searching in males, at some expense to sperm production. In species that copulate, the ejaculate mass necessary to ensure fertilisation will usually be much smaller than the mass of the batch of ova to be fertilised, unless the female produces few ova and invests a high degree of parental care in each zygote. Even in the latter case, the total direct investment by a female in a given zygote will generally exceed that of the male except in a few monogamous species or those showing sex role reversal (e.g. see Trivers 1972). Ultimately, because of their high investment in each zygote, females form a limiting resource for male reproductive success in the vast majority of species (those with promiscuity and polygamy).

This limitation will usually be manifest as a highly male-biased operational sex ratio. The operational sex ratio (Emlen 1976b, Emlen & Oring 1977) is the ratio of fertilisable females to sexually active males; it is quite distinct from the actual sex ratio which should (unless special circumstances operate) approach unity at the end of parental care (see 6.3). Highly male-biased operational sex ratios will be generated by (i) high value for:

$$\frac{\text{expenditure on ovum production plus female parental investment}}{\text{expenditure on sperm production plus male parental investment}}$$

(ii) much asynchrony in the onset of sexual receptivity in the female population (see also Emlen & Oring 1977), e.g. as in continuous-breeding species. Intense intra-sexual competition will occur when these two features are coupled with a high degree of clumping of the resources used by receptive females.

Both sexual strategies (male and female) will be under selection to maximise contribution to succeeding generations, hence any reduction in male gametic investment (as would accompany the change from external to internal fertilisation) will be offset by increased expenditure on some other male strategy such as enhanced mobility, territoriality, harem defence, etc. As inescapable conclusion for many species is that female reproductive success is limited mainly by the successful rate of conversion of environmental energy into offspring, whereas male reproductive success is limited by the number of matings a male can achieve with different females (see also Bateman 1948, Trivers 1972, and 7.1).

This chapter will be concerned almost exclusively with problems that males face when searching for mates, and in particular how evolution may be expected to operate when several male competitors are searching simultaneously in the same area. Evolutionary 'games' in which the best strategy depends upon the strategies 'played' by competitors are common in animal behaviour. For each problem we shall seek an 'evolutionarily stable strategy' or ESS, i.e. the strategy that (once fixated in the population) cannot be invaded by a mutant showing any alternative strategy (see also 1.2 and 10.3.1). John Maynard Smith's ESS concept (see 1974b, 1976c) is probably the most important recent development in the study of behaviour and evolution; classically many interpretations of behaviour have assumed an 'advantage to the species' which would not be stable against invasion by mutants with more selfish strategies. In many cases, the best strategy for a given individual is strongly dependent upon the strategy played by his opponent(s); this is true for competitive search behaviour.

8.2 The dung fly mating system

Theories benefit much from empirical substantiation. Studying flies mating on cow pats may seem a rather extreme perversion. Perhaps I can hope it provided quantitative data that enabled fairly precise empirical tests of models for mate searching under intra-sexual competition.

Scatophaga stercoraria L. is the common yellow dung fly so ubiquitous in cattle pastures in spring and autumn. They arrive often in large numbers to mate around and oviposit in the cow dung, where the larvae develop. Because they are discrete, observable units, cow pats constitute a unique example of a patchy environment; theoretical models for optimal searching rely much on the idea of scattered distribution of resource patches. The dung fly mating system is relatively simple in that there is no evidence that individual males hold territories. Females arrive quickly to the fresh dropping and lay all their mature eggs at one visit, then they stay away from the dung until the next batch is mature, and so on. Sexually active males arrive even faster than the females and in much larger numbers. The sex ratio is around 4–5 males to each female, thus intra-male competition for mating opportunities is intense (Parker 1970a). Flies are attracted (almost certainly by olfaction) to the dropping from downwind. In their search for females, males may move back and forth between the dung surface and the surrounding grass on its upwind side; as they arrive females usually fly over the dropping to the upwind side before walking cautiously to the dung surface. A male sighting a female approaches, leaps, mounts, orientates and then mates immediately; sometimes females flee during the approach.

Fig. 8.1. A passive phase *Scatophaga* male standing as an attacker (top) encounters the pair. The female is ovipositing. (From Parker 1970b.)

Fig. 8.2. A struggle in *Scatophaga* for the possession of a female. The attacker male (left) is attempting to push the paired male (right) away from the female. (From Parker 1970b.)

Although most females arriving to oviposit contain an adequate supply of sperm from a mating at a previous oviposition, they are fully receptive and during their lifetimes undergo many more matings than are necessary to ensure fertility. During mating, it appears (from sperm competition experiments 8.3.3) that the male displaces most of the previously-stored sperm contained by the female; the last male to mate thus predominates in the fertilisation of a given egg batch. Copulation can occur either on the dropping or in the surrounding grass, and afterwards the male does not dismount immediately from the female. He withdraws genital contact but remains mounted in a 'passive phase' while the female oviposits. Soon after the last egg is laid the female sways from side to side several times, the male dismounts, and the female flies upwards and away from the area.

Throughout mating and oviposition, pairs are continually harassed by searching males and a specialised series of defensive reactions have evolved by which the paired male dispels attackers. Commonly the paired male rises up on its first and third pairs of legs ('standing') and at the same time raises the middle legs, deflecting the attacker away from any contact with the female (see Fig. 8.1). If an attacker manages to gain a grasp on the female, there is usually a fierce struggle (Fig. 8.2) in which the attacker may eventually succeed in displacing the original male (a 'take-over'). The new male than invariably mates with the female before she continues oviposition, with the new male in attendance. Struggles can be prolonged and may involve several males clustered around one female; the female may sustain damage and become badly smeared with liquid dung (see also Hammer 1941).

8.3 Spatial equilibria: where to search

8.3.1 *Ideal free theory*

Habitats are seldom if ever uniform in their resource value and distribution of receptive females within a habitat will always be non-random. Since in promiscuous species selection will act on males to maximise their mating frequency, then the female distribution will exert a major effect on the male distribution and search strategy. It has been suggested (Parker 1970a, 1974a) that where there is no territoriality, selection will act on males to favour spatial and temporal distributions such that all males experience exactly equal fitness gains, expressed as mating rates (female quotas) or fertilisation rates. This equilibrium distribution would be achieved by adjustments in competitor densities in the various parts of the habitat, or in the lengths of time spent by the various individuals in given, transient patches within the habitat. If any part of the habitat is underexploited relative to the rest, selection favours variants that exploit it until the stable equilibrium is restored. This is a form of frequency-dependent selection.

The problem of obtaining matings is parallel to that of extracting any other fitness-related resource from a habitat, and (independent of the above work) Fretwell and Lucas (1970) and Fretwell (1972) proposed a similar but more general theorem for habitat use in the absence of despotic, resource-guarding behaviour (see also 11.3 and 12.6.3). They suggested that where there are several habitats, each varying in value, the distribution of individuals should be such that each individual obtained the same fitness prospects; again the equality would be achieved by adjustments in competitor densities in the habitats. It is assumed that resource value decreases with increased competition, thus good habitats are filled first and contain more competitors than poorer habitats. Some of the poorest habitats may remain uncolonised, only to become occupied gradually if population size increases. They termed the equilibrium spacing an 'ideal free distribution', i.e. individuals are 'free' to enter or leave habitats and selection favours various forms of competitor density assessment ('conspecific cueing', see Kiester and Slatkin 1974) and habitat selection so that all individuals experience equal gains (see 11.3).

Clearly, ideal free distributions can be obtained only when the expected gains of individuals are reduced by competitor density. Allee's principle (Allee *et al.* 1949) is that individual benefits increase as population density increases, up to some maximum, then decrease. This phenomenon can theoretically generate an advantage in moving from an apparently more suitable habitat into a less suitable one, as population density increases (Fretwell 1972). However, the models of mate searching examined in the present chapter are all concerned with cases where gains decrease as competitor density increases; it seems likely that male mating prospects will commonly decline as an inverse function of competitor density.

In addition to the work mentioned above, there is a considerable literature on theory of feeding strategy that is directly relevant to the models we shall consider, and also theory on the host-searching of insect parasites (see Chapter 2). Other general treatments of the problem of resource utilisation within habitats are given by Svärdson (1949), Kiester and Slatkin (1974), Grant (1975) and Baker (1978).

8.3.2 *The first twenty minutes*

A searching male dung fly can obtain a mating either by catching a female as she arrives at the dropping, or by a take-over of a female during mating or oviposition. The pay-offs are not equal, especially if the female has laid several eggs at the time of take-over. The three types of female (arriving, copulating and ovipositing) have distinct and non-random incidences in time and space at the oviposition site and this ultimately establishes a complex set of spatial and temporal equilibria for the male search strategy.

During the first 20 min. after the dropping is deposited, the male's strategy must be determined almost exclusively by the availability of newly-arriving females; copulation takes around 35 min. and the prospects of gains from take-overs of the few mating pairs available initially are relatively insignificant. Counts were made of the total searching males present in each of a series of zones and on around the droppings, and simultaneously counts were made of the total female captures in each zone (Fig. 8.3). If ideal free theory holds, males should

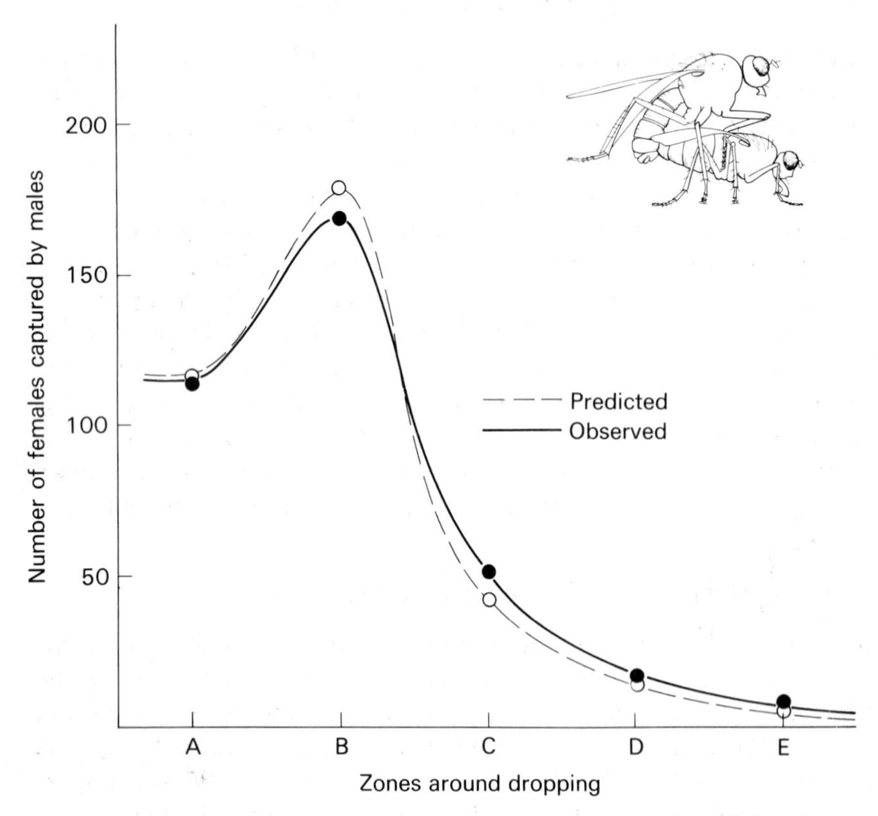

Fig. 8.3. Comparison of observed and predicted number of female captures by male *Scatophaga* in each of a series of zones on and around cattle droppings (aged 0–20 min.). Zone A = the dung surface, zone B = the peripheral band of surrounding grass up to a distance of 20 cm from the dropping edge, zone C = the band 20–40 cm from the edge, zone D = the 40–60 cm band, and zone E = the 60–80 cm band. The predicted numbers of captures assume an ideal free distribution of males. (From Parker 1974a.)

be distributed between zones in such a way that all individuals experience equal expectations of gain. Hence the proportion of females captured in a given zone should equal the proportion of males searching there, assuming that all females arriving are equally valuable irrespective of where they are caught. This prediction concurs with the data in Fig. 8.3; the difference between observed and expected distributions is not significant.

8.3.3 *After the first twenty minutes*

After the first 20 min. or so it becomes more difficult to calculate the equilibrium distribution because take-overs offer significant gain prospects and their effect cannot be excluded. Oviposition must occur on the dung surface, and the peak of egg-laying occurs at around 50–60 min. after deposition. Copulation, on the other hand, commonly occurs some 1–2 m from the dung and peaks at around 40 min. A model was constructed which calculated the ideal free equilibrium strategy, through time, for the alternatives of searching on the dung surface versus searching in the grass (all the grass zones in Fig. 8.3 are grouped into a single alternative strategy to dung searching). At equilibrium, the gain rates (measured as eggs fertilised per male per min.) must be equal for each alternative, and the stability is achieved by varying the competition density in the two localities. Exact details of this model can be found elsewhere (Parker 1974a); the following is a summary.

Gain rates from matings with incoming females

As mentioned, many incoming females eventually arrive at the dung by walking from an upwind position, after their initial flight from downwind. However, a proportion of arrivals are always direct—the female lands on the dung surface. If there are very few competitors around the mating site the best strategy to obtain incoming females would seem to be to wait on the dung—this would yield immediate access to females that fly there direct, and eventual access to indirectly-arriving females. But as the number of competitors increases, those males searching on the extreme upwind periphery would win most of the matings with indirectly-arriving females. As male density around the mating site increases, there will thus be an increasing premium on invading the upwind surrounding grass. These predictions were borne out by data (Parker 1974a); the probability of capture of an incoming female in the grass (P_g) increases in relation to the number of males searching in the grass zones (m_g) up to an asymptotic value, where presumably all females arriving indirectly are captured before they reach the dung. Where $P_g(m_g)$=the probability that an incoming female will be captured before reaching the dung in relation to the number of males searching in the grass, then the probability of capture on the dropping is obviously $1-P_g(m_g)$. Thus where there are $f_i(t)$ females arriving per minute at time t after deposition, the average rate of capture of incoming females by a male searching in the grass will be $[P_g(m_g) \times f_i(t)]/m_g$ females/male/min. The average search time per male will be the reciprocal of this rate. Let the handling time (to mate and guard a female during oviposition)$=c_i$. Hence a gain rate of

$$r_{gi}(t) = g_i \Bigg/ \left\{ \left[\frac{m_g}{P_g(m_g) . f_i(t)} \right] + c_i \right\}$$

eggs/male/min. is achieved by each male searching in the surrounding grass, and a gain rate of

$$r_{di}(t) = g_i \Bigg/ \left\{ \left[\frac{m_p}{(1 - P_g(m_g)) . f_i(t)} \right] + c_i \right\}$$

eggs/male/min. is achieved by searching on the pat, where m_p=the number of males on the pat and g_i=the number of eggs that will be fertilised by a male mating with an incoming female.

Gain rates from take-overs

The attack rate sustained by pairs on the dropping can be approximated to (0·07 m_p+0·1) attacks/pair/min. (Parker 1970a, 1974a). Take-overs occur on average once every 57·2 attacks of ovipositing pairs and once every 154 attacks of pairs *in copula*. We can calculate by a similar method as before that the expected gain rate from oviposition take-overs of a male searching on the dropping becomes

$$r_{do}(t) = g_o \Bigg/ \left\{ \left[\frac{57{\cdot}2\ m_p}{(0{\cdot}07\ m_p + 0{\cdot}1) . f_o(t)} \right] + c_o \right\}$$

eggs/male/min.; where g_o=the eggs fertilised from a mating with an ovipositing female, c_o=the handling time, and $f_o(t)$ is the mean number of ovipositing females present during the tth min. after deposition.

Potential gain rates on the dung surface from take-overs of copulating pairs are very small compared to those from ovipositing pairs. This is because at all but the lowest male densities, pairs meeting on the dung emigrate to the downwind surrounding grass for most of the duration of copulation. The probability $P_e(m_p)$ of emigrating in relation to the number of m_p of males present on the pat is known (Parker 1971). Where $f_c(t)$ is the total number of copulating pairs at time t, as an approximation it can be taken that the number which met and remained on the dropping will be $f_c(t).(1-P_g(m_g)).(1-P_e(m_p))$. Hence the gain rate from take-overs of copulating females to a male on the dropping will be

$$r_{dc}(t) = g_c \bigg/ \left\{ \left[\frac{154\, m_p}{(0{\cdot}07\, m_p + 0{\cdot}1).f_c(t).(1-P_g(m_g)).(1-P_e(m_p))} \right] + c_c \right\}$$

eggs/male/min., where g_c=the eggs fertilised by a male that takes over a copulating female, and again c_c is the appropriate handling time.

Potential gains from take-overs that arise by searching in the grass are not worth including in the model. Ovipositing pairs do not occur there (unless driven from the dung), and the attack rate in the surrounding grass of the copulating pairs is virtually negligible, especially since most emigrate to a downwind position.

Comparison between predicted and observed search profiles

The equilibrium numbers of males searching in the two localities through time t after dropping deposition can now be found by setting

$$\begin{array}{ccc} \text{total gain rate on the dung} & & \text{gain rate in grass} \\ r_{di}(t)+r_{do}(t)+r_{dc}(t) & = & r_{gi}(t) \end{array} \qquad (1)$$

The only variables at a given time t are m_p and m_g; where m_t=the total number of searching males, then $m_t=m_p+m_g$. Hence equation (1) predicts m_p and m_g, for any chosen density of searching males. There is good evidence from field data that at any given pat age, a direct relationship exists between m_t and the various female availabilities (f_i, f_o, f_c) and so the female availabilities can be estimated for any m_t.

The only remaining values needed are the egg gains; g_i, g_o, and g_c. These can be determined only by laboratory experiments in which the sperm from a given male are 'labelled' by exposing the male to high doses of gamma irradiation; this has the effect of inducing a very high chance of death in the stages of development before hatching. Thus though irradiated males are able to mate and their sperm able to fertilise ova, very few of the eggs they fertilise will hatch if a high enough radiation dose is given. Hence females can be multiple-mated with normal and 'labelled' males, allowed to oviposit, and the number of fertilisations ('egg gains') achieved by the labelled male calculated from records of the number of eggs that fail to hatch. This sort of experiment (see Parker 1970c) showed that for *Scatophaga*, the last male to mate fertilises about 80% of the batch to be laid, and that the

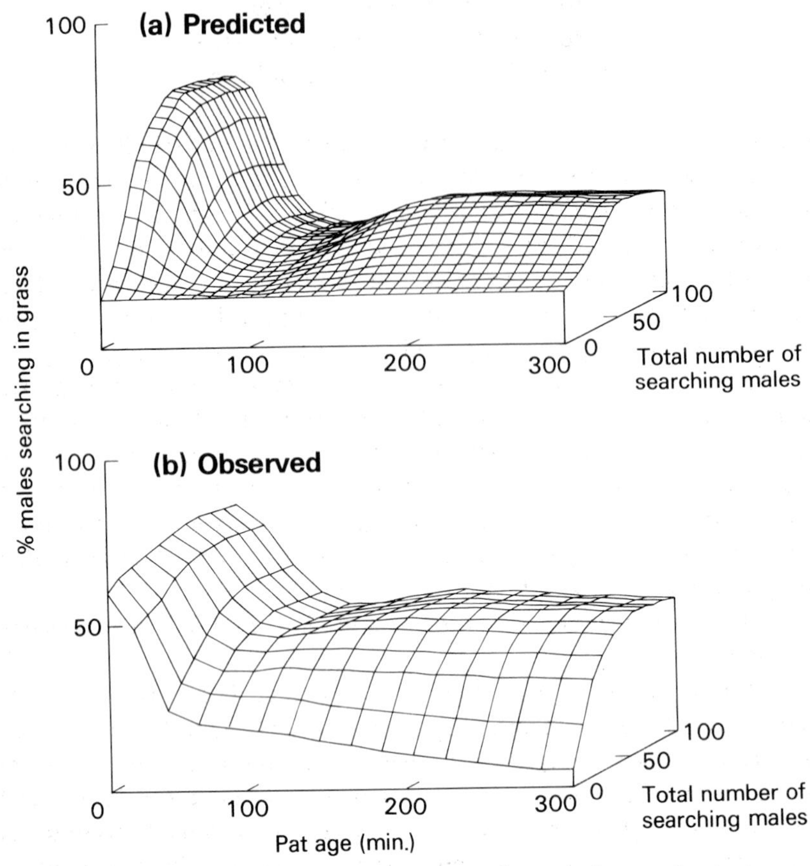

Fig. 8.4. **a** = Predicted profile for *Scatophaga* male search strategy (percentage males searching in the grass ($= 100\ m_g/m_t$) plotted against dropping age and total searching males (m_t)), assuming that fertilisation rates achieved by searching on the dung equal those achieved by searching in the grass. For calculation, see text. **b** = Observed profile for male search strategy. This is derived from the observed data which has been averaged for certain groupings of dropping age which would therefore tend to be smoothed out to some extent. To emphasise that this profile must be regarded as approximate only, half the grid lines from each axis are omitted as compared with the predicted profile. (From Parker 1974a.)

sperm from the previous males compete over the remaining 20% in the same proportions as they did at the last batch (i.e. before the last mating). For example, where there have been two previous ovipositions (and two previous matings) egg gains at the third oviposition (i.e. after the third mating) will be ranked as follows:

third (last) male $= \cdot 8$ of batch
second male $= \cdot 8 \times \cdot 2$ of batch
first male $= \cdot 2 \times \cdot 2$ of batch

This sort of data can be linked with field data on probabilities of take-over and survival to compute an expected eventual egg gain arising from a given mating. For take-overs during oviposition it is assumed that on average, half the eggs will have been laid and hence fertilised at the time of take-over (fertilisation occurs as the egg passes down the oviduct during oviposition). The computations give egg gain estimates as $g_i = 34{\cdot}9$, $g_o = 18{\cdot}5$, and $g_c = 33{\cdot}9$ (Parker 1970c).

Figure 8.4a shows the equilibrium percentage of the searching male population which should be found in the grass, in relation to dropping age t and total male density m_t, predicted from equation (1). The observed profile for the male search strategy is given in Fig. 8.4b; it is derived from field data taken from groups of droppings of similar ages but the averaging procedures used must necessarily have smoothed the profile to some extent. Both profiles show:

(1) the proportion of males in the grass is lowest at low male densities (m_t low),

(2) the equilibrium proportion of males searching in the grass falls sharply with time from an initially high value (when almost all available females are newly-arriving ones) towards the onset of the oviposition peak at 50–60 min. (when oviposition take-overs become very important), especially at high male densities. There is then a trough around 100 min., rather later than the oviposition peak, followed by a plateau.

The only notable disagreement between the two profiles is that early after deposition and at low m_t, many more males are observed searching in the grass than is predicted by the model. There are several possible interpretations of this discrepancy, but perhaps the most plausible explanation relates to the problem of obtaining olfactory information about new droppings. Once a skin forms over the pat surface, it may be possible for a fly on the dropping to perceive the presence of a new dropping upwind. This would not be possible early on, before the skin forms, because of the strong smell of the dropping currently colonised. However, this information about new droppings may be obtained by spending time in the grass upwind.

All in all, these studies offer some validation of the ideal free theory. In essence, it shows that within a part of a habitat in which there exist regular gradients of some fitness-related resource, competitor distribution can evolve to adjust individual benefits to equality. Note that we have assumed a negligible cost in moving from one part of the habitat to another; this is probably valid for dung flies since a shift between grass and dung surface can be achieved almost instantly. Yet we have been concerned throughout with the rates of gain experienced by individuals, the assumption being that selection will favour strategies that maximise resource uptake/time. Where the moving or transit times become significant, then this must be incorporated, as we shall discuss later (see 8.4). Also, though we anticipate that the equilibrium distributions will be achieved by some process of conspecific cueing, we have made no predictions about what makes one particular individual depart rather than any other individual; this can also become a complex problem. In *Scatophaga*, males move repeatedly between grass and dung, but it is not known whether certain genotypes have a higher tendency for one or other search strategy. Nothing is known of the mechanism by which the spatial ESS is achieved.

Given that selection may act to minimise time costs for given gains, two other features of the dungfly mating system deserve a mention. Firstly, why do pairs that meet on the dung emigrate to the grass to copulate? Generally the male flies there, carrying the female. The obvious solution would be that emigration reduces the probability of take-over, and this is so (Parker 1971). However, because it is hotter on the dung than the ambient temperature, copulation is faster there and thus emigration has a significant time cost. This probably explains why relatively few pairs emigrate when the density m_p of males on the pat is low, but almost all ($>90\%$) emigrate at high m_p. When the costs and benefits are analysed in a quantitative model for emigration, the prediction is that males should stay on the dung to copulate if $m_p < 5$, otherwise emigrate (Parker 1971). Though the probability of emigration in relation to male density is a continuous rather than a step function, half the males show emigration at $m_p = 5$, below this value there is an increasing tendency to stay and vice-versa above it.

Secondly, why do males remain with their females during oviposition? A mutant which showed no passive phase behaviour would save some 17 min. oviposition time. However, remember that the last male to mate fertilises approximately 80% of the egg batch to be laid by a mated female. The passive phase male thus guards his rights to precedence at fertilisation of the mature eggs, and it can be calculated that the benefits to the male of this behaviour far outweigh the costs (Parker 1970d). The female also benefits slightly from the male's passive phase by avoiding rejection delays and other costs of harassment, even though it must usually be preceded by what is (from the female's point of view) a supernumerary mating. But an interesting paradox arises. Most Diptera (and many other insects) show unreceptivity to further matings when they have had one successful mating. If *Scatophaga* females were initially 100% unreceptive after mating, then the passive phase (a sex-limited male character) could not evolve by immediate selection processes, even if it yielded a selective advantage to the female as a means of speeding up her oviposition. Spread of the passive phase characteristic is obviously dependent on an advantage to males possessing it. However, if there is some chance that a mated female can be remated (as little as probability ·1 is enough), then sex-limited genes for the passive phase behaviour would be favoured and as they spread, so would sex-limited genes for full receptivity in the female (see Parker 1970d). Thus if as many as 1 in 10 females would be remated, then it pays a male to 'spend' 9 time-expensive passive phases for the sake of being able to prevent fertilisation losses due to his mate's infidelity on the tenth. The passive-phase evolved to reduce losses due to sperm competition. Males of several other animals show parallel adaptations to reduce sperm competition; in addition to pre- and post-copulatory passive phases there are copulatory plugs that function as chastity belts and various forms of territorial guarding (e.g. Devine 1977, Ross & Crews 1977, Parker 1970e) (see also 7.10).

8.4 Temporal equilibria: how long to stay in a patch

8.4.1 *No competition in the patches*

Theory: the marginal value model

We shall now consider some rather more rigorous models in which the effects of a significant search or transit cost can be assessed. Though specifically interested here in the problem of mate searching, the models can be expressed in a general form for fitness-related resource accrual, and will in some cases be exactly equivalent to those discussed by Chapter 2 for optimal exploitation of food patches.

Consider a series of resource patches discontinuously-distributed within a large habitat. Fitness-related gains can accrue only within one of the patches, and each patch gradually decreases in value with time of exploitation. How long should a single competitor remain in a patch when the quality of the patch cannot be predicted in advance of its encounter? For the present purposes the resource to be gained is mating opportunities and the patches are (generally) areas containing or attracting receptive females. That mating prospects have a patchy distribution within the habitat is evident for most species; a review of specific types of sexual encounter sites and their nature is available for insects (Parker 1978). The following theoretical analysis is based on the works of Charnov (1976b) and especially Parker and Stuart (1976).

Let the mean cost in terms of units of fitness of travelling between patches $=S$; search cost S is related to the density of utilisable patches within the habitat. All patches are of similar quality or value (this assumption is to be relaxed later). Once in a patch, the male invests I units of fitness (again measured through time and energy losses) in resource uptake, i.e. on obtaining matings. I is the absolute cost of resource exploitation and is not weighted by the benefit that accrues during I. Throughout I there is resource accrual and the cumulative fitness gain obtained by I will be defined by the function $G(I)$. Again this is an absolute gain rather than net gain; it excludes cost I. Now as before, we anticipate that selection will act to maximise the individual's gain rate, measured as benefit/cost. Throughout, we shall call this maximum rate SL_{max}. We seek the optimal investment, I_{opt}, for a male to spend in a given patch before leaving to search for a new, unexploited patch. Stated verbally, the emigration threshold I_{opt} occurs at equilibrium where

(Expected future fitness due to continued investment with an existing resource)	=	(Expected future fitness due to withdrawal from the existing resource to start a new search phase)
A		B

i.e. when $A > B$ stay; when $A < B$ emigrate.

This rule (termed the 'marginal value theorem' by Charnov 1976b) is demonstrated graphically in Fig. 8.5. If the abscissa is produced a distance S beyond the origin to point P, then I_{opt} is given by the tangent to $G(I)$ drawn from P. This is obviously the maximum slope (SL_{max}) that can be obtained from the system. Where $G(I)$ is not monotonic, then I_{opt} is given by the steepest tangent obtainable to $G(I)$; there may be several possible tangents.

For most practical purposes, it is rather difficult to measure G and I in terms of direct fitness units. In fact this does not matter, provided that each is directly related to fitness; and there is no necessity for the

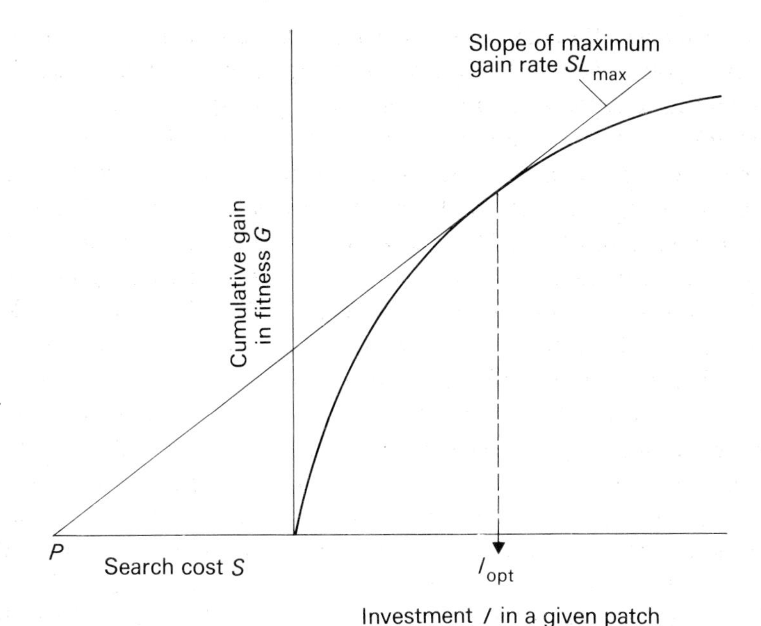

Fig. 8.5. Graphical model for solving optimal emigration threshold (I_{opt}) in habitat where animal searches for single resource-patch type and with no competition in resource. Probable total fitness gain (G) assuming that its extraction costs nothing, is plotted against extraction investment cost I. Mean search cost is plotted by producing abscissa beyond origin distance S to point P. I_{opt} is given by tangent shown at which point $(dG/dI)_{I=I_{opt}} = SL_{max}$. Selection will favour assessment mechanisms which ensure emigration at I_{opt}. (From Parker & Stuart 1976.)

two conversion constants to be equal (I_{opt} is the same even when different scales are used for G and I). For sexual selection and mate searching, G will generally be probable numbers of inseminations, or probable number of ova fertilised, and I and S will generally be time. Both these measures are likely to be directly related to fitness, because time spent away from mate searching can usually be converted directly into missed mating opportunities.

Data: copulation time fits the model

There is an example in the search strategy of male dung flies that fits the model in Fig. 8.5 fairly well, assuming the exponential form for $G(I)$. It concerns the optimal copulation time, and the resource patch is a single female. The longer a male spends copulating with a non-virgin female, the more sperm from previous matings he displaces, and the more eggs he will fertilise himself. However, a male that continues to copulate for a prolonged period misses opportunities to mate with new females. The curve in Fig. 8.6 was obtained by sperm labelling experiments in which females were mated with both a normal and an irradiated ('labelled') male (Parker 1970c, Parker & Stuart 1976). The first mating was of normal duration, but the second was terminated at

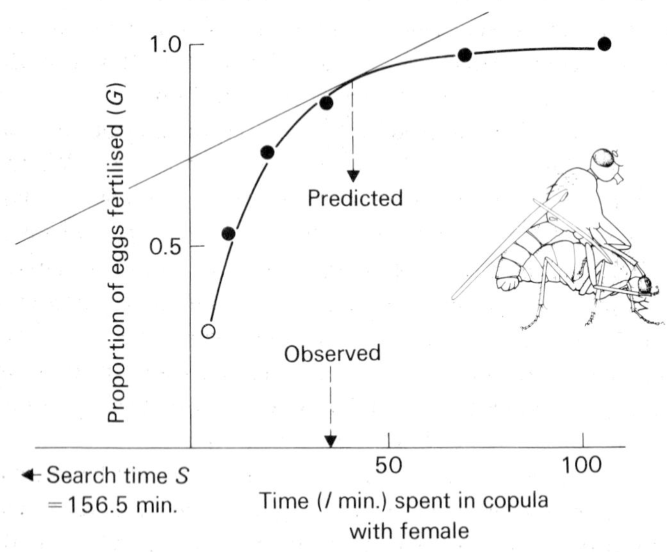

Fig. 8.6. Optimal copula duration in *Scatophaga* (reanalysed from Parker 1970c). Observed points were obtained experimentally from sperm-competition experiments but converted to allow for a ratio of 1 virgin: 4 mated females. G is thus proportion of eggs which will be fertilised by male if it spends time I on copulation. With virgins it is assumed that any copulation time would give full fertilisation of the batch. This may invalidate first point but is unlikely to cause errors for the rest. Curve gives best exponential fit to data ($= 1 - \exp - I/16$; $I_{opt} = 41{\cdot}4$ min.). Observed $I_{opt} = 35{\cdot}5$ min. (From Parker & Stuart 1976.)

a variable interval. To obtain data about extremely long matings, the second male was sometimes allowed to mate twice or even three times. It is unlikely that males can differentiate between virgins and non-virgin females, and so the curve for $G(I)$ (see Fig. 8.6) can be calculated assuming that females occur around droppings in a ratio of 1 virgin to 4 non-virgins (i.e. that likely in nature). Search cost S was calculated from field data by counts of the numbers of each sex present with time after deposition of the dropping. The search cost is, for the 'lifetime' of a dropping, the total male minutes spent searching divided by the total females which arrive, i.e. the mean time for one male to find one female. To this must be added the oviposition-guarding time.

Curve $G(I)$ fits closely to an exponential form. This enables a predicted I_{opt} to be computed, which can then be compared with the observed duration for I_{opt}. The fit between the predicted and the observed copula durations is quite close, suggesting that mating duration and hence degree of sperm displacement may be optimised in response to sexual selection as envisaged. The model assumes that sperm costs nothing; this may account for the fact that the observed mating duration is less than the predicted duration.

In nature, whether patches are related to feeding or to mating, it seems certain that the resource patch values will vary. What is the optimal response when the environment contains several patch types each varying in $G(I)$? We now need solutions for the optimal emigration thresholds for each patch type. It turns out that marginal value theorem applies exactly as before. Thus a male should persist with a given resource j until the gain rate in j reduces to become equal to the maximum mean gain rate obtainable from the total environment (all I optimised). (Further details and implications of this type of model can be found in 2.3 and in Parker & Stuart 1976.)

For mate searching, patch variation can occur in at least two ways. Firstly, where the resource patch is a single female (as in the *Scatophaga* copula duration example), it is easy to find reasons why females will vary in their reproductive value to males. For instance, fecundity often increases with the size of the female (e.g. in *Pieris*, Baker 1968; *Scatophaga*, Parker 1970c; *Asellus*, Manning 1975; *Bufo*, Davies & Halliday 1977). Thus the optimal copula duration (where there is stored sperm to displace), or the optimal female-guarding duration (where there is a precopulatory stage of amplexus in which the male holds on to the female until mating) may be modified by the reproductive value of the female. Reasons why a female guarded during amplexus should offer gain prospects that diminish with time are somewhat tenuous; it seems possible, however, that the probability that the guarded female will allow mating in the next unit of time initially increases up to some maximum, then decreases (see Parker 1974b). Do males show optimal emigration thresholds in response to variance in female reproductive value? There is evidence in *Scatophaga* that this is so. As we have seen, the copula duration of males mating with gravid females is close to an optimum predicted in relation to sperm displacement. Rather rarely, a male on the dropping captures a female that has just finished laying her eggs. (This is an unusual event because the female flies away from the dung within seconds of the male's dismounting at the end of oviposition.) Now a mating with a post-oviposition female can yield gains only through future ovipositions, and clearly the optimal sperm displacement (and hence copula duration) will differ. Because the probable egg gains are much less, it turns out that the predicted optimal copula duration with post-oviposition females is 17·5 min, which is close to the observed duration of 15·1 min. (Parker 1970c). This is less than half the time for gravid females, and indicates that males may be monitoring female egg content when their copula duration decision is being determined.

Other applications of the marginal value model

A second source of patch variation can arise when resource patches are sexual encounter sites rather than single females. Encounter sites are commonly (but not always) areas that have some resource value to females such as a feeding, nesting, or oviposition site. Hence as the primary resource decays in value to females, the number of females exploiting or arriving into the patch declines. $G(I)$ is thus now the number of matings G achieved by a male through time I in the patch. Patch value variation will be extensive (and dependent on variations in the primary resource), stipulating a wide array of forms for $G(I)$.

Remember that we are presently concerned with the case where there is only one male in a patch, i.e. where there is no competition within patches. The above sort of model may apply therefore when patches (encounter sites) are widely spaced and exploited quickly so that the chances of two males occurring together in the same patch are small. More importantly, perhaps, it would also apply when there is despotic guarding with the convention that only one male (the owner) exploits an encounter site at any one time. Such mating systems are quite common. In this case, $G(I)$ is likely to be a probability density function in which G is the cumulative probability that a female will have arrived by time I. Search costs may be high, because many patches may be encountered before an unoccupied one is found.

In summary, we may expect that animals searching for and singly exploiting discontinuously distributed resources will show 'resource assessment strategies', i.e. selection will favour that they monitor various cues correlated with current resource quality. These cues will be used in accordance with marginal value theorem to determine the decision whether to stay in the resource patch, or to leave to search for a new one i.e. to determine the emigration threshold. There will be a unique pure ESS (optimal emigration threshold) for each resource patch type, which will depend on the overall habitat quality. For a fine adjustment (rather than a response based on an average habitat) the animal should modify its set of optimal emigration thresholds in relation to the habitat's current condition. Where learning capacities allow these fine adjustments, this poses the fascinating problem of an optimal memory window (Cowie 1977) i.e. how much of the previous resource accrual history should be monitored for use as the sliding average to predict SL_{max}? A second form of fine adjustment concerns the individual's phenotype. Phenotype features such as size, age, or current state of physiological deficits may exert important effects on the optimal emigration thresholds. We may look to the motivation mechanism as the process by which strategy is shifted in relation to deficit. For features such as size, colour or age variation, it seems possible that genes for the appropriate strategy could be 'switched on' by a particular phenotypic or ontogenic state. Though there has been no work on phenotypic variation in emigration threshold, it has been found for several groups that there are gross differences between phenotypes in observed mate-searching strategy (e.g. Alcock 1978).

8.4.2 *Competition within patches: continuous resource input*

For mate searching, the main application of the models where resource patches are exploited singly may be the case of despotic guarding of territories or single females. What evolutionary rules will apply for ideal free systems, where several competitors can invade a patch and exploit it simultaneously? As we shall see, the answer to this question is influenced markedly by whether resource input continues after the competitors arrive.

Theory

Consider again the dung fly mating system. Males arrive very quickly to the fresh dropping (usually within 4 min. of deposition) and few if any leave until the dung is 15–30 min. old. The arrival rate of females shows a gradual negative exponential decrease as the dung ages. Because there is such a predominance of males, it is obvious that a given male's fertilisation gain rate at the dropping will be dependent upon both the number of competitors and on the input rate of females. Many mating systems will be similar, and we need a model with an appropriate set of assumptions. These will be (i) the competitors (males) arrive very quickly relative to the resource (female) input; (ii) resource input is continuous with monotonically decreasing rate; (iii) the resource is exploited *immediately* on input by the competitors. The following account is taken from Parker and Stuart (1976).

Let $G(I)$ be the total females which will have arrived by time I after deposition. The resource is exploited immediately on input, and so $G(I)$ is also the theoretical gain of one individual with no competitors. Where there are n male competitors, then the gain rate per male will be $(1/n)\ (dG/dI)$, which we will call dg/dI. The gain at time I of one individual is thus $g(I)$. Suppose that all males arrive simultaneously at $I=0$, and that they sustain a search cost S due to moving from one patch to another. Initially, imagine that all resources have similar $G(I)$. It can be deduced that evolution will favour the following rules:

(i) All competitors should remain in the resource patch initially,

(ii) As I increases, a value of I is achieved where, assuming all other competitors stay, the fitness value of leaving becomes greater than that of staying for any one given competitor. We shall call this point I_{1crit}, the first emigration threshold; at I_{1crit} one of the competitors should leave to search for a new patch. There should then follow a phased withdrawal of competitors until eventually the last competitor leaves at I_{ncrit}. The range of I between I_{1crit} and I_{ncrit} is termed the emigration range.

Such a pattern of behaviour is a mixed ESS not a unique pure strategy, as for the non-competitive case with I_{opt}. The population will show a distribution of emigration thresholds for departure, determined by the resource input rate; probably each individual will 'play' the ESS distribution when examined over a given long time period. However, the fundamental principle of a mixed ESS is that all its component pure strategies will experience equal payoffs. For our case, this means that all competitors must experience the same gain rate from the system, irrespective of when they leave. Thus the gain rate must be constant and independent of I within the emigration range. Hence for $I_{1crit} < I < I_{ncrit}$, let

$$\frac{dg(I)}{dI} = \frac{g(I)}{S+I} = \text{constant } SL_{\max} \tag{2}$$

and since $\dfrac{dg(I)}{dI} = \dfrac{1}{n(I)}\dfrac{dG(I)}{dI}$, then from (2) above the number of competitors present at a given I in the emigration range must be

$$n(I) = \frac{1}{SL_{max}} \cdot \frac{dG(I)}{dI} \tag{3}$$

This can be termed for convenience the 'input matching theorem'; i.e. the number of competitors should continually match or track the resource input rate divided by a constant equal to the expected gain rate for the habitat as a whole. The theorem thus stated applies only when each individual gains $1/n$ of the input at any given time; different forms of input matching will operate when the individual's gains do not obey a strict inverse proportionality.

Figure 8.7 gives a graphical demonstration of the principle on which rule (3) is based. The upper curve is $G(I)$, i.e. the cumulative input of resource to the patch through time I. The lower curve is $g(I)$, the cumulative uptake of a single competitor where there are n competitors. Up to the first emigration threshold (I_{1crit}), the gain of a single competitor is simply $1/n$ that of the total input, $G(I)$, and n is constant as n_{max} because all of the competitors should stay. Now the first emigration threshold will be equal to the emigration threshold for an individual exploiting the patch singly (no competition); i.e. $I_{1crit}=I_{opt}$ for the non-competitive case. The reason for this is that I_{opt} is independent of the scaling of $G(I)$—hence dividing $G(I)$ by a constant (n_{max}) does not affect the optimal emigration threshold. The reason that only one competitor should leave at I_{1crit} can be seen as follows. Imagine that all competitors stayed. Each would now begin to experience a gain rate less than the slope ($=SL_{max}$) of the tangent to $g(I)$. Hence it would have been better for any one of them to have left earlier, when $dg/dI=SL_{max}$. However, if several should leave when I_{1crit} occurs, then there is an advantage in staying, because the gain rate (dg/dI) of the remaining competitors would be raised above SL_{max}. Hence the only way that gain rates can be maintained constant ($=SL_{max}$) is for individuals to show a gradual withdrawal from the patch in a manner defined by the input matching theorem (3). The maximum gain rate from the non-competitive system (slope of tangent to $G(I)$ curve) will be a factor n_{max} greater than that of the competitive case (slope SL_{max} of tangent to $g(I)$ curve). The end of the exploitation period occurs with the departure of the last competitor. This last emigration threshold (I_{ncrit}) can be found from (3) quite simply, because at this stage there is only one individual left, hence $n(I)=1$.

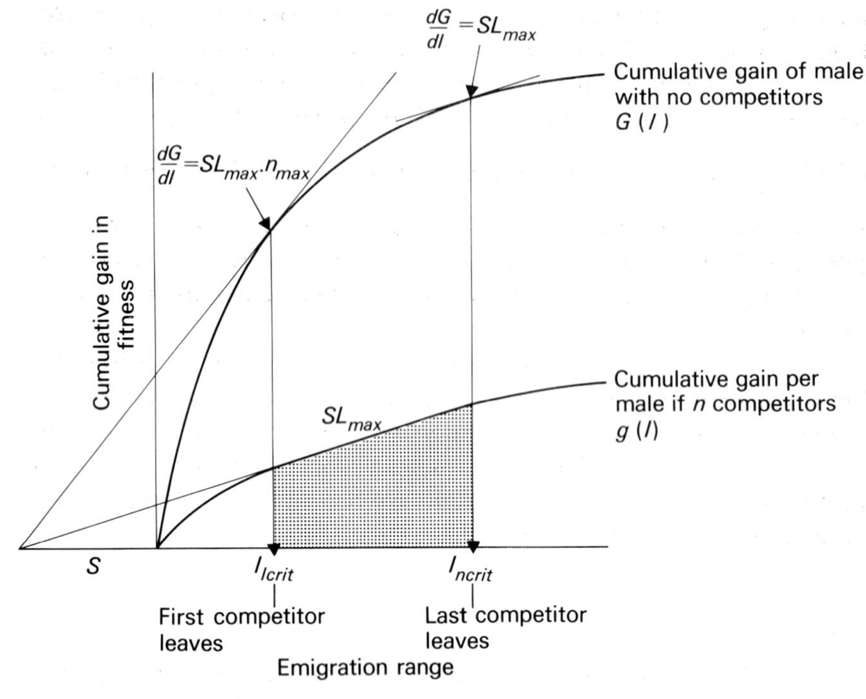

Fig. 8.7. Range of mixed emigration strategy where there is resource sharing, continuous input of resource to a patch, and immediate uptake of resource. All competitors arrive at $I=0$, or randomly before I_{1crit}. Mean cumulative gains per individual are $g(I)$; $G(I)$ is gain in noncompetitive situations. Thus maximum slop $SL_{\max}$ of competitive situations corresponds to $SL_{\max} \cdot n_{\max}$ where resource is exploited by one individual. Between $I=0$ and I_{1crit}, all individuals should stay; the last individual should leave at I_{ncrit}. Within the emigration range there should be a graded departure at a rate set by (3). Note that $I_{1crit}=I_{opt}$, where $dG/dI_{opt}=SL_{\max} \cdot n_{\max}$; $I_{ncrit}=I$ of noncompetitive situation where $dG/dI=SL_{\max}$. (From Parker & Stuart 1976.)

Hence by substituting $n(I)=1$ into (3), it is evident that this last individual must leave if $dG/dI<SL_{\max}$.

The data fit the predicted mixed ESS

Do male dung flies show input matching? It is easy to test this from field data (Parker 1970a, Parker & Stuart 1976). The cumulative resource input measured as mean total females that will have arrived at a dropping approximates to $G(I)=22[1-\exp-(I/90)]$. The overall mean capture rate during the observations was roughly $SL_{max}=0{\cdot}005$ females/males/min. The average number of males around at time I was measured directly, and hence $n(I)$ observed can be compared with $n(I)$ predicted from input matching theory (3) in which $SL_{max}={\cdot}005$ and $dG/dI=\frac{22}{90}\exp-(I/90)$. These are plotted in Fig. 8.8. Though the fit between the predicted and observed curves is quite close, a better fit (see legend to Fig. 8.8) can be obtained if the predicted curve is corrected to allow for the effects of handling times (copulation and passive phases).

This sort of model is relevant to many species (particularly invertebrates) in addition to *Scatophaga*. There is evidence for an exactly parallel mixed ESS for input matching in *Sepsis cynipsea*, another fly that uses cattle droppings as a mating and oviposition site; it is probable that it will be found in the mate searching of many other insect species (see Parker 1978).

A bit more theory

The mechanism by which the mixed ESS of emigration thresholds is maintained in *Scatophaga* and *Sepsis* is not known. Perhaps the most likely possibility is that the population is genetically monomorphic, and each individual is programmed to select (before I_{1crit}) a value of I within the emigration range with a probability distribution that satisfies the ESS requirements defined by (3). This is to say it may select an emigration threshold randomly within a probability distribution $p(I)$ that is fixed by the requirements of input matching.

An alternative explanation could be that the population is genetically polymorphic for emigration thresholds, though this poses genetic difficulties about maintaining the ESS distribution (see Maynard Smith & Parker 1976).

Though the simple input matching model outlined in the previous section can be useful biologically, we need now to seek a greater degree of realism by relaxing some of its more restrictive assumptions. Firstly, what happens when there is non-synchronous arrival of competitors? That all competitors will arrive simultaneously must be a very unlikely contingency for most systems. If all the competitors which are likely to arrive will have done so before the first emigration threshold then rule (3) will apply exactly as before, provided that arrival order is random with respect to the emigration threshold chosen. Early arrivers will experience a higher gain curve $g(I)$ than late arrivers because of

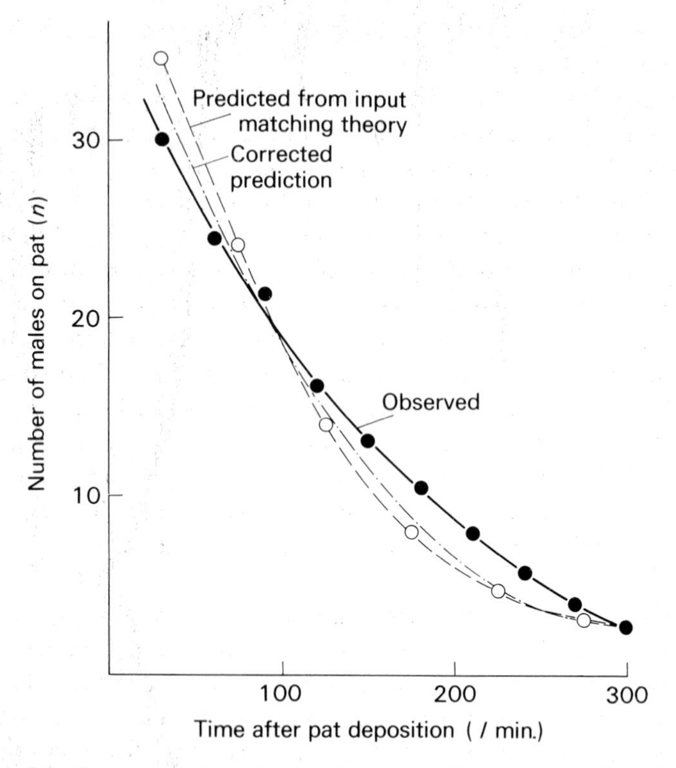

Fig. 8.8. Input matching during the emigration range of *Scatophaga*. Males build up to a peak number at a dropping around 15 min., then begin to depart as shown by $n(I)$ observed. Assuming that emigration thresholds exist in mixed ESS predicted by (3), then predicted $n(I) = (dG/dI)(1/SL_{max})$. Closer fit is obtained if this predicted $n(I)$ is corrected to allow for effect that copulation and oviposition guarding will have on time at which males will leave. This was done by finding mean number p of males paired to females in each 50-min. period (= copulation + guarding time) after dropping is deposited ($I = 0$). I also found number d of males leaving out of mean total number n of males present during same period. Predicted $n(I)$ is thus altered by subtracting $f_1 = (p_1 d_1)/n_1$ from predicted $n(I)$ for first period, $f_2 = \{[p_2(d_2 - f_1)]/n_2\} - f_1$ from that for second period, and so on. f initially lowers predicted $n(I)$ but later raises it. This calculation is based on assumption that a male paired to a female will not leave area even if its emigration threshold has been exceeded during pairing phases—though it will leave soon after. (Taken from Parker & Stuart 1976.)

the initially reduced competition, but over many cycles of searching, such effects will even out and the mean $g(I)$ curve becomes the same for all competitors and again equal to $1/n$ of the input curve, $G(I)$. Strictly, the dung fly case is of the non-synchronous type though virtually all competitors have arrived by 4–5 min. I_{1crit} is between 15–30 min. (Parker & Stuart 1976).

If competitors continue to arrive throughout the emigration range, the rate of departure will be higher during the emigration period, to compensate for the continuing input of competitors.

The effect of variations in patch quality and number of competitors is to alter the width of the emigration range in Fig. 8.7. For a given patch quality, if the number of competitors arriving at an early stage increases, the first emigration threshold will be reached sooner. This is because the greater the number of competitors within the patch, the lower the expected gain from staying. The position of final emigration threshold, which occurs when only one animal is left in the patch, will not be changed by varying the initial number of competitors, so that the overall effect of increasing the number of competitors is to increase the width of the emigration range. Similarly, when the quality of a particular patch is low, the first emigration threshold will be reached sooner.

In summary, the input matching theorem is a logical product of the combination of the marginal value and ideal free theories, and is supported by field data on mate-searching strategies of male dung flies. Relaxing certain of the more restrictive assumptions used to construct the simple model outlined above (page 234) does little to affect the theorem. It seems likely that selection will favour forms of input matching that rely on assessment of resource quality (resource assessment) in conjunction with assessment of competitor density (conspecific cueing), though this remains to be demonstrated empirically.

8.4.3 *Competition within patches: no continuous input*

I will now outline a type of model with a major similarity to the last one in that resource patches gradually deplete in value throughout exploitation. However, there will be a very fundamental difference in that it will have no further *input* of resource into a patch from the time that exploitation begins. Patches decay in value since, as resource items are used up, these items become increasingly difficult to find because of their reduced density in the patch. Thus there are now two components of searching: searching for patches and searching for resource items within a patch. This sort of model can yield a solution radically different from the continuous input model.

If there are no competitors in the patch, the optimal emigration threshold is given simply by the marginal value model (Fig. 8.5); all patches are reduced to the same threshold density and the individual male leaves each patch when he could expect to do better by travelling to another one. What happens if there is competition within patches?

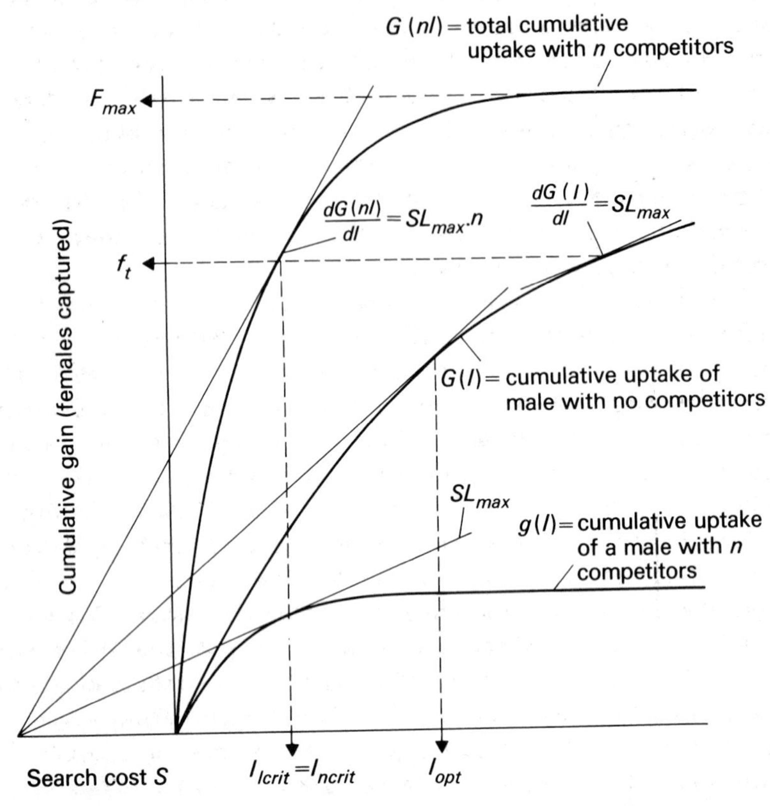

Fig. 8.9. Optimal resource utilization in a model where there is no further input after the competitors have arrived and the competitors must search randomly for resource items within the patch. $G(I)$ is the cumulative captures experienced by one individual with no competitors, I_{opt} is found as in Fig. 8.5. But when there are n competitors from $I = 0$, then the total captures which will have occurred in the patch will follow $G(nI)$. There are F_{max} females in the patch. Thus at any level of resource depletion, e.g. f_t, the rate $dG(nI)/dI$ is n times the rate $dG(I)/dI$. Curve $g(I)$ is the cumulative gain per competitor, this $= \left(\frac{1}{n}\right) G(nI)$. All individuals should leave at $I_{1crit} = I_{ncrit}$, i.e. when there are f_t females left in the patch. It is not advantageous for one individual to stay beyond $I_{1crit} = I_{ncrit}$, because when there are only f_t females left $dG(I)/dI$ for the case of solitary exploitation would in any case be falling below SL_{max}.

Figure 8.9 shows the cumulative uptake of resource by one male searching alone; the total uptake by n competitors (the rate of uptake is n times as fast as with a single male); and the cumulative uptake of a single male in a group of n competitors. The important point is that there is only one optimal departure threshold for a male in a group of n competitors. All individuals should leave at this point (shown by f_t in Fig. 8.9). It is not advantageous for any one given individual to stay beyond this threshold level, since its gain rate would still fall below the maximum level. At any given time, the gain rate per competitor is dependent *only* on the density of available females, *not* on the number of competitors. Hence individual gain rates cannot be held constant at SL_{max} by departures.

Thus where all resource input ceases before the competitors arrive, the ESS can be a unique, optimal departure time defined by a threshold density of resource items within the patch and there need be no emigration range. However, suppose that the gain rate per individual competitor at a given value for F_i is reduced by competitor density. This could be because of time spent in interactions between competitors (known as the 'interference effect') or because of increased avoidance behaviour shown by the resource individuals. Such conditions will now make individual gain rates dependent on n, and thus it is possible to obtain mixed emigration strategies (emigration ranges) even in the absence of continuous resource input. One effect which must make gain rates for mate searching to some extent dependent on n concerns a 'proximity' effect. When two or more males perceive an unmated female, then obviously only one of them can mate. The chances of this happening may be negligible; e.g. when mating occurs rapidly after encounter and the density of males within a patch is low so that the mean time between two successive encounters by males of the same female is short. However, if males must sometimes compete with one another after having encountered the same female, then this effect can be significant and will be dependent on competitor density n. It should then lead to input matching.

8.5 Who stays, who leaves? Emigration decisions

We have been concerned exclusively with deterministic analyses, i.e. with continuous functions for input, uptake, and departure. Individuals depart as discrete events, one by one, and hence when a competitor leaves during the emigration range he may temporarily raise the gain rate of each of the remaining competitors *above* SL_{max}. This effect increases as the number of competitors becomes small. When rule (3) requires, say, that $n(I)=4\cdot36298$ males, then where there are 5 competitors, if all others will stay it pays a given male to leave. If this male does leave, then the 4 left achieve $dg/dI > SL_{max}$. Thus in circumstances where others may leave, there can be an advantage in staying. Overall, individuals that show tendencies for 'hanging on' when dg/dI begins to reduce below SL_{max} must experience equal payoffs to those with opposite tendencies.

This problem is almost equivalent to an N-competitor version of the 'war of attrition' game analysed by Maynard Smith (1974b). In the war of attrition, two opponents display (expending time and energy) in contesting a single resource; the winner is the one that displays longer. There are two sorts of ESS solution. Where there is no observable asymmetry associated with the contest, then the ESS is a mixed one in which the population plays a negative exponential distribution of display times. A mixed ESS can apply for emigration thresholds (Parker & Stuart 1976). For example, using the example above (where the required value for $n(I)=4\cdot36298$), then stability can be achieved by an ESS ratio of individuals that play a high threshold for emigration (emigrate only if $n(I)>5$) to those with a low threshold (emigrate only if $n(I)>4$). In reality, for ideal free distributions it seems likely that the ESS may often be a mixed one that includes the range of emigration thresholds above the critical density $N_{crit}=4\cdot36298$. This will arise if commonly many more competitors arrive at a given habitat than the critical density, N_{crit}. As individuals depart, there will thus be a change in their composition. If low threshold individuals have a tendency to go first, the mean threshold of those remaining will gradually increase. This will continue until all remaining individuals have a higher threshold than the number left in the patch, at which point further emigration ceases. Thus the mean of the ESS distribution will always exceed N_{crit}, and will be closer to N_{crit} when the mean number of competitors arriving at the start of the game is high (rendering high thresholds relatively more disadvantageous).

A second type of solution to conflicts of the war of attrition type is the 'conventional settlement' (Maynard Smith & Parker 1976), which can occur when there is some recognisable asymmetry associated with the dispute (e.g. one opponent is an 'owner', the other an 'interloper'; or one is bigger, etc.). Here the ESS can (for instance) be a pure strategy for: when owner be prepared to display to a cost in excess of the resource value; when interloper retreat without display (see 10.3.2). It is possible that conventional settlements may apply to some of the N-person conflicts posed by deterministic demands of input matching and ideal free theory. Also, conventional cues could be used to settle emigration decisions in a more general way. Rather than individuals choosing before I_{1crit} a value of I within the emigration range with probability $p(I)$ as discussed in 8.4.2, it could be (say) that the last individual to have arrived departs at the next emigration threshold value for I. Some investigation of the factors controlling individual emigration decisions in a real system is most desirable.

8.6 Some concluding remarks

Territoriality and resource guarding is discussed elsewhere in this volume (Chapter 11); we are concerned in the present chapter mainly with ideal free solutions and resource sharing. From the previous section, it is evident that there must be an element of conflict in all practical applications of ideal free theory. The gains to a 'despotic mutant' able (at negligible cost to himself) to exclude all competitors from his patch are considerable. For instance, in the continuous input model the gain rate of such a mutant leaving at I_{opt} would change from SL_{max} to $n_{max} \cdot SL_{max}$ (see Fig. 8.7) provided it can assimilate all the input immediately. In the zero input model the increase in gain rate of a despot which guards the resource until it leaves at I_{opt} is not quite as great (see Fig. 8.9) because a single competitor cannot here achieve the uptake rate of n competitors.

A second problem left unexplored is that of searching for unwilling mates. Supposing that mated females congregate in different patches from virgins. A male may be capable of achieving matings by rape in such a patch, and this will have effects on ideal free stability. A second type of unwilling mate that may be encountered is one that exerts mate choice. Males will be selected to be far less discriminate than females over their choice of partners (see Chapter 7) and this can lead to sexual conflict concerning the optimal choice. This can apply to matings between closely-related species. Some models involving mate searching in conditions of sexual conflict are discussed elsewhere (Parker 1974b, 1978).

Finally, perhaps we should consider the complexity of the individual's 'response surface' required by the models. If individuals have adapted so as to obey ideal free and input matching theories, they must be able to monitor (i) current resource patch quality and to 'predict' the way it will change, (ii) the current level of competitor density in the patch, and (iii) the current level of the competitor population in relation to the current general habitat quality, i.e. they must have their response surface modified by SL_{max}. Whether evolution has favoured response surfaces of such adaptive complexity remains largely to be investigated; however, the empirical support for some of the models from field data given here and elsewhere in this volume leads one to proceed with optimism.

Chapter 9
The Evolution of Cooperative Breeding in Birds

STEPHEN T. EMLEN

9.1 Introduction

Of the many questions facing sociobiologists, one of the most important is to better understand the evolution and functioning of apparent altruistic behaviour. Altruism is here defined as any behaviour performed by an individual that benefits a recipient while incurring a cost to the donor (Hamilton 1964, Wilson 1975). Such behaviour is not only of general biological and anthropological interest; it also poses an evolutionary dilemma since it appears to contradict the fundamental theorem of individual natural selection (see 1.5).

Altruism, outside of a parent–infant context, is presumed to be rare in the animal kingdom. It is commonplace only among the social insects, and reappears spottily among the vertebrates. Here it reaches its highest development among primates, social carnivores, and certain tropical, cooperatively breeding birds.

Most current genetic theories pertaining to the evolutionary 'hows' of altruism have developed from studies of social insects (see 1.5); most of the hypotheses on the ecological 'whys' of communal and cooperative living stem from field studies of birds. Birds are excellent subjects for longitudinal studies of cooperative breeding systems because they combine elements of simplicity (sufficient stereotypy of displays to allow behavioural interpretations; short generation time and reproductive periods to allow collection of demographic data) with sufficient complexity of social organisation to make the results applicable as models of similar behaviour in higher forms.

Since fitness is measured in terms of genetic contribution to succeeding generations, the ultimate in seemingly altruistic behaviour (outside of sacrificial death) is to forego breeding and, instead, to aid other individuals in their reproductive efforts. Cases of such behaviour have been reported in over 150 species of birds (see compilations by Skutch

1961, Rowley 1968, 1976, Harrison 1969, Fry 1972a, Grimes 1976, and Woolfenden 1976a) and, until recently, were lumped under the designation of 'helpers at the nest'. 'Helping' is known to be well developed in about 80 species of which fewer than 20 have been studied in detail. The list of 'helper' species cuts widely across taxonomic, geographical, and ecological lines.

Early workers speculated that helpers (or 'auxiliaries' as I will frequently call them) were primarily young birds, gaining experience before setting out to breed on their own. Since most cooperative breeders (defined here as species with helpers/auxiliaries) are relatively long-lived, an individual sacrifices only minimal loss of reproductive potential by spending a year or two as a non-breeding helper. But fitness of an individual must be measured relative to the fitness of others in the population. The critical question is not whether the individual gains experience by remaining a helper, but whether this experience, expressed in terms of increased life-time fitness, exceeds the life-time fitness it could have had if it had foregone helping and bred on its own earlier in life.

The majority of cases involving helpers at the nest among birds involve young remaining on the parental territory and aiding their parents in producing later broods. Studies on a variety of species indicate both (1) that helpers are often young or siblings of the breeding birds, and (2) that helpers themselves become breeders later in life.

9.2 Case history I: the florida scrub jay system

These points are well exemplified by the Florida scrub jay (*Aphelocoma coerulescens*). The scrub jay is a member of the family Corvidae with a range extending widely in the American southwest and with a small population inhabiting the oak scrub habitat of peninsular Florida. Suitable habitat is scarce and patchy in Florida, leading to disjunct and isolated small populations of jays. In suitable habitat in Florida, the shortage of nesting territories is a limiting factor of great importance to scrub jay biology.

Scrub jays live in stable groups comprised of a breeding pair and their young of the previous one or more years (Woolfenden 1973, 1975). The groups maintain permanent, year-round territories that are defended by all members of the group. In early spring, the breeding pair selects a nest site and builds a nest, unaided; the breeding female alone incubates and is fed on the nest by her mate. Other members of the group attempt to feed the incubating female but they are driven off by the paternal male. After the young have hatched, however, the non-breeding members of the group play a significant role in bringing food to the young (often transferred to the young via the mother), in providing nest vigilance, and in mobbing potential predators.

Woolfenden's data allow a detailed examination of the kin-status of the individual helpers. Of 199 helpers recorded between 1969 and 1977*, 118 were assisting both genetic parents, 49 were helping one genetic parent plus a step-parent, and only 32 were aiding other breeding combinations of siblings, grandparents, and non-kin.

The tendency to help was also age-related. Virtually all one year olds were helpers, but the proportion of individuals attempting to breed on their own increased steadily with advancing age. By three and five years of age, respectively, almost all females and males were breeding on their own.

Between 1969 and 1977, 124 out of 234 (or 53 per cent) of all breeding attempts had at least one helper in attendance. The number of helpers varied considerably, ranging from 0 to 6 (with an average of 1·9 for groups with helpers). The size of the helper contingent bore a direct relationship with the reproductive success of the previous year, adding even greater support to the hypothesis that helpers normally come from the ranks of young that remain on their natal territories (Woolfenden 1975, and personal communication).

An analysis of the reproductive success of groups with and without helpers clearly demonstrates the positive contribution made by auxiliary birds. Table 9.1 shows scrub jay fledging success analysed as a function of the presence or absence of helpers on the one hand, and as a function of the previous experience of the breeders on the other. Even with experience factored out as a variable, groups with helpers succeeded in fledging significantly more young. These differences are enhanced if one considers the survival of young to the age of full independence (see Woolfenden 1975).

The contribution of the helpers lies partly in easing the energetic burden of bringing food for the new nestlings. Studies by Stallcup and Woolfenden (in press) reveal that helpers provide approximately 30 per cent of the food. Starvation is rare among scrub jay young

*From unpublished data generously provided by Glen Woolfenden (personal communication). The numbers of helpers might better be termed 'helper-seasons', since all instances of helpers occurring during the nine year study are included. Since the same individual bird might be a helper in successive seasons, the value does not represent a number of independent, individual birds.

Table 9.1. The effect of helpers on reproductive output in Florida scrub jays (measured as number of fledglings produced per year). Analysed from unpublished data kindly provided by Glen Woolfenden.

	Fledglings produced by groups		
	Without helpers	With helpers	Average number of helpers in groups with helpers
One or both members of breeding pair inexperienced	1·03 (37)*	2·06 (18)	1·7
Both members of breeding pair experienced	1·62 (45)	2·20 (81)	1·9

*Numbers in parentheses refer to sample size of nesting attempts.

(Woolfenden in prep.) and the total amount of food brought to nests does not differ between groups with and without helpers. Rather, the effect of auxiliary birds is to enable the parents to reduce their number of foraging trips and thus lessen the degree of parental care necessary for successful rearing of the young. The selective advantage (to the parent) of this energetic reduction may relate to an improved physical and physiological condition of the parent bird. This, in turn, could reduce stress-related mortality among the breeders. Preliminary data support this contention. Analyses of survival of breeding scrub jays from one year to the next revealed a mortality of 13 per cent among breeders that had helpers the previous year compared with 20 per cent for breeders that did not (Stallcup & Woolfenden in press).

An additional, although difficult to quantify, contribution of auxiliary scrub jays relates to the increased vigilance and anti-predator mobbing protection that comes with having a large group residing on the same territory. This may be the main factor contributing to the differences in Table 9.1.

To better understand the helper situation among scrub jays, we must separately examine the reproductive options available to non-breeding males and females. While remaining on the parental territory, a dominance hierarchy develops which is of importance in predicting these options (Woolfenden & Fitzpatrick 1978a). In this hierarchy, the breeding male is dominant, followed by older non-breeding males, younger males, and then females (in turn, age related). As a group increases in numbers, its territory often expands. This can lead to subdivision or budding off of a portion of the parental territory. This new territory is then occupied by the dominant male helper (Woolfenden & Fitzpatrick 1978b). Other means of obtaining a territory and breeding include occupying vacancies left by the death of nearby male breeders or directly displacing them; the latter is rare in scrub jays. In all instances, a young male must remain on its parental territory until it attains sufficient age, experience, and status to enable it to compete successfully for breeding status via one of these routes.

Females, on the other hand, are subordinate to males and thus stand virtually no chance of achieving breeding status on their own natal territory. Incest has virtually never been reported by Woolfenden. If a son accedes to breeding status on the parental territory, it generally mates with a new female who immigrates from outside. (A male has never been recorded to mate with his sister.) Thus a female's strategy must lie in surveying different territories, searching for vacancies where a breeding female has succumbed or finding a new territory. If such a vacancy is found, a female can compete to become a breeder on the new territory; if no vacancy is found she is tolerated back on the parental territory until another opening presents itself. The only chance for successful breeding on her natal territory would come if the male breeder (usually her father) died, or was challenged and defeated, and his vacancy filled by a new breeding male. Even then the female breeder of the group would generally be her mother, the mate of the previous male and the dominant individual among the females. There is thus little tendency for females to remain on as breeders on their natal territories.

In keeping with predictions from these 'options', females tend to disperse more often and at an earlier age than males. Woolfenden suspects that this differential dispersal incurs a heavier mortality upon the females, leading to a slight skew in the adult sex ratio of scrub jays.

Interestingly, the relative contributions made by different age and sex categories of helpers closely parallel their options for taking over, and successfully breeding in a portion of the parental territory. When feeding visits made by all helpers were analysed by Stallcup and Woolfenden (in press) it was found that older, more dominant male helpers contributed considerably more than younger males and equally or more than the contribution of the male parent itself. First year males (that stand virtually no chance of breeding or attaining dominant status in the near future) made only minimal contributions, as did all females, regardless of age.

9·3 Common characteristics of cooperative breeders

A review of the literature suggests that most species of cooperatively breeding birds share several general features of the Florida scrub jay 'system'.

First, they share the ecological problems of severe habitat-, territory-, or nest-site limitation. Most are tropical or sub-tropical in distribution, are sedentary, and inhabit relatively stable environments. In the absence of large seasonal changes in the availability of unoccupied territories, suitable habitat becomes filled or 'saturated', and a portion of the population remain non-breeders. The hypothesis that habitat saturation might provide an impetus for the evolution of group territoriality and cooperative breeding (by both increasing the survival of young that remain on the natal territory, and reducing the chance that an emigrating individual will successfully establish a territory) was first suggested by Selander (1964), and greatly expanded by Brown (1974) and Gaston (1976). It has been echoed by workers studying such diverse cooperative breeders as flightless gallinules in Tasmania (Ridpath 1972a, b, c), wood-hoopoes in East Africa (Ligon & Ligon unpublished observations), acorn woodpeckers in California (MacRoberts & MacRoberts 1976), Mexican and scrub jays in North America (Brown 1974, Woolfenden 1975, 1976a), malurid wrens in Australia (Rowley 1965a) babblers in Israel (Zahavi 1974, 1976), and in India (Gaston 1976), and troglodytid wrens in Central America (Selander 1964).

Certain population parameters are also held in common. A 'demographic portrait' of the typical cooperative breeder was painted by Brown (1974) and included: low fecundity, deferred maturity, high survival (long life span), and low dispersal. These traits typify a population with a relatively old age structure (Brown 1974). They are also the classic diagnostic features of *K*-selected species (see Chapter 14), whose population densities are continually near or at the carrying capacity of their environments. This demographic profile reinforces the notions of habitat saturation and strong competition for available space.

Finally, most species of cooperatively breeding birds appear to recruit helpers via retention of the young on the parental territory. The donors (helpers) and recipients (breeders) of any helping contribution thus are close kin.

One might conclude, therefore, that the 'dilemma' posed by cooperative breeding is resolved. Most known cases of helpers at the nest fit the models of ecological habitat saturation and K-selection, as well as the genetic constraint of helpers being close kin (generally siblings or offspring). Remaining with the parents on the natal, group territory can be viewed as a strategy to maximise the fitness of the helper by increasing its probability of survival while allowing it to gain experience; fitness might be reduced by attempting to breed independently because of the low probability of finding and defending a suitable, unoccupied territory.

9.4 Do helpers really help?

Not all cases of cooperative breeding are so easy to explain. Suppose a young auxiliary did benefit by remaining on the natal, group territory. It might gain from an increased security resulting from group membership (see Chapter 3) (e.g. via increased alertness to, mobbing of, or defence against potential predators); it might benefit from the groups' more efficient localisation and utilisation of food resources; it might acquire access to group 'possessions' such as communal nests or roosting sites, or group maintained food stores (Selander 1964, MacLean 1973a, b, MacRoberts & MacRoberts 1976, Brown in press). But these benefits do not explain why a non-breeder should contribute to the care of additional young being produced on the group territory. The benefits can be realised without incurring the cost, in terms of time and energy, of being a 'helper'. In fact, by contributing to the successful rearing of additional young, a helper would be increasing the pool of individuals with which it must compete for social dominance and for access to vacant territories that might become available. We are left with the question of why helpers should help.

How might a breeder view the costs and benefits of retaining auxiliaries on the group territory? If the non-breeders make a significant contribution to the rearing of future young, the benefit is obvious. But there are numerous anecdotal reports of young birds being incompetent or downright disruptive in their helping attempts, and of parents aggressively chasing such would-be helpers from active nests. In these instances it might be advantageous for a naive individual to gain experience through helping; but it might not always be advantageous for the parent to risk the well-being of its current offspring to the ministrations of such helpers. An increase in the number of individuals

Table 9.2. Do helpers really help? A tabulation of available data on reproductive success comparing groups with and without helpers. Numbers in parentheses denote sample sizes.

Species	Reproductive output: Pairs alone (or small groups)	Reproductive output: Groups with helpers (or large groups)	Measure of output	Source
Accipitridae				
Parabuteo unicinctus (Harris' hawk)	1·3 (27)	2·0 (23)	Advanced nestlings/nest	Mader, 1975
Rallidae				
Tribonyx mortierii (Tasmanian native hen)				
Inexperienced breeders	1·1 (15)	3·1 (7)	Independent young/group	Ridpath, 1972b
Experienced breeders	5·5 (22)	6·5 (24)	,,	,,
Alcedinidae				
Dacelo gigas (kookaburra)	1·2 (9)	2·3 (10)	Fledglings/nest	Parry, 1973
Meropidae				
Merops bulocki (red-throated bee-eater)	2·3 (36)	2·7 (10)	Fledglings/nest	Fry, 1972a, b
Merops bulockoides (white-fronted bee-eater)[1]				
1973	1·0 (1)	1·2 (20)	Fledglings/nest	Emlen, Demong and Hegner, unpubl. data
1975	0·27 (11)	0·38 (37)	,,	
1977	1·31 (16)	1·94 (18)	,,	
Phoeniculidae				
Phoeniculus purpurens[2] (green wood-hoopoe)	0·75 (28)	2·05 (19)	Independent young/year	Ligon and Ligon, unpubl. data

Aphelocoma coerulescens (Florida scrub jay)						Woolfenden,
Inexperienced breeders	1·0	(37)	2·1	(18)	Fledglings/nest	unpubl. data
Experienced breeders	1·6	(45)	2·2	(81)	,,	,,
Malurinae						
Malurus cyaenus (superb blue wren)	1·5	(16)	2·8	(12)	Independent young/year	Rowley, 1965a
Timilinae						
Turdoides squamiceps[3] (Arabian babbler)	1·8	(14)	1·9	(18)	Independent young/year	Zahavi, 1974
Turdoides caudatus[4] (common babbler)	2·7	(13)	3·8	(12)	Young fledged/season	Gaston, 1976
Turdoides striatus (jungle babbler)					Independent young/year	
Favourable Habitat (Type 1)[5]	1·6	(7)	2·3	(9)	,,	Gaston, 1976
Less Favourable Habitats (Types 2 and 3)[6]	0·3	(7)	1·0	(10)	,,	,,
Icteridae						
Pseudoleistes virescens (brown & yellow marshbird)	1·3	(4)	2·0	(11)	Fledglings/nest	Orians *et al.*, 1977

[1]Excludes nests destroyed by catastrophic events unrelated to group size (including Safari ant predation, cave-ins, and flooding of entire colonies).

[2]Comparison of small (<5) versus large (>5) groups.

[3]Comparison of small (2 to 4) versus large (5 to 7)groups.

[4]Comparison of small (4 to 7) versus large (8 or greater) groups.

[5]Comparison of small (2 to 7) versus large (8 or greater) groups: group size defined relative to the mean group size present in that habitat ($\bar{x}=7{\cdot}8$).

[6]Comparison of small (2 to 4) versus large (5 or greater) groups: group size defined relative to the mean group size occurring in those habitats ($\bar{x}=4{\cdot}7$).

residing on the territory should also lead to depletions of the food resources and increased conspicuousness of the nest to predators. Why should breeders tolerate such 'helping' at all?

To resolve questions such as these, we must raise the fundamental question of whether helpers really do help. Is it safe to generalise from scrub jays to cooperative breeders in general?

Only a handful of studies have obtained sufficient quantitative data to examine whether production of young is greater in groups with helpers than in pairs breeding alone. These are analysed and summarised in Table 9.2. Although there is a tendency for groups with helpers to fare better, this effect is by no means universal.

Even these data are difficult to interpret. Some studies showed that the feeding contribution of helpers reduced the time and energy load on the parents, but did not result in significantly increased numbers of young produced. Such benefits could be manifest in terms of improved physiological condition or greater survival of breeders with helpers (see Stallcup & Woolfenden in press), but this would not be apparent from the measurements of reproductive success shown in Table 9.2.

Alternatively, in some species, the number of young produced per group was larger than that for pairs, but when recalculated as the number of young produced per adult individual, no such helper contribution was evident.

Finally, among Tasmanian native hens, kookaburras, green woodhoopoes, arabian babblers, and jungle babblers, pairs with helpers (or larger groups) occupied larger and/or better quality territories than breeding pairs alone (or small groups). Is the observed correlation between presence of helpers and higher reproductive success due to the 'helping' contributions provided by auxiliary birds, or to the occupancy of better areas for rearing young? If the latter, did helpers join or remain with the breeders on large territories because the breeders were more experienced and/or held a better quality territory? Or was the presence of the helpers the reason for their occupancy of these areas? Gaston (1976) and Brown and Balda (1977) have begun asking these important questions. Their initial results suggest that the importance of territory quality as a determiner of flock size and, thus, indirectly of reproductive success, is greater than previously realised.

9.5 Does helping benefit the helper?

But the critical evolutionary question should not be whether helpers benefit the breeders but rather whether helping benefits the helpers. To answer this we must thoroughly understand the options available to a potential helper. What are its chances of establishing an independent territory? What are its chances of obtaining a mate and, if successful, how many young might they produce? Most importantly, how would this reproductive output compare with the contribution the helper would make if it delayed breeding and assisted another individual in a 'cooperative' breeding attempt?

At the beginning of its first potential breeding season, an individual has two alternative strategies. It can stay at home and help its parents or it can go off and attempt to rear its own offspring.

Let N_0 = No. of young produced by a novice individual during its initial breeding attempt,
N_1 = No. of young produced by an experienced member of a pair breeding without the assistance of a helper,
and N_2 = No. of young produced by a member of an experienced pair aided by a helper.
Further, let r_p = the coefficient of relatedness (see 1.5) between a breeder and its own offspring ($r_p = 0{\cdot}5$),
and r_h = the coefficient of relatedness between a helper and the offspring of the breeding pair that it helps.

We can now make a straightforward prediction: a potential helper will maximise its inclusive fitness by remaining a helper *only* so long as

$$(N_2 - N_1) \times r_h > N_0 \times r_p$$

We rarely have sufficient data from the field to derive accurate estimates of N_2, N_1 and N_0. Furthermore, even if estimates are available, they are based on the production of young in a given year or breeding season. They represent only one portion of the life-time inclusive fitness of the animal, which is the actual parameter maximised by natural selection. In order to calculate the lifetime fitness from various reproductive strategies, we would have to take into account the probability of surviving from one year to the next for helpers and non-helpers as well as summing the above calculation over successive seasons. Nevertheless, attempting to apply this formula (derived from

Brown 1975b) and testing the above prediction can be an informative exercise.

Fully aware of the simplifying assumptions and shortcomings of this quasi-quantitative approach, I present three examples to show

Table 9.3. Comparison of the relative genetic contributions of a Scrub Jay that helps in the reproductive efforts of its parents vs. one that fills a vacancy and becomes a novice breeder.* (See text.)

	Fledglings produced/year	No. of nests in sample	Category
N_2	2·20	81	Pairs where both members have prior breeding experience, and where helpers were present
N_1	1·62	45	Pairs where both members have prior experience, but no helpers present
$\bar{H}$	1·86	81	Average number of helpers present in groups with helpers
N_0	1·36	55	Pairs where at least one member is breeding for the first time (=novice)
N_0^1	1·03	37	Subsample of above in which no helpers were present
N_0^2	1·50	6	Subsample of N_0 where both members are novice breeders and no helpers were present

$r_h = r_p = 0{\cdot}5$

$$\text{**Case 1:}\ \frac{N_2 - N_1}{\bar{H}} \times r_h < N_0 \times r_p;\quad 0{\cdot}16 < 0{\cdot}68$$

$$\text{Case 2:}\ \frac{N_2 - N_1}{\bar{H}} \times r_h < N_0^1 \times r_p;\ 0{\cdot}16 < 0{\cdot}51$$

$$\text{Case 3:}\ \frac{N_2 - N_1}{\bar{H}} \times r_h < N_0^2 \times r_p;\ 0{\cdot}16 < 0{\cdot}75$$

*Analysed from unpublished data (1969–1977), kindly provided by Glen Woolfenden.

**In order to calculate the fitness gain to one helper, we divide by the number of helpers.

how the formula can be used to better understand the options actually available to a helper or novice breeder and how it can lead to the formulation of new questions for future study.

For scrub jays, the helping prediction is difficult to test. It is impossible to obtain a meaningful estimate of N_0, since novice scrub jays virtually never strike out on their own, establish a new territory, and proceed to breed. Rather, the strategy of the birds is to remain on the parental territory until a breeding vacancy occurs. Generally, this comes about following the disappearance of a previously breeding, mated bird. Thus, when a novice scrub jay breeds for the first time, it generally 'marries into' an established territory, and pairs with an experienced mate. (In only 9 of 48 known cases (19 per cent) did a novice breeder first breed with another novice.*) Frequently, the novice breeder obtains a full set of helpers as well, non-breeders that are the previous offspring of the experienced mate. (This occurred in 18 of 55, or 33 per cent, of the cases.*) No valid measure of N_0 can be obtained, and our estimates therefore will be inflated.

Table 9.3 shows that the annual contribution to inclusive fitness achieved by remaining a helper scrub jay can be estimated as 0·16 genetic equivalents. This is considerably lower than the fitness gain achieved by filling a vacancy and breeding on its own. The same is true in Cases 2 and 3 where we estimate N_0 using a subset of the data in which no helpers aid the novice, and both members of the pair are novices, respectively. We should predict that non-breeding scrub jays will cease being helpers *whenever* they can successfully compete for a vacancy left by the disappearance of a previously breeding individual. Within the severe constraints imposed by the ecological saturation of territories and by their system of behavioural dominance hierarchies, scrub jay helpers behave as predicted.

Table 9.4 presents quantitative data for the superb blue wren (*Malurus cyaneus*), a cooperative breeder from Australia. A completely marked population of these birds was studied for five years by Rowley (1965a) and Table 9.4 is drawn from his results. Superb blue wrens also live on permanent territories, and of 43 breeding groups studied, 14 (33 per cent) consisted of the breeding pair plus a supernumerary helper that generally was a male offspring from the previous year's breeding effort. As seen from Table 9.4, the contribution of the male helper was considerable. N_1 can be used as a fair estimate of N_0 since age and experience did not influence reproductive success (Rowley 1965a,

*Calculated from unpublished data kindly provided by Glen Woolfenden.

p. 288). Since N_0 is slightly greater than $N_2 - N_1$, we might predict that adult superb blue wrens should not remain as helpers on their parental territories. How then can we explain the large percentage of groups with non-breeding, auxiliary males? Part of the answer may lie in the fact that most dispersal in this species is carried out by females that suffer a higher mortality, giving rise to an unbalanced sex ratio. Adult males outnumber females 1·4:1, resulting in a shortage of female breeding partners. Competition should exist among males for access to potential mates and not all males will be successful in obtaining partners. Thus the value for N_0 is inflated since it is calculated only

Table 9.4. Relative contribution (in terms of inclusive fitness measured in genetic equivalents) of superb blue wrens that breed on their own vs. remain as helpers. (Data from Rowley 1965a.)

	Independent young produced/year	No. of breeding groups in sample
N_2	2·83	12
N_1	1·50	16
$\bar{H}$	1·08	12
$N_0 = N_1$	1·50	—*
$N_0^1 = \cdot 7\, N_0$	1·05	See text
$r_h = r_p = 0{\cdot}5$		

Case 1: $\frac{N_2 - N_1}{\bar{H}} \times r_h < N_0 \times r_p$; $0{\cdot}62 < 0{\cdot}75$

Case 2: $\frac{N_2 - N_1}{\bar{H}} \times r_h > N_0^1 \times r_p$; $0{\cdot}62 > 0{\cdot}52$

*For justification of using N_1 as an estimate of N_0, see text.

from breeding attempts in which the novice male does obtain a mate. A more accurate approximation of N_0 might be calculated by multiplying N_0 by the probability that a new male would obtain a mate (estimated as the ratio of females per male=0·7). With the addition of this correction factor, $N_2 - N_1 > N_0$ yielding the prediction that the inclusive fitness of male superb blue wrens would be maximised by remaining as helpers.

The gains from the two strategies of breeding versus helping are similar enough that we might expect the occurrence of helpers to vary from year to year in relation to the skew in the sex ratio (which measures the 'shortage' of female partners). This prediction is upheld in Fig. 9.1 which plots the proportion of breeding units containing helpers

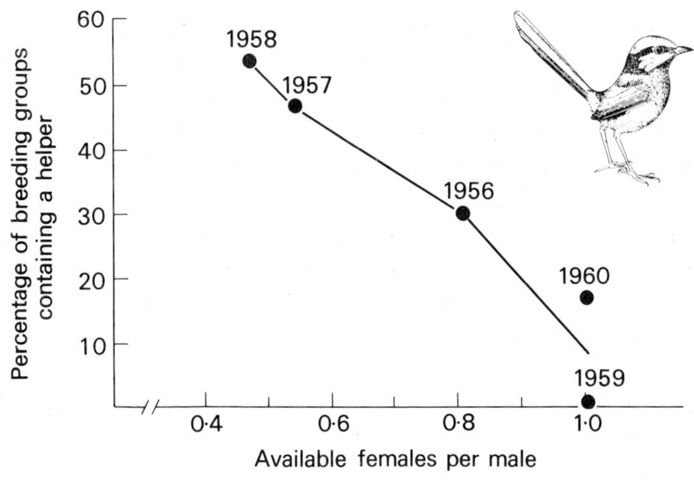

Fig. 9.1. The frequency of occurrence of male helpers among superb blue wrens plotted against the availability of female mates (analysed from data of Rowley 1965a).

against the ratio of females available per male for each of the five years of Rowley's study.

The Tasmanian native hen, an Australian member of the rail family was studied in depth by Ridpath (1972a, b, c). He describes a social system in which two males frequently pair with the same female. The trio defend a permanent territory and cooperate in the incubation of eggs and care of the young. The typical trio is composed of two genetic brothers and an unrelated female (Ridpath 1972b, Maynard-Smith & Ridpath 1972). Both the age of the breeding birds and the presence or absence of the helper were shown to be important determinants of reproductive success. Young birds normally disperse or 'rove' prior to their first breeding season. Whether an individual becomes a breeder in a *pair* or a breeder (or helper) in a *trio* depends in large part on whether it disperses solitarily or together with its siblings. Three options thus are available to a young male. First it might disperse together with a brother to whom it is dominant, tolerate (or encourage?) the continued presence of the brother, and become a breeder in a new trio (N_2 in Table 9.5). Second, it might disperse solitarily (or drive off its subordinate sibling following dispersal) and become a breeder in a pair (N_1). Or it might disperse together with a sibling to whom it is subordinate, remain on the group territory and contribute as a helper. Since these disperal options occur prior to the first breeding season, only data from first year breeders are presented in Table 9.5. In this absence of prior experience, N_0 can be considered equivalent to N_1.

Table 9.5. Relative contribution (in terms of inclusive fitness) of Tasmanian native hen males that breed on their own, breed as dominant members of a trio, and help as subordinate members of a trio. (Data are for first year birds. From Ridpath 1972b.)

	Surviving young	No. of breeding groups in sample
N_2	3·1	7
N_1	1·1	15
$N_0 = N_1$	1·1	See text
$r_p = 0{\cdot}5$		
$r_h = 0{\cdot}25$		
	Case 1: $N_2 - N_1 \times r_h = N_0 \times r_p$;	$0{\cdot}50 \simeq 0{\cdot}55$
	$r_\alpha = 0{\cdot}42$	
	$r_\beta = 0{\cdot}33$	
	Case 2: $N_2 - N_1 \times r_\beta > N_0 \times r_p$;	$0{\cdot}66 > 0{\cdot}55$
	Case 3: $N_2 - N_1 \times r_\alpha > N_0 \times r_p$;	$0{\cdot}84 > 0{\cdot}55$

Since the helper is contributing to rearing the offspring of its brother (the helper's nieces and nephews), r_h=0·25. The calculations (Case 1) show that the annual inclusive fitness realised by contributing as a helper closely approximates that achieved by breeding in a pair.

The Tasmanian native hen is one of the few species of cooperative breeders where the auxiliary male frequently copulates with the female and is not aggressively excluded from doing so. In fact, Ridpath's (1972b) data reveal that the beta male achieves approximately one third of all copulations. Information concerning the actual split of paternity of the offspring is lacking; but accepting the simplified assumption that all observed copulations have an equal probability of leading to fertilisation, we can recalculate r_h for the beta male as 0·33 and r_p for the alpha male as 0·42. Shared paternity of the clutch obviously increases the inclusive fitness realised by the helper (Case 2).

Should the dominant male tolerate his brother as a helper if this means having to share the paternity of the clutch? The calculations in Case 3 show that the greater reproductive success of the trio more than outweighs the slight decrease in relatedness to the offspring.

9.6 Breeder-helper conflict

The above examples are, of necessity, oversimplified. Nevertheless such analyses are often valuable in allowing a better understanding of the selective pressures operating on different individuals in cooperatively breeding groups, and in leading to the formulation of further hypotheses.

As long as N_2 exceeds N_1, and $N_2 - N_1$ exceeds N_0 (assuming $r_h = r_p$), we should expect harmonious interactions between members of the cooperative breeding group. The 'best interests', in terms of inclusive fitness, of all members of a group are maximised when the auxiliary bird remains with the parental group. Similarly, when N_0 exceeds $N_2 - N_1$, and N_1 equals or exceeds N_2, the self-interests of both auxiliary and parent are again in agreement—this time favouring the strategy of the auxiliary leaving the group and initiating breeding on its own.

However, in situations where the success of an auxiliary as a novice breeder exceeds its contributions as a helper, yet the parental group fares better when helpers are present, conflict is to be expected. The interests of breeder and helper should be in direct conflict whenever

$$N_0 > N_2 - N_1 > 0.$$

If the auxiliaries are not direct offspring of the breeders (if r_h does not equal r_p), conflict will occur when

$$N_0 r_p > (N_2 - N_1) r_h,$$

and

$$(N_2 - N_1) r_p > 0.$$

Trivers (1974) has discussed the reasons for expecting an analogous conflict between parent and offspring and mentions some of the behavioural ploys or strategies that each might use to maximise its own fitness at the expense of the other. I wish to stress here that similar conflicts are to be fully expected in cases of seemingly cooperatively breeding birds. Alexander (1974) carries this idea even further and considers several instances of cooperative breeding as extreme forms of parental manipulation of offspring by forced retention as helpers (but see 1.5.1).

Several predictions can be made concerning the intensity of this conflict if we examine (1) the ecological opportunities available for novice breeding, and (2) changes in the degree of relatedness between auxiliaries and breeders in the cooperative group.

With increasing age, experience, and dominance, the chance of an individual obtaining a mate and a territory increases—hence N_0 increases. N_0 should also increase when favourable ecological conditions 'open' new territories or lead to vacancies in old ones. It might also increase in times of abundant food supply and low predation, when experience (in the form of parental expertise) is not so essential for rearing young. Note that the intensity of the breeder-helper conflict, and the balance in inclusive fitness payoffs to the different individuals, will shift from year to year as ecological conditions change. All of the factors listed above increase the value of N_0 relative to $(N_2 - N_1)$ and suggest conditions when auxiliaries should show interesting tendencies either to leave the parental group and attempt to breed on their own, or to enter into increasing conflict with breeding members of their groups. If the habitat is saturated and new territories are few, the conflict could take the form of the auxiliary challenging the breeder and taking over (or subdividing) the parental territory. Zahavi (1974, 1976) has proposed that helpers go further than this. He hypothesises that they minimise their contribution to the success of the breeding effort of the group and, instead, actively disrupt or undermine the parental breeding attempts, thus speeding up the process of territory take-over (but see Brown 1975a). While all of this must remain speculative, it is interesting that many workers describe serious challenges between breeders and helpers (e.g. Skutch 1953, 1961, Brackbill 1958, Koster 1971, Gaston 1976, Woolfenden personal communication, Emlen & Demong personal communication), and both Zahavi (1974, studying Arabian babblers) and Woolfenden (1973, 1976b, studying Florida scrub jays) report cases where non-breeding birds have destroyed eggs in their own groups' nests or in nests of adjacent territories.

An increased conflict is also to be expected when the breeder and the helper view the contribution by the auxiliary differently. This will be the case whenever r_h drops significantly below 0·5.

In calculations in Tables 9.3 to 9.5, our starting assumption was that auxiliary birds were offspring from the breeding pair and thus were related to all future offspring as full sibs. But if either member of the breeding pair died or disappeared and was replaced by a new partner, then future offspring would only be *half-sibs* of the auxiliary. Thus although the helper would still be assisting one of its parents, r_h between the donor (helper) and recipient (future offspring) would drop to 0·25. Under such situations, the inclusive fitness gain by remaining a helper decreases (regardless of any changes in N_0), and we should expect auxiliaries to show tendencies to (1) make a lesser contribution to the group breeding efforts, (2) leave the territory and initiate nesting on their own, or (3) strive to attain dominant status on the parental territory itself. Since the 'best interests' of any former helpers and the new breeder may differ, the step-parent might be less tolerant of the presence of helpers (to which it is *unrelated*) near the nest. Although I know of no hard data available pertaining to these predictions, there have been anecdotal reports of an increased tendency among helpers to disperse following the death and replacement of a member of the breeding pair, as well as tendencies for new (non-related) breeders to aggressively chase off former helpers after moving in and mating with a former breeder (Brackbill 1958, Woolfenden personal communication). A somewhat analogous finding, but from a non-cooperative breeder, was reported by Power (1975). Studying mountain bluebirds, *Sialia currucoides*, he reported that after the artificial removal of a male member of breeding pairs, new replacement mates arrived rapidly and formed pair bonds with the breeding female. However, the new mate provided minimal, if any, parental care for the current offspring that had been fathered by the previous male and hence were not genetically his own (but see also Emlen 1976a).

Changes in behaviour and in group composition are known to occur among certain cooperative mammals coincident with a change-over of the dominant or breeding male. Among both African lions and Indian langurs, *Presbytis entellus* (see Chapter 3), the takeover of a group by a new male (or group of males) is frequently followed by the killing of infant individuals by the new male(s). The presumed adaptiveness of this behaviour lies in ridding the group of genetically unrelated individuals, and in causing the females to reproductively recycle, hence becoming sexually receptive more rapidly to produce young for which the new male is assured paternity.

This discussion should serve to emphasise an extremely important point: that what appear to be cooperative and harmonious societies on the surface are often expected to be extremely competitive underneath. Part of our current state of ignorance about cooperative social systems stems from our frequent inability to see beneath this surface. Individual members of 'cooperative' groups usually are tied together by complex kin-relationships; they may also benefit from group defence of important resources, group anti-predator behaviours, etc. But the potential conflicts between helper and breeder, between parent and offspring, between sibling and sibling, and between kin and non-kin, should be great. The agonistic interactions frequently are subtle and often have been missed by observers who have not had their populations individually colourmarked. But who breeds, and who will breed in the future, remain crucial questions for individual members of any society, no matter how seemingly cooperative. And the various agonistic interactions, social bondings, or influential 'friendships', regardless of of how subtle or sophisticated, determine the dominance relationships of the individuals and thus, ultimately, access to breeding status. Long-term, in-depth studies of individually marked populations of co-operative breeders thus remain an important and yet critically under-studied aspect of sociobiology.

9.7 Case history II: the groove-billed ani

This subtle competition is well exemplified in the groove-billed ani, *Crotophaga sulcirostris*, originally thought to represent a pinnacle in the evolution of true cooperation. Anis are neotropical members of the cuckoo family (Cuculidae). They live in small groups with fairly closed membership and maintain year-round territories that are defended by all. A group is composed of from one to four monogamous pairs (plus an occasional unpaired bird that functions as a helper). All members of a group contribute to the building of a single nest in which all females communally lay their eggs. Incubation and care of the young is shared by the different members of the group.

But all is not as cooperative and as harmonious as it first appears. There is a limit to the number of eggs that can be efficiently incubated—beyond a certain clutch size, eggs tend to be buried, fail to be turned regularly, and are unable to receive sufficient heating from incubation to guarantee normal growth and development (Koster 1971). Thus the per cent of eggs that hatch decreases as the communal clutch size increases. We should therefore expect competition among the various females to ensure that *their* eggs are among those successfully incubated.

It has been known for some time that one often finds eggs strewn on the ground under active nests of anis (Davis 1940, Skutch 1959). Several hypotheses have been advanced to explain this occurrence: improperly built nests; the results of overly anxious predators; and even the hypothesis that since anis are related to cuckoos, they might be partially brood parasitic and thus not have developed strong or efficient tendencies toward parental care.

The groove-billed ani was recently studied in Costa Rica by Sandra Vehrencamp (1976, 1977). By individually marking her birds and closely monitoring nests, she was able to record that eggs were deliberately rolled out of active nests by the females themselves. No eggs were tossed when the group was comprised of a single pair, but the proportion of the total eggs laid that were tossed increased with increasing group size. Continuous observations at the nests revealed that the females in a group have a linear dominance hierarchy. Interestingly, the timing of the onset of laying of eggs was inversely correlated with dominance, such that the most subordinate female initiated laying first, while the most dominant was the last to lay. Vehrencamp (1976) found that an individual female visited the nest on several different days prior to actually laying her first egg. During these vists she frequently tossed out an egg. Once she had initiated laying herself, however, she ceased removing eggs. This makes intuitive sense since a female that continued to toss eggs would run the risk of removing her own.

The outcome of these behaviours is that late-laying (dominant) females have fewer of their own eggs tossed and end up with a larger fraction of the eggs in the communal basket.

But the matter does not end here. Vehrencamp outlined several strategies used by the subordinate females to partially overcome this unfavourable skew in the ownership of the final clutch of eggs. First: as the number of females in a group increased, the average number of eggs produced per female increased. Further, this increase occurred disproportionately among the early laying females (Fig. 9.2). Second: subordinate females frequently produced a 'late' egg, laid several days after the completion of the rest of their clutch. Such late eggs were more apt to be exempt from tossing (since the last female might already have initiated laying). Third: subordinate females frequently prolonged the interval between the laying of successive eggs. Data from one group under intensive observation yielded an inter-egg interval of 2·6 days for the first female, 1·8 days for the second, and 1·5 for the last, or dominant, female.

There is a limit to how long one can delay or stretch out the egg-laying process. This is determined by the timing of the initiation of incubation. Eggs that are laid after incubation has begun will hatch later, the young will be smaller, and they will suffer a competitive disadvantage relative to their nest mates (not all of whom, remember, are siblings). When the communal anis are viewed in this competitive framework, it is not surprising to learn that the subordinate females were the ones that initiated incubation, thus exerting a pressure on the dominant females to produce their full clutch in a shortened period of time.

In spite of these conflicting female strategies, Vehrencamp (1977) found a definite skew in the final egg ownership, with the dominant, last-laying, female obtaining an increasing number of eggs in the communal clutch as the number of laying females increased (Fig. 9.2).

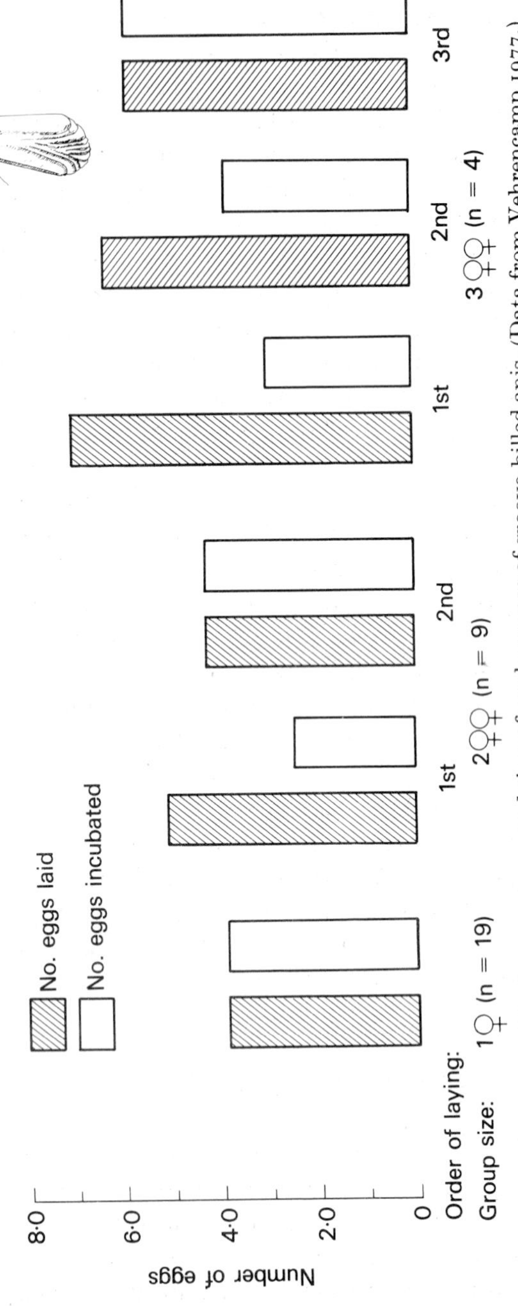

Fig. 9.2. Egg ownership for one-, two-, and three-female groups of groove-billed anis. (Data from Vehrencamp 1977.)

One final point is that the contribution to parental care made by different members of the communal group correlated with their degree of egg ownership. For males, incubation effort increased with increasing dominance status. For females, on the other hand, the trend was the opposite. 'Female behaviour . . . is reminiscent of parasitism, in which the alpha individual replaces the eggs of low status individuals with her own and greatly reduces the incubation effort relative to the other females. Males on the other hand, incubate in direct proportion to their presumed gamete contribution . . . The position of an individual in the communal hierarchy thus has important implications for its reproductive success' (Vehrencamp 1977 p. 405).

Other species known to be communal nesters (in which several females lay their eggs in the same nest) also have been reported to strew their eggs about. This is true of both rheas (Bruning 1974a) and ostriches (Sauer & Sauer 1966, L. Hurxthal personal communication). In these species, a group of females visits the territory of a male, lay their eggs in the nest, and then move on, frequently laying for a second (and in the rhea, a third or even fourth) male. In ostriches, a dominant female (called the major hen) remains and together they 'rearrange' the eggs prior to the initiation of incubation. In rheas, females also display an interest in the nest, returning and rolling eggs about when males take feeding breaks from incubation. The communal clutch in both these species often exceeds the number of eggs that can be incubated efficiently, and eggs that are in a peripheral position frequently fail to hatch (Bruning 1974a, b, L. Hurxthal personal communication). The 'manipulations' occurring when the eggs are shuffled could be interpreted as a strategy resulting from female-female competition which is adaptive to the dominant hen by assuring her a favourable skew in the final ownership of the clutch. This hypothesis, which could partially explain the strewn eggs and the low hatching success, demands additional investigation.

9.8 Case history III: the white-fronted bee-eater

In a paper entitled 'Alternate Routes to Sociality in Jays', Jerram Brown (1974) traces two major evolutionary pathways to complex avian societies. Starting with the classic all-purpose territory, one route leads to communality and cooperative breeding via the retention of young with the parents. Group membership is relatively closed, and group territories are defended. The ecological correlates of environmental stability (small fluctuations in K) and habitat saturation already have been mentioned.

An alternate route leads to coloniality. Certain ecological conditions, such as limited but clumped nesting areas, can lead to colonial living. So too can group advantages stemming from increased predator alertness or defence, or from increased abilities to locate and exploit unpredictable, ephemeral, and patchily distributed food resources (see Chapter 3).

Only a handful of species among the swifts, kingfishers, and bee-eaters are known to be *both* colonial and cooperative. Chimney swifts (*Chaetura pelagica*), as the name suggests, build stick and saliva nests on the insides of chimneys. Third individuals frequently are found tending the nestlings (Dexter 1952, Fischer 1958). Pied kingfishers (*Ceryle rudis*) in Uganda occasionally breed in colonies and up to four auxiliaries may help certain breeding pairs (Douthwaite 1970). Fry (1972a, 1977) studied red-throated bee-eaters (*Merops bulocki*) for many years in West Africa. Colonies generally ranged from 30 to 100 individuals and roughly one third of the breeding nests had a helper present. Emlen, Demong and Hegner (unpublished observations) in ongoing studies of its sibling species, the white-fronted bee-eater (*M. bulockoides*) in East Africa, find that over 70 per cent of all breeding nests are tended by helpers and from 40 to 65 per cent of the sexually mature population are non-breeders in any given year.

Since our own bee-eater studies are still in progress, the following discussion is based upon preliminary data. This leaves me free to speculate on various selective pressures that might be operating on these birds, as well as on possible behavioural strategies that bee-eater helpers might be employing. Some of the ideas presented below are purposefully extremist; many undoubtedly will be proven incorrect as our studies continue. But such speculation can help to formulate specific hypotheses and to suggest the types of data that must be obtained to test them.

White-fronted bee-eaters are common birds inhabiting the savannahs of east and southern Africa. They are social throughout the year and breed in dense colonies ranging from 20 to 120 nests. Each nest is dug deep into the ground or into vertical cliffs along riverbanks or ravines.

Social interactions in a bee-easter colony are intricate and involve at least three levels of social bonding between individuals. The primary social unit is a cooperatively breeding group of from two to seven individuals (average group size=3·3). Members greet one another with a series of stereotyped mutual displays, and they share a common roost chamber at night. During the breeding season, they share in most aspects of the nesting cycle, including nest excavation, incubation (by both sexes), and feeding and defending of the young.

Many individuals also have close friendships with one or two other birds residing in other holes and belonging to other breeding units. These secondary social ties grant immunity from attack and allow frequent visitations to be made to specific other nesting chambers. Superimposed on these close social alliances, is a looser bond with most, if not all, other bee-eaters in the colony.

Most breeding units appear to maintain ownership of additional roosting holes at other colony locations. Birds frequently change nesting sites between breeding attempts, and may also shift their roosting locations in response to natural disturbance or human harassment. Such a complex network of social interactions, involving several layers of social bonding with a large number of other individuals, and involving several spatial locations at different colony sites, is rare among avian species studied to date.

Normally we think in terms of the social advantages that stem from colonial living. However, the presence of large numbers of interacting individuals, and the close proximity of large numbers of active nests, can also increase the opportunities for various subtle forms of behavioural manipulation and cheating (Alexander 1974, Hoogland & Sherman 1976). The availability of large numbers of potential sexual partners might afford increased opportunities for cuckoldry among males (males inseminating females other than their own mates) and promiscuity among females. Possible access to nearby nests might provide opportunities for nest parasitism among females (females laying eggs in nest chambers other than their own). To counteract such possible exploitation, individuals might be expected to recognise friends and kin, males to closely guard their mates, and groups to defend access to their nest chambers. A very complicated network of subtle, competitive, interactions thus should be expected in any society that is both colonial and cooperative.

As an example of this complexity, consider the possible functions of the visitations that occur outside of the confines of the cooperative breeding units. Who is really visiting whom, and why?

One possible interpretation might be that kinship bonds are strong among bee-eaters, and parent-young, or sibling-sibling alliances continue long after the birds have matured, dispersed, and joined separate breeding units of their own. Visitations might then be expected between kin. Preliminary data are consistent with this idea. Many visitations are either between older parents and matured and relocated offspring, or between individuals of unknown genetic relationship that roosted or nested together (in the same group) in prior breeding seasons.

But what function might such friendships serve? Suppose a male breeder has been cuckolded. A male parent is related to his offspring by 0·5. But if that male has been cuckolded (and assuming that the population is not highly inbred), the degree of relatedness between male and offspring approaches 0. If his mate has been promiscuous, the male might actually increase his inclusive fitness by leaving the breeding group. He could attempt to find a new mate and breed again, or he could change groups and provide 'parental' care for another set of offspring. The second strategy could be adaptive in terms of inclusive fitness provided that he direct his attention to close relatives that may be breeding elsewhere in the colony. In short, when the certainty of paternity is low, a male might maximise inclusive fitness by joining ranks as a helper with his mother or his sisters (to whose offspring he is related by 0·25). He might be less prone to aid his male kin since their certainty of paternity could also be in question (Alexander 1974).

Individual bee-eaters that are unsuccessful in a breeding attempt might adopt a similar strategy. An individual could 'recoup' some inclusive fitness by transferring its services to breeding groups containing close kin. (Obviously the costs and benefits of such behaviour must be weighed against the costs and benefits of remaining with the mate and initiating a new nesting attempt.)

Interestingly, there is a high degree of fluidity of members between groups in the white-fronted bee-eater (see later) and many of the instances of regrouping do involve the re-establishment of bonds between individuals that shared membership in some group in the past. For example, in 1977 the reproductive efforts of one small colony failed when rising river waters flooded the nesting holes. Many individuals left their breeding units at this time and six birds gained access to and joined other groups at an adjacent colony where breeding was already in progress. When the young fledged from this new colony, all six of the transfer birds left their 'foster' units, and four of them rejoined with individuals from their original units. Although the kinship bonds between these late-joining helpers and the breeders of their new 'foster' units is unknown, these observations at least document that temporary transferral of 'services' between groups can occur following reproductive failures.

The occurrence of prolonged social ties between parents (especially mothers) and young, and also between young and their 'aunts', is well-known in primates and social carnivores (Rowell *et al.* 1964, van Lawick-Goodall 1968, 1969, Spencer-Booth 1970, Lancaster 1971, Jolly 1972, Hrdy 1976, Estes & Goddard 1967, Schaller 1972, Bertram 1975). It now appears that analogous bonds, possibly serving similar functions, may also occur among bee-eaters.

An alternative (and more selfish) explanation of visitations might be that non-breeding helpers are keeping a close watch on the breeding activities of other units in the colony. In bee-eaters, as in many other cooperatively breeding birds, the adult sex ratio is skewed with an excess of males (Fry 1972a, Emlen, Demong and Hegner unpublished data). This means that males compete for female partners. Preliminary observations indicate that groups may break up if a breeding attempt fails. Thus it might behove an individual to be alert to the potential availability of new mates. According to this hypothesis, one function of visitations could be to increase the likelihood of locating prospective breeding partners for new nesting attempts. Both sexes might be involved in such activities, but males should predominate as visitors since females are the limiting sex. Female visits, however, might serve an additional selfish function of locating nests in early stages of incubation where extra eggs could be added parasitically.

Both Fry (1972a) and Emlen *et al.* (unpublished data) have observed fairly stable dominance relationships between the members of a cooperative breeding unit. Presumably one important advantage of dominant status among males is preferential access to the sexually receptive female in the group. Yet our preliminary observations indicate that females may be promiscuous and accept copulations from many males during the period one to three months *prior* to breeding. The ovaries are not enlarged at this time (Emlen *et al.* unpublished data) and we assume that the copulations have a social rather than a gametic function. Immediately prior to breeding, the situation changes as each female consorts with a single male that aggressively keeps others at bay. However, male helpers continue feeding the female long after egg-laying. What is the function of such a pseudo-sexual bond between helper and breeding female? Does such feeding during incubation merely serve to replenish the female's energy reserves? Or could it be a selfish ploy to cement a social bond that might lead to cuckoldry or a change in paternity of later clutches? How assured is a dominant male of paternity of the young? We do not yet know.

A general corollary to existing models of the evolution of altruistic behaviour via kin selection is that altruistic acts be directed preferentially to close relatives (nepotism). Situations where a donor and recipient would be closely related are generally met in 'closed' societies where dispersal is low. Observations on group stability and dispersal from a large number of cooperative species of both birds and mammals conform to this prediction.

White-fronted bee-eaters, however, may possess a more open form of social organisation than most avian species studied to date. Two types of data (both indirect) suggest that membership in the co-operative breeding units is somewhat fluid and that helpers may frequently be other than immediate kin.

First, when breeding groups failed in a reproductive attempt, members frequently deserted or abandoned the breeding effort. To our surprise, many groups split up at this time. Individuals began exhibiting behaviour typical of 'unattached' birds, roaming the colony, 'inspecting' numerous nest cavities where breeding was in progress, and roosting in unused holes at the periphery of the colony. After variable periods of time, most joined with others in already existent groups or formed groups de-novo.

Table 9.6 presents preliminary data on the post-breeding membership changes of marked birds that were active members of 19 breeding holes whose eggs or young were destroyed. Only two nesting units remained fully intact, although 27 birds (68 per cent, excluding birds not seen again) remained with at least one other member of their original group (Columns A+B+C). However, 19 individuals (48 per cent) split off alone or in diads and later joined with others to form new units (Columns C+D). At the time of this writing, most of these birds have been followed for only three to six months since splitting from their original groups, and their new groups have not reinitiated breeding. Hence it is not known whether group composition will shift again (possibly even back towards the original membership) prior to the next breeding season.

Second, in several breeding colonies, nesting was highly asynchronous with some groups fledging young when others were still in early incubation. In 4 of 6 such colonies, we observed birds in juvenile plumage playing an active role in bringing food to nestlings. In one highly asynchronous colony sampled at the very end of the breeding season, 68 per cent of the auxiliaries (21 of 31 helpers from 20 nesting holes) were juveniles. These observations are of interest since Fry (1972b) reports that bee-eaters are single-brooded and do not renest

Table 9.6. Open vs. closed membership in white-fronted bee-eater groups. Group fluidity of 62 adult bee-eaters from known groups followed during the 3 to 6 months following the failure of their breeding effort. Numbers represent totals of individual birds falling into each category.

	A Full group remained intact	B Subgroup split off from original unit but remained intact (did not merge with others)	C Subgroup split off and joined with other birds to form a new group	D Single bird left original group and joined with others to form a new group	E Not seen again
Colony 1	2	4	0	3	7
Colony 2	4	11	6	10	15
Total	6	15	6	13	22

following a successful breeding effort. If this proved true of the white-fronted bee-eater, then these juveniles could not be interpreted as aiding the renesting efforts of their own parents. The alternative hypothesis would assume that they joined with genetically unrelated groups and were actively contributing to the rearing of foster (non-kin) young.

Our behavioural observations suggest that many juveniles ceased roosting communally with their parents shortly after fledging. As in the case of adults that split off from breeding units, these juveniles roamed widely throughout the colony, inspecting numerous active and inactive nesting chambers. Frequently, siblings remained together and, occasionally, they remained with a non-breeding helper who formally also had belonged to the 'parental' group.

Thus it appears that there is a fair degree of fluidity in the membership of bee-eater breeding units. Certain individuals are regularly accepted as visitors. Others, including juveniles, may split off and join other units. Groups break up, disperse, and occasionally regroup for later nesting attempts. And frequently individuals reunite with others with whom they shared some prior membership at some point in the past.

Such behaviour may be fascinating, but it is not in keeping with the idea of closed genetic kin groups between whose members cooperative or altruistic acts are performed but outside of which they are withheld. Unravelling the possible patterns of genetic (kin) relationships in this mobile and fluid society poses one of the greatest challenges of our current and future research.

This discussion has been largely speculative, and has extended far beyond the limits of our preliminary data. But although we still do not understand the evolutionary adaptiveness of cooperative breeding in the white-fronted bee-eater, it is obvious that many aspects of the breeding system are fundamentally different from either the Florida scrub jay or the groove-billed ani societies described earlier in this chapter. Bee-eaters do not maintain group nesting territories and their social organisation is not 'closed'. Furthermore, the principal ecological correlates mentioned earlier are absent for bee-eaters. Suitable habitat is not limiting, the birds are not territorial, and nest sites occur in profusion. The white-fronted bee-eater does not appear to fit the 'ecological saturation' model that is of such importance in explaining the evolution of cooperative breeding among many other species.

This reinforces the belief that several, different, evolutionary routes have lead to social systems involving cooperative breeding (Fry 1972a,

Brown 1974, Ricklefs 1975). 'Helping' at the nest is not a unitary solution to a specific set of ecological conditions. Rather, cooperative and communal breeding have evolved independently many times, in a variety of ecological contexts, and have assumed a wide range of forms.

9.9 The evolution of cooperative breeding

9.9.1 *Cost-benefit considerations*

What, then, are emerging as the basic models or hypotheses for the evolution of cooperative nesting? Perhaps the best approach to this question is to re-examine the potential costs and benefits accruing first to breeders, and then to non-breeding auxiliaries, as a result of membership or helping in a group.

Breeders

For breeders, at least five potential advantages can be listed.

(1) They may obtain important assistance in the care and rearing of their current offspring. This may be especially crucial in harsh or unpredictable years, when the parents alone would have difficulty meeting the nutritive and/or protective demands of the young.

(2) The helpers could function as an insurance back-up, increasing the probability that offspring would fledge and survive in the event of the death of one of the parents.

(3) The experience gained by the helpers in rearing young should improve their ability to become successful parents in the future. If the helpers are close relatives of the breeders (e.g. previous offspring), their future success could increase the inclusive fitness of the original breeders.

Note that under severe conditions of habitat saturation, the chance for younger birds to establish themselves and breed successfully might be sufficiently low that the parents are 'forced' to tolerate their continued presence in order to minimise any loss of inclusive fitness resulting from the higher mortality of dispersing individuals.

(4) Inter-group competition might favour large group size if an increased number of individuals could better defend a high-quality territory or secure a better nest site. Retention or recruitment of helpers is a primary means of increasing group size.

(5) Other advantages accrue merely from living in groups rather than solitarily. As mentioned previously, these include an increased potential for predator detection and avoidance, and increased abilities to localise and efficiently exploit sparce and unpredictable food supplies (see Chapter 3). These, in turn, increase the probability of survival for all members of the group.

Against these gains, must be weighed several potential costs and risks incurred by breeders that tolerate helpers in their group.

(1) The additional birds require food and thus might deplete the available food resources. If food was limiting, this could lead to a reduction of reproductive success of the breeders.

(2) The activity of the helpers could draw the attention of potential predators to the group and increase the conspicuousness of the nest site.

(3) If experience is important to providing proper offspring care, then the inability or incompetence of novice helpers could be detrimental to the survival and well-being of current offspring.

(4) Finally, helpers are potential breeders in their own right and thus are competitors both for breeding partners and for territorial space. Although anthropomorphic, the question of 'trust' in a helper (particularly a non-kin helper) must be raised. The potential danger of disruptive interference by auxiliaries has already been mentioned (see Zahavi 1974, 1976).

Non-breeders

Non-breeders also can benefit in several ways by becoming members of breeding groups.

(1) They gain experience in caring for young.

(2) They gain access to the resources and security of the group (including access to any communally stored resources that might be present).

(3) They benefit from the same advantages of group living as do breeders.

(4) They may gain a kin selected advantage through rearing relatives.

The chance of surviving to the age and status necessary for independent breeding thus can be enhanced by remaining with, or joining, a group as an auxiliary.

The non-breeder must weigh these advantages against the chances of successfully breeding on its own. Generally, such chances are low in cooperatively breeding species as a result of:

(1) The difficulty in establishing an independent territory (because the habitat is saturated, and population density is at the environmental carrying capacity),

(2) the scarcity of female mates (resulting from differential dispersal strategies among the sexes, with females dispersing earlier and more often than males, and suffering a higher resultant mortality), and

(3) the low success of novice breeding attempts (resulting from environmental unpredictability or harshness of breeding conditions).

It is the *balance*—the trade-offs—of these different factors that should determine when and where cooperative breeding will occur.

9.9.2 *Hypotheses to explain the evolution of cooperative breeding*

From these cost-benefit considerations, a series of possible hypotheses for the evolution of cooperative breeding can be formulated. In the remainder of this chapter, I list six such hypotheses, and discuss each within the definitional framework of 'altruistic', 'cooperative', 'selfish', and 'spiteful' behaviours as defined by Hamilton (1964) in terms of individual (not inclusive) fitness and as shown schematically in Fig. 9.3.

Hypothesis 1 (Donor –, Recipient + or 0). The occurrence of non-breeding helpers serves to regulate the population size. Helpers behave altruistically, foregoing breeding in years when the population density is high relative to the available resource base, but providing a pool of potential breeders that can rapidly increase the population size in good years or following decimation in harsh years. This hypothesis has been

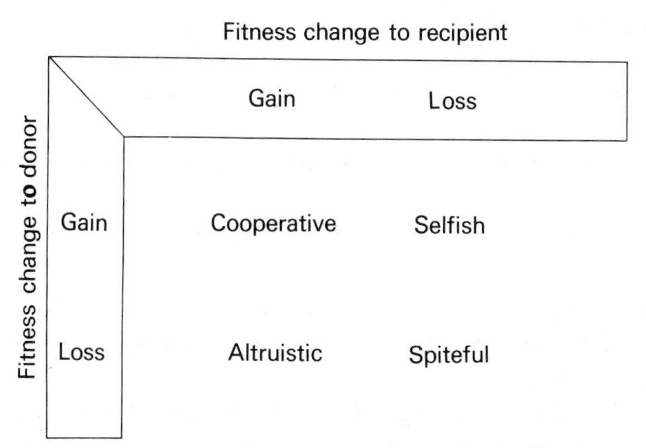

Fig. 9.3. Schematic representation of types of behavioural interactions (modified from Hamilton 1964).

championed by Wynne-Edwards (1962) and echoed by several recent workers studying cooperative breeders (e.g. Skutch 1961, Fry 1972a, 1975, Parry 1973, 1975). This model is basically group-selectionist (section 1.4 discuss the difficulties of group selection). If helping behaviour evolved for the purpose of population regulation, it would be truly 'altruistic'.

Hypothesis 2 (Donor− or 0, Recipient+). Helpers significantly improve the reproductive success of breeding members of their groups, but do *not* benefit themselves. This could occur in at least three, different, circumstances.

A. An auxiliary could increase its individual fitness by leaving the group and breeding on its own, but it is constrained from doing so. At first such behaviour might seem altruistic but could be explained as *forced* altruism imposed on the young through parental manipulation (and thus be selfish from the point of view of the parent). (For a fuller discussion, the reader is referred to Alexander 1974, and 1.5.1.)

B. A mature offspring might have little choice but to remain on the natal territory. Habitat saturation and/or harsh breeding conditions could make it ecologically imperative that auxiliaries remain with the group, and put pressure on the breeders to tolerate their continued presence. Contributing to the current breeding effort could even be viewed as a strategy by an auxiliary to increase the probability that it will be tolerated in the group (see also Hypothesis 3).

C. A genetic relative of the breeding pair might return and contribute to a reproductive effort after failing in a breeding attempt of its own. Such behaviour might also appear to be altruistic, but as long as the helper preferentially aided its relatives, such behaviour could represent the best option available for increasing its own inclusive fitness.

Hypothesis 3 (Donor+, Recipient+). Helpers significantly improve the reproductive success and/or the survival of the breeding members of the group, but they also benefit themselves. The real advantage of being an auxiliary bird might lie in increased survivorship of individuals in groups, as opposed to solitary individuals. To the degree that a group functions more efficiently than a solitary individual in holding a higher quality territory, in more efficient predator detection and avoidance, in better food localisation and utilisation during harsh periods, or perhaps in terms of some form of social manipulation of non-members, membership per se could lead to increased survival. While benefiting through group membership, the auxiliary could gain valuable experience, obtain a higher social status, improve its competitive ability and, later, scout for available mates and vacant territories. This alone might be sufficient to explain the presence of helpers in many species. Gaston (1976) champions the view that contributing to offspring care might be a requisite or 'payment' for acceptance of an auxiliary into a group (a form of reciprocal altruism, Trivers 1971, see 1.6).

The advantages of group living might be greatest during the harsh non-breeding seasons. The benefit to a helper, then, might lie not so much in experience gained *during* a breeding season, but in the increased probability of survival *between* breeding opportunities. We might be focusing our attention on the wrong aspect of cooperative breeding if we limit our studies to the role of helpers in providing care to young.

Note that many of the advantages of group membership should extend to breeders as well as auxiliaries. Additionally, any contribution of the auxiliary to offspring care would benefit the breeders. Thus, in the classification scheme shown, helping would be termed 'cooperative' since both the recipient and the donor gain.

Hypothesis 4 (Donor+, Recipient+, 0, or −). Helping is really a behavioural strategy not to benefit the present breeders but to improve the long-term success of the helper. Again, this could occur in at least four different ways.

A. The auxiliary might remain because of the security achieved through group living and the access afforded to the resources of the group. Helping would be a short-term strategy adopted while waiting for nesting opportunities to 'open up' (see Hypothesis 3).

B. The helper might significantly increase the ability of the breeding group to hold and defend a larger or better quality territory. This, in turn, could increase the chances of later territorial subdivision and thus territorial establishment by the helper (e.g. Woolfenden & Fitzpatrick 1978b).

C. In several species of cooperative breeders, sex ratios are skewed towards males, and females are a limiting resource. If group membership leads to the formation of alliances or bonds that increase the probability of future mate location, mate takeover, or mate acceptance (depending on the process of mate selection), then helping could be a behavioural ploy with a delayed, selfish, 'pay-off'.

D. Helping could also be a selfish female strategy to gain access to a nest chamber in which she communally or parasitically lays eggs. In this way a low ranking female might succeed in having her young reared while contributing only minimally or not at all to their care. Extending this idea to colonial species, membership in a breeding group might enhance acceptance by others in the colony thus providing opportunities for parasitically laying eggs in additional nests in the colony.

Each of these strategies is basically selfish from the point of view of the helper. Whether they should be classified in our scheme as 'selfish' or 'cooperative' would depend upon the importance of additional birds to the group (territory defence, predator alertness, resource localisation, etc.) and on the magnitude of the contribution of the helpers to the reproductive effort of the breeders.

Hypothesis 5 (Donor+, Recipient−). Helpers might actually *decrease* the reproductive success of the breeders but benefit themselves by gaining experience or status through group membership. Novice helpers frequently are inefficient (e.g. Skutch 1969, Brackbill 1958, Rowley 1965b) and their bungling efforts convievably could decrease the reproductive success of the unit. The activities of auxiliary birds also could deplete food resources near the nest area and increase conspicuousness to predators. In such cases, why should breeders tolerate the presence of auxiliary individuals? Although 'selfish' in our scheme, this form of offspring manipulation of parents might be explainable by kin-selection. Provided that members of cooperative units are close kin, the well-being and survival of the helpers could be of direct genetic concern to the breeders. To paraphrase Wilson (1975: p. 351), 'The parents can be expected to lose fitness by turning their offspring over to the ministrations of incompetent helpers. One reason for taking the risk could be kin-selection. By permitting close relatives to practice with their offspring, parents can improve their inclusive fitness through the proliferation of additional kin when the relatives bear their own first offspring.'

Hypothesis 6 (Donor−, 0, or +; Recipient−). Finally, it is plausible that helpers actively, but subtly, undermine the reproductive efforts of the breeders, thus lowering the fitness of the 'recipients'. Recall that in many species, interactions between helpers and breeders are antagonistic early in the breeding cycle. Helpers often are aggressively excluded at the times of courtship, copulation, egg laying, and incubation. One probable reason is that helpers provide a threat to the breeders. The problems of cuckoldry and uncertainty of paternity already have been discussed. Helpers could also lead predators to the nest, actively damage or injure the eggs or nestlings, or fail to provide food for hungry young. Such behaviour might seem 'spiteful' in the short-run, but could lead to an overthrow of the current breeders and thus a takeover of the territory (or, conceivably the mate) by an auxiliary in the long-run. Such a model has been proposed by Zahavi (1974, 1976).

This list is by no means exhaustive. Neither are the hypotheses mentioned above mutually exclusive. The combinations and permutations of these six viewpoints lead to innumerable predictions touching virtually all aspects of the lives of cooperative breeders.

Altruism and spite, behaviours that decrease the fitness of the individual performing the acts, pose an evolutionary dilemma since they are not expected to evolve through the process of natural selection. The expansion of the theory of natural selection to include kin selection partially eleviates this paradox. Many cases of apparent altruistic behaviour 'become' cooperative if the donor and recipient are close kin. Similarly, many apparently spiteful acts 'become' selfish (see 1.5).

With few exceptions, studies of cooperatively behaving birds have concentrated on looking for differences in reproductive success of parental groups alone compared with groups containing one or more helpers. Hopefully this chapter has pointed out that such simplistic data collection, while extremely important, is not sufficient in itself to unravel the complex and subtle behavioural conflicts and interrelationships that form the network of social behaviour in cooperative breeders. Demographic information on natality, mortality, or dispersal often have been calculated as *average* values. Unfortunately, these types of data are of minimal use when searching for, or testing, sociobiological models. Future studies instead of concentrating only on the breeding season will have to deal with individually identifiable animals, and follow cohorts marked as young through their entire life cycles.

Studies of cooperative behaviour remain in their infancy. Data bases sufficient to test the six basic hypotheses listed above are practically nonexistent. But the importance of such studies is great, if for no other reason than the general interest in understanding complex societies and in explaining the phenomena of cooperation and altruism. Hopefully, a decade from now, we will be in a much better position to provide concrete answers to such questions as 'What ecological conditions favour or promote the expression of cooperative tendencies?' and 'What behavioural strategies are followed by individuals that live within such seemingly 'cooperative' societies?'.

Chapter 10
Animal Signals: Information or Manipulation?

RICHARD DAWKINS AND JOHN R. KREBS

10.1 Introduction

10.1.1 *The cynical gene*

Because of the way natural selection works, it is reasonable for us to picture an animal as a machine designed to preserve and propagate the genes which ride inside it (Dawkins 1976). As a means to this end it will often manipulate objects in its world, pushing them around to its own advantage. Some of these objects will themselves be living creatures —mates, parents, prey, rivals—each one a machine designed to propagate its own genes in similar ways. When an animal seeks to manipulate an inanimate object, it has only one recourse—physical power. A dung beetle can move a ball of dung only by forcibly pushing it. But when the object it seeks to manipulate is itself another live animal there is an alternative way. It can exploit the senses and muscles of the animal it is trying to control, sense organs and behaviour machinery which are themselves designed to preserve the genes of that other animal. A male cricket does not physically roll a female along the ground and into his burrow. He sits and sings, and the female comes to him under her own power. From his point of view this *communication* is energetically more efficient than trying to take her by force.

Cullen (1972) uses a human analogy to illustrate the distinction between force and communication. '. . . to a man the command "Go jump in the lake" is a signal, the push which precipitates him is not'. Making a similar point Wilson (1975) cites J. B. S. Haldane's remark that 'a general property of communication is the pronounced energetic efficiency of signalling: a small effort put into the signal typically elicits an energetically greater response'. This is reminiscent of electronic amplification. A transistor or valve in an amplifier receives a low energy fluctuating signal, and uses it to control a high energy signal so that its fluctuations, with more or less fidelity, follow the original. A man's muscles are too feeble to pull a plough, but by a judicious mixture of direct sensory stimulation, reward and punishment, he can manipulate the behaviour of a horse so that it pulls the plough for him. When the man gently tugs the horse's left rein, the horse pulls massively to the left, a high power low fidelity amplifier of the man's weak leftward movement. A male cricket has the physical strength to walk about looking for females, but he can apparently use his muscles to greater advantage by sitting in one place and singing. Communication, which we use interchangeably with 'signalling', could be characterized as a means by which one animal makes use of another animal's muscle power. This can be developed into a definition, although the definition leads us so far from the spirit of what is conveyed by the ordinary English usage of the word that we are tempted to abandon the word communication altogether.

10.1.2 *Definition*

Call the two animals actor and reactor. Natural selection in the past has worked on individuals of the class to which the actor belongs, to improve their power to manipulate the behaviour of individuals of the class to which the reactor belongs. Statements of this kind may be shortened for convenience, using the phrase 'is selected to'. In this case our short form is that the actor is selected to manipulate the behaviour of the reactor—male crickets are selected to manipulate the behaviour of female crickets. Communication is said to occur when an animal, the actor, does something which appears to be the result of selection to influence the sense organs of another animal, the reactor, so that the reactor's behaviour changes to the advantage of the actor.

10.1.3 *The actor*

Of course the actor does not benefit every time it communicates. A cricket may spend its whole life singing out of range of any females. Its song may attract a parasite rather than a female. We believe it is fruitful to interpret the attributes of animals in terms of the selection pressures which may have shaped them, but perfectionism is no part of this belief. When you watch a particular animal doing a particular action, the chances are good that on this occasion the action will turn out to be a mistake. Many an animal dies as a direct result of its own behaviour, even if that behaviour is well adapted to average statistical circumstances. On average, male crickets who sing propagate their genes more efficiently than male crickets who do not, even though some crickets fail as a direct result of singing. This is no paradox. Motorists who wear seat belts are less likely to be killed than those who do not, yet some individuals die *because* they are wearing a seat belt and it traps them. As selectionists we are concerned with average statistical benefits.

10.1.4 *The reactor*

Our definition stipulates that communication results in a net average benefit to the actor, but it says nothing on whether the reactor benefits. The point is irrelevant to the definition. Female crickets may benefit from their tendency to approach male song. Small fish do not benefit from their tendency to approach angler fish lures. Foster mothers do not benefit from their tendency to push food into the coloured gapes of baby cuckoos. But all are examples of communication. The actor in all three cases, male cricket, angler fish, and cuckoo nestling respectively, 'is selected to' manipulate the behaviour of the reactor. Then why do reactors respond, if they only harm themselves by doing so? The answer is that in a sense, even in extreme cases like angler fish prey, they do benefit *on average* from their tendency to respond.

It is the lesson of the seat belt over again. Small fish benefit from their tendency to approach wriggling worm-like objects, because the majority of such objects are good to eat. It is true that a minority turn out to be anglers' lures, but this is not sufficient to reverse the net average benefit. All sensory discrimination involves some generalization. To a stickleback the definition of a conspecific male is anything red. This is on average a serviceable definition, even if it occasionally leads to wasteful attacks on harmless mail vans (Tinbergen 1953). To a taxonomist, 'anything small and wriggling' is scarcely an adequate definition of a worm, but to a hungry fish it is normally good enough. The existence of anglers' lures doubtless leads to selection pressure to change the definition or sharpen up the generalization gradient, and some such evolutionary improvement may well have occurred. But this is only one of many selection pressures bearing on the matter, and in any case selection is also acting on the angler fishes to improve the quality of their deception. Like men who wear seat belts, fish who approach worm-like objects sometimes die as a result, but still are on average more likely to survive than those who do not.

10.1.5 *Who benefits?*

To summarize the point of view we are adopting: as an inevitable by-product of the fact that animals are selected to respond to their environment in ways that are on average beneficial to themselves, other animals can be selected to subvert this responsiveness for their own benefit. This is communication. It may happen that both parties benefit by the arrangement, in which case the word subvert will seem inappropriate. But as far as our definition of communication is concerned, whether the reactor benefits or not is incidental.

Many authors, on the other hand, use the term communication only when the reactor, as well as the actor benefits. The following quotations are representative:

> 'Central to any definition of communication is the reception by an organism of information conveyed by a stimulus from the external world . . . In stimulus exchanges with the environment, or exchanges between an animal and its prey, the relationship between sender and receiver is one-sided; while one participant tries to maximize the efficiency of the stimulus exchange, the other is at best neutral and often seeks to minimize it. In true communication, however, both participants seek to maximize the efficiency of information transfer.' (Marler 1968.)

> 'One party—the actor—emits a signal, to which the other party—the reactor—responds in such a way that the welfare of the species is promoted.' (Tinbergen 1964.)

> 'One of the basic functions of a display . . . is to make the behavior of the communicator more predictable to a recipient by making available some information about the internal state of the communicator.' (Smith 1968.)

> 'Displays are acts specialized to make information available'. (Smith 1977.)

Tinbergen (1952), Morris (1956), Marler (1959) and other ethologists have built upon the ideas of Darwin on the expression of the emotions (1872) to produce an elegant account of how communication systems might have evolved, assuming mutual benefit to both actor and reactor. We here call this the classical ethological approach.

10.2 The classical ethological approach

10.2.1 *Ritualization*

Each signal is supposed to have been derived in evolution from another behaviour pattern which earlier was used for something else. The evolutionary process whereby an incidental movement becomes built up into an effective signal is called ritualization, defined by Huxley (1966) as '. . . the adaptive formalization or canalization of emotionally motivated behaviour, under the teleonomic pressure of natural selection so as: (a) to promote better and more unambiguous signal function, both intra- and inter-specifically; (b) to serve as more efficient stimulators or releasers of more efficient patterns of action in other individuals; (c) to reduce intra-specific damage; and (d) to serve as sexual or social bonding mechanisms'.

10.2.2 *Information*

The emphasis on reduction of ambiguity (see also Cullen 1966) clearly makes sense only in the context of an exchange of information, and is not necessarily compatible with the cynical view given at the beginning of this chapter. The idea of an exchange of information is a carry-over from human language, where the end result of communication is that the receiver learns something which he did not know before, from the sender. In the case of animal signals, what is the 'information' supposed to be 'about'? In some cases, such as the celebrated bee dance, discovered in the classic research of von Frisch and ingeniously confirmed by Gould (1976), we can regard the information as being about the outside world. In the terms of information theory, each receiving bee's prior uncertainty about the location of food is reduced when she reads the dance of a successful returning forager. Here one cannot doubt that the benefit is, in a sense, mutual, but we would still prefer to avoid information terminology and would instead think of the dancing bee as a manipulator, making efficient use of the muscle power of her sisters Wilson (1975) says: 'The straight run represents, quite simply, a miniaturized version of the flight from the hive to the target'. A similar point was made by Haldane and Spurway (1954) in their famous paper on the information-theoretic analysis of the dance. It follows that the receiving bees can be regarded as amplifiers of the dance in two senses. Firstly, each one of them amplifies the 'miniaturized' dance into a full distance flight. Secondly, one dancer may recruit a large number of new foragers simultaneously, and there is thus an amplification in terms of numbers.

Other examples where animals can be said to be communicating information about the outside world are not numerous. Songbirds have an alarm call that 'means' 'aerial predator'. It is reasonable here to regard information as flowing from actor to reactor, but it is no less reasonable to eschew the ideas of information and of meaning and to think instead of the caller as 'manipulating' the behaviour of its companions. (Charnov & Krebs 1975, Dawkins 1976 pp 181–183.)

Most 'informational' interpretations of animal communication have concentrated on information about the actor's internal state rather than about events in the outside world. This is the significance of the phrase 'emotionally motivated behaviour' in Huxley's definition, and of Darwin's title 'The Expression of the Emotions in Man and Animals'. Even an alarm call can be interpreted as meaning 'I am afraid' rather than 'There is a hawk', though we would, of course, add that there is no need to think of signals 'meaning' anything at all.

10.2.3 *Origins*

The classical theory that signals evolve from involuntary expressions of the emotions is a powerful one, whether or not informational concepts are invoked. It is compatible with the 'cynical gene' view we are adopting. The basic idea is as follows. The behavioural acts in an animal's repertoire occur non-randomly in time with respect to one another. In some cases the sequential or temporal connectedness is obvious, and in others statistical analysis is required to show it up (Nelson 1964). Either way, what it means is that the animal's future behaviour is, at least statistically, predictable from its past behaviour. Tooth-baring in a dog is a practical preparation for biting, and a dog who has just bared his teeth is statistically more likely to bite than a dog whose teeth are covered. We can, if we wish, see this as an expression of emotion: the dog with bared teeth is 'angry'. What is more important is that if an ethologist, with or without a computer, and whether or not he uses words like angry, is capable of predicting what an animal is likely to do next, then so, probably, is another animal. This other animal does not have a push-button event recorder and a computer, but he has the great inductive technique known as learning, and he inherits the genes of a long line of successful ancestors. Between them, these two equip him for the same kinds of feats of induction as are achieved by the ethologist with his computer. If an animal can benefit by 'predicting' the behaviour of other animals, he will tend to do so. Needless to say, there is no implication of conscious prediction. Predicting means, here, behaving as if in anticipation of another animal's future behaviour. If it is the case that an animal who has bared his teeth is statistically likely to bite, successful rivals will be those who behave in a way appropriate to a future bite, for instance by running away. So selection favours heightened responsiveness in the reactor.

The fact that other animals are responding to their behaviour induces new selection pressures on actors. If a dog can cause rivals to flee simply by baring his teeth, selection will favour dogs who exploit this power. Tooth-baring will become ritualized, exaggerated for increased power to frighten, and the lips may be pulled back further than is strictly necessary merely to get them out of the way. Over evolutionary time teeth may get larger, even if this makes them less efficient for eating.

Signals are thought to evolve from any incidental movements which happen to be perceptible to other individuals, and which happen to have been 'informative' even before they became ritualized. Their name suggests that 'intention movements' might be good predictors of future behaviour, and they do indeed appear to have been often ritualized (Tinbergen 1952). It is a little less obvious why 'conflict' movements seem to have been so favoured as primordial signals (Tinbergen 1964) but an 'information' enthusiast might suggest that it is because they tend to occur at moments of transition between one motivational state and another, i.e. moments of high 'surprise value' or uncertainty, which is another way of saying high information content (Dawkins & Dawkins 1973).

Byproducts of autonomic system activity are such effective indicators of internal emotional state that they are the basis of police lie-detector tests. Animals cannot strap electrodes to each other, but their sense organs are in any case sensitive to some external manifestations of sympathetic and parasympathetic activity. Morris (1956) has suggested that a large number of animal signals can be traced back to changes in systems involved in thermoregulation (hair and feather erection, surface blood-vessel dilation, sweating), excretion (e.g. urine-marking of territory in dogs, urination over female rabbits by males) and respiration. Darwin himself pointed out that changes in breathing are indicative of strong emotion, and suggested that this was the origin of vocalization. The heart beat seems to have one of the two qualifications necessary in a prime candidate for ritualization—it is a good indicator of emotional state. It does not appear to have been obviously ritualized in fact, perhaps because it lacked the other essential qualification—detectability by another's sense organs before ritualization began. An imaginative classical ethologist might speculate that a population of animals, experimentally fitted with amplifying stethoscopes over hundreds of generations, might evolve heartbeats so loud that the stethoscopes would eventually become superfluous and the ritualization process would take off on its own.

To summarize what we are calling the classical ethological view of the evolution of animal communication, reactors are supposed to be selected to behave as if predicting the future behaviour of actors. Actors in their turn are selected to 'inform' reactors of their internal state, to make it easy for reactors to predict their behaviour. According to this view, it is to the advantage of both parties that signals should be efficient, unambiguous and informative. Communication is seen as a vehicle of inter-individual cooperation, and its evolution is mutual co-evolution.

But a consideration of the fundamentals of how natural selection actually works (Williams 1966) leads to the more cynical view of the interactions between individuals which we gave in the first part of this chapter. Cooperation, if it occurs, should be regarded as something surprising, demanding special explanation, rather than as something automatically to be expected. Even mates (Trivers 1972), and parents and offspring (Trivers 1974, Trivers & Hare 1976) often have divergent genetic interests (see also 1.5), and must be expected to conflict with each other rather than to cooperate. Returning to the question of why the heart-beat has not been ritualized, the real reason may be neither the lack of a stethoscopic bridge over the initial audibility gap, nor the potential danger to the heart's vital function of pumping blood. It may be that the heart-beat is such a true and unfakeable informer of internal state that it had to be hushed up! For every case of ritualized exaggeration of an external indicator of internal state, there could be many cases of systematic suppression, of negative ritualization.

10.3 Ritualization and combat

10.3.1 *Ritualized fighting as an evolutionarily stable strategy*

In the case of signals used in fights, the word ritualization has special connotations which raise particular theoretical problems. When two animals contest a piece of food, a mate, a nest or some other resource, the winner clearly benefits, yet even animals with dangerous weapons often settle such disputes by conventional displays. The loser gives up without a struggle, and even in the moment of victory the winner does not go all out for the kill (Lorenz 1966). Like many generalizations, this one has exceptions (Geist 1971) but there does seem to be a problem here for the 'selfish gene' approach to communication. Ritualized combat is obviously 'good for the species' because it saves lives and prevents injury, but the crucial question is whether it is good for the genes which cause individuals to indulge in it. Ethologists have often argued in a qualitative way that individuals benefit from ritualized contests because they themselves avoid injury (Tinbergen 1951), but a more searching analysis has only recently been made (Maynard Smith & Price 1973, Maynard Smith & Parker 1976, Maynard Smith 1976c). The essence of Maynard Smith and Price's argument, which they developed from earlier ideas on sex ratios (Fisher 1930, Hamilton 1967), was that the best strategy (e.g. 'fight dangerously' or 'fight conventionally') for an individual to adopt depends on what all the others are doing. Suppose, for example, that in a hypothetical population everyone uses only ritualized or conventional signals, retreating from a contest at the first sign of escalated, dangerous, fighting. A new mutant, called 'hawk', which fights viciously in every contest would prosper because it would always win, and would suffer no risk of injury since its opponents always retreat. Now imagine that after a few generations of this prosperity the hawkish mutant has spread and replaced the ritualized signaller. Most contests now involve two hawks, and on average each hawk has an even chance of losing the fight and getting seriously injured in the process. The average benefit from contests is no longer obviously higher for hawks than for ritualized signallers: if the advantage of winning is less than the cost of serious injury the retiring conventional competitor does better on average than a hawk in a population dominated by hawks. The critical conclusion is that both fighters and signallers do well when they are rare, but can be outdone by the other when common. Table 10.1 shows how this hypothetical example can be formalized as a payoff matrix. The ritualized signaller and escalated fighter are named 'hawk' and 'mouse' and the formulae in the cells of the matrix are the payoffs from a contest (in fitness units) to the two types of individual when fighting against each type of opponent. When hawk meets mouse, its payoff is V, the value of the resource, since mouse always retreats. Similarly, mouse gets nothing from a contest against hawk, but gets on average $\frac{1}{2}V-T$ against

Table 10.1. A payoff matrix for a simple 'mouse' and 'hawk' contest. A 'hawk' always fights viciously and risks injury, while 'mouse' only uses conventional displays and retreats at once if attacked by a hawk. The payoffs assume that in a contest between two hawks each has an equal risk of injury and an equal chance of winning. Mouse-mouse contests are similarly equally likely to be won by either individual. The fitness scores are: V = benefit from winning, W = cost of injury, T = cost of time wasted in conventional displays (Maynard Smith 1976c).

		In a contest against:	
		Hawk	Mouse
Payoff to:	Hawk	$\frac{1}{2}(V-W)$	V
	Mouse	0	$\frac{1}{2}V-T$

Note

If $W > V$, the ESS is when the proportion of hawks (p) is such that the average payoff to hawk = average payoff to mouse or:

$$p[\tfrac{1}{2}(V-W)]+(1-p)V=p(0)+(1-p)(\tfrac{1}{2}V-T)$$

which gives:

$$p=\frac{V+2T}{W+2T}$$

another mouse: $\frac{1}{2}V$ because each contestant wins $\frac{1}{2}$ of the contests and $-T$ because in every contest, whether it wins or loses, mouse has to waste time displaying, which is represented by the fitness cost T. The important point to note is that, as implied by our verbal argument, if the cost of injury is bigger than the value of victory ($W > V$) then hawk does worse than mouse in a population of hawks [i.e. $\frac{1}{2}(V-W)<0$], while mouse does worse in a population of mice (i.e. $\frac{1}{2}V-T<V$). Because of this frequency dependence of benefit, neither strategy when common is resistant to invasion by the other or in other words neither is an *evolutionarily stable strategy* (*ESS*). An ESS is a strategy such that if most members of a population adopt it there is no rare strategy that would give higher reproductive fitness. There is, however, an ESS for the matrix in Table 10.1. It consists of the particular mixture of hawks and mice in which the payoffs for the two strategies are equal; if either hawks or mice became commoner, they would start to lose ground. As shown in Table 10.1 this *mixed ESS* can be expressed as a proportion of hawks (p) in terms of V, W, & T. The mixed ESS could also be realised if each individual played hawk with probability p and dove with probability $(1-p)$.

We have dwelt on this very simple example at some length to show the technique of analysing an ESS but, to return to our original point, the simple model shows that, as long as $W > V$, neither pure conventional display nor pure escalated fighting is an ESS. If the benefit of victory is enormous ($W < V$) as for example in the case of two elephant seals fighting over a harem where the winner may obtain a huge number of copulations, hawk can be an ESS, and escalated fighting with injuries should be common.

If we extend the model to include three strategies, hawk, mouse, and retaliator, the latter being an individual who displays conventionally against mouse, but escalates in retaliation against hawk, it turns out that retaliator can be an ESS (Table 10.2), again assuming $V < W$. The

Table 10.2. A payoff matrix similar to that of Table 10.1, but incorporating the additional strategy 'retaliator' after Maynard Smith (1976c) except that a mouse, before it flees, incurs a risk of injury S when it meets a hawk. Retaliator is an ESS if $S > \frac{1}{2}(V - W)$ and $V < W$.

		In a contest against:		
		Hawk	Mouse	Retaliator
Payoff to:	Hawk	$\frac{1}{2}(V - W)$	V	$\frac{1}{2}(V - W)$
	Mouse	$-S$	$\frac{1}{2}V - T$	$\frac{1}{2}V - T$
	Retaliator	$\frac{1}{2}(V - W)$	$\frac{1}{2}V - T$	$\frac{1}{2}V - T$

general conclusion is that on the basis of payoffs to individual genotypes we would not expect to observe animals using purely escalated fighting, if costs of injury are high relative to benefits from winning. They should use predominantly conventional displays, and escalate only in retaliation.

Some animals, however, possess no weapons with which to escalate. When contests are settled purely by conventional means, it seems likely that the one who persists longer wins, in which case one can ask for how long an individual should persist. If all individuals always choose to persist for m minutes, a mutant persisting for just a little longer than m would always win contests, the upper limit to m being set by the cost of displaying in relation to the value of winning. The ESS cannot therefore consist of a single strategy, and in fact the mixed ESS is a random distribution of persistence times. In other words, each individual displays for an unpredictable length of time so that its opponent cannot anticipate how long the contest will last and decide to hang on just a little longer to be sure of winning. As with the hawk-mouse ESS, the mixture could be realised either by an appropriate proportion of individuals using each alternative, or by every individual sometimes using one strategy, sometimes another, and making a random choice according the appropriate probabilities. One example of a random distribution of persistence times is that of male dung flies waiting on dungpats to mate with incoming females (see Chapter 8.7). Although this does not involve a display it is an analogous problem because the best persistence time for any one male depends on how long the others persist.

10.3.2 *Asymmetric contests and assessment*

An important difference between the models discussed so far and the real world is that in contests between real animals there are usually asymmetries (Parker 1974c). These are of three types (Maynard Smith & Parker 1976): (a) the two contestants differ in strength or fighting potential; (b) they differ in their expected benefit from winning the contest (for example, a hungry animal benefits more than a satiated competitor from winning a fight over food); (c) they differ in some way which is unrelated either to fighting potential or expected benefit, but the difference could be used as an arbitrary one to settle contests, somewhat as humans settle a dispute by tossing a coin.

Differences in fighting potential and assessment

Suppose the two individuals in a contest differ in fighting ability. The weaker individual should withdraw as soon as it assesses its relative strength, since it could not win the contest by outlasting the opponent in conventional display, or injuring it in an escalated fight. At the same time, if the stronger individual can win a contest by means of a simple signal such as a raised crest or a loud shriek, bluff by weak individuals should evolve. What sort of cues would be good to use in assessment of an opponent's fighting ability? Clearly the cues used should be closely linked to fighting ability and give reliable information. Assessment signals which are easily mimicked by weak individuals will, in the course of evolution, soon come to be ignored in ritualized disputes, while reliable cues will become established as displays to the benefit of both sender and receiver of the signal. Cues such as size, which are obviously linked to fighting ability will tend to be resistant to bluff and should be used to settle contests. For example if two hermit crabs (*Clibariarius vitatus*) differ appreciably in size, the smaller one retreats from a contest immediately (Hazlett 1968). Similarly, ritualized contests are often literally trials of strength: adult male African buffalo (*Syncerus caffer*) charge at each other and collide head on (Sinclair 1977); vixen foxes (*Vulpes vulpes*) stand on their hind legs and try to push each other over (Macdonald 1977); male bullfrogs wrestle with one another (Howard in press b); and cichlid fish may be so exhausted after winning a ritualized contest that they cannot start again until after a rest (Baerends & Baerends von Roon 1950). Ritualized trials of strength may also be more indirect. Siamese fighting fish (*Betta splendens*) settle contests by a series of display movements involving alternation between head on and sideways postures (Simpson 1968). These ritualized swimming movements probably allow each rival to assess the other's strength and fighting ability. It is especially interesting that the movements of the eventual loser of a contest closely parallel those of the winner until a few moments before giving up, just as one would expect if the contest involves both bluff and assessment. A similar effect was observed in red deer stags (*Cervus elephus*) by Clutton-Brock (in prep.); the stags compete for hinds to add to their harems, and contests consist of prolonged roaring duels. Escalated contests are rare, and they are costly because of the high risk of injury and because subordinate males, known as sneaky fuckers, may steal matings during a prolonged fight. Contests are settled by roaring: the two males roar at each other with a gradually increasing tempo until one suddenly gives up (Fig. 10.1). Clutton-Brock's interpretation is that roaring is a form of assessment and is hard to bluff because roaring contests are so exhausting. A stag will increase its roaring rate in response to an accelerating tape recording but it gives up if the tape accelerates too fast.

Another much more general link between assessment, threat signals, and vocalisations has been discussed by Morton (1977). He points out that the vocal threat signals of many birds and mammals are low pitched harsh shounds. The pitch of a call depends in part on the tension, length and thickness of the vibrating membrane and on the size of the resonating chamber (this second factor is important in mammals but may not be so crucial in birds (Greenwalt 1968). This means that larger animals are capable of making lower pitched sounds so that pitch is a reliable cue for assessment of body size and hence fighting ability of an opponent. Therefore it is not surprising that low-pitched sounds have become ritualized as threat signals. The harshness of threat sounds is probably a byproduct of their low pitch, since a vibrating membrane under low tension tends to produce harmonically unrelated tones which sound harsh.

It seems, therefore, that displays used in assessment are often hard to bluff because they are direct or indirect trials of strength and hence

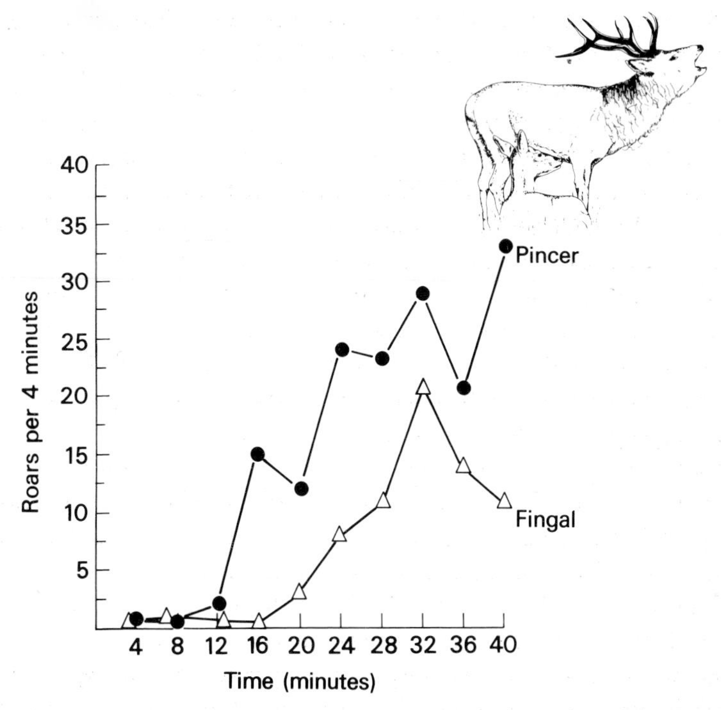

Fig. 10.1. Roaring as a means of assessment in red deer stags. This graph shows a contest between two stags called Pincer and Fingal. As the contest proceeds, both stags roar at a progressively faster rate until one (Fingal) appears to 'give up'. His roaring rate drops off sharply while Pincer's continues to rise. Pincer won the contest which was over a harem. (Clutton Brock in prep.)

costly to perform. Can we turn the argument the other way round and say that an animal can signal its dominance by the degree of cost it is willing to incur? Zahavi (1977b) argues along just these lines. He notes, for example, that a dominant bird in a flock of Arabian babblers (*Turdoides squamiceps*) gives food to others in the group, apparently altruistically, and he suggests that the dominant is in effect saying 'I am strong enough to be able to afford to give up food so don't come and fight against me'. There is a striking parallel between Zahavi's idea and the Potlatch tradition of the Kwakiutl Indians of the Pacific Northwest. The tradition was for local chiefs to invite rivals to their village, and lavish on them gifts or food, and even destroy valuable houses in the village to show how much they could afford to spare. The bigger the destruction of resources, the more effective was the ritual at impressing rival chiefs (Harris 1976). The Potlatch ritual underlines a problem with Zahavi's idea: if strong individuals indicate their fighting ability by simply throwing away resources or taking risks, then the way is open for weak individuals to 'save up' until they are capable of successfully outcompeting the stronger ones: the 'Potlatch strategy' is not evolutionarily stable. Referring back to Zahavi's example, if a subordinate young babbler ate enough food provided by dominants it might eventually grow sufficiently to be able to beat the dominant in a fight. This could not happen, however, when a co involves the same cost to the winner and the loser, as in red roaring competitions. Our conclusion, then, is that assessment s usually have a cost, but the incurring of a cost itself cannot be said to have selective value. Rather the cost is a consequence of assessment.

Our contention that ritualized contests are usually settled by displays which indicate fighting ability seems to be contradicted by many aggressive signals that appear to be easy to fake. It is important to distinguish assessment signals such as the roaring of red deer which are used to settle contests by ritualized displays, and aggressive signals such as the red breast of a robin which merely enhance a posture by making the displays more conspicuous; but nevertheless there are many examples of signals used to announce fighting ability which would appear to be easily faked. Two points can be made about these signals First, the limit to cheating is set by probing and escalation. This is well illustrated by the Harris's sparrow (*Zonotrichia guerula*), in which there is a correlation between dominance status in winter flocks and the size of a black bib of feathers under the chin. This black bib is, in effect, a badge of status. When Rohwer (1977) tried to create cheaters by enlarging the bibs of subordinate birds with black dye, he found that the experimental cheaters did not win more contests, but instead they were involved in more escalated disputes in which they were defeated by true dominants. One interpretation of these results is that contests are not settled by bib size alone, and that assessment also involves escalated fights. This is supported by an experiment of Rohwer and Rohwer (1978). They implanted subordinate birds with testosterone at the same time as painting their bibs. These birds successfully increased their status, while controls implanted but not painted failed to win more contests, even though they fought more. The conclusion is that both bib size and escalated contests are used in status assessment. The second point about fakeable signals is that contests are more likely to be settled without probing and escalation when the payoffs for winning are valued low by the contestants. The ringtailed lemur (*Lemur catta*) settles disputes over pieces of food (which are not highly valued) by easily faked signals such as staring, calls, and feinting blows. In the breeding season, however, when the stakes are high, things are different; biting, chasing and tearing out of fur are common during escalated contests between males over the chance to mate with receptive females (Jolly 1966).

Asymmetry in benefit

Even when two contestants are equally matched in fighting ability, one of them might be willing to escalate further because it has more to gain (i.e. $V - W$ will remain positive for higher levels of escalation). As a general rule, an individual should be willing to put more into a fight, the more it can get out of winning. For example, female iguanas (*Iguana iguana*) try to steal from each other the burrows which they dig for the purpose of egg laying. Both a resident and an intruder are more likely to escalate a fight if the burrow is deep than if it is shallow. A deeper burrow represents a bigger payoff because it requires less future digging before it is ready for egg laying (Rand & Rand 1976). In an escalated contest over a deep burrow the iguanas use high cost displays such as biting, lungeing and rapid approach, while they settle disputes for shallow burrows by milder displays such as opening the mouth and head swinging. Both intruder and resident may gradually escalate the contest, but the correlation between hole depth and tendency to escalate is better for residents than intruders. This is perhaps because the resident has the more accurate assessment of the depth of the hole and can adjust its investment in displays appropriately. This difference in knowledge about the depth of hole probably also explains why residents are more likely to win contests over deep holes.

In the iguanas, the value of the hole is similar for both resident and intruder, the only asymmetry in benefit resulting from the degree of certainty about the state of the hole. Very often, however, the resident defender of a resource such as a territory stands to gain more than the intruder because, having learned the good feeding and hiding places, it can better exploit the territory in the future. An example of this is referred to in section 2.4.3: territorial Hawaiian honeycreepers (*Loxops virens*) can gain more than an intruder out of their own territory because the territory holder systematically avoids revisiting flowers from which nectar has been taken while intruders do not. With this clear asymmetry in payoff, it would benefit both the resident and intruder to save time and energy by using a cue correlated with the asymmetry to settle disputes. If the resident has more to gain, an obvious cue would be prior residency. In fact it is well known that prior residency is used as a cue to settle contests in fish and at least some birds (Phillips 1971, De Boer & Heuts 1973, Zayan 1975, Krebs 1977a). Some fish such as the blenny (*Blennius fluviatilis*) change colour when they establish a territory, and the colour cue is probably used to settle contests (Wickler 1957): the benefit to the prior resident is obvious, and the intruder benefits because it saves time in a contest which it could not expect to win.

Uncorrelated asymmetries

Even if there is no difference in fighting ability or expected gain, a totally arbitrary asymmetry between contests could theoretically be used to settle disputes (Maynard Smith & Parker 1976). If the resource which is being contested is not in very short supply, it could be advantageous for both contestants to save time by settling the dispute with a totally arbitrary convention, just as two men may toss a coin. Table 10.3 illustrates how the strategy of settling contests between hawks and mice by an arbitrary asymmetry such as 'first come first served' could be resistant to invasion by an alternative strategy of

Table 10.3. An example to show how the acceptance of an arbitrary asymmetry to settle contests may, by saving time, increase the payoff to both participants. Consider the hawks and mice game in Table 10.1 with the arbitrary values of $V=60$, $W=100$, $T=10$. The payoffs for Table 10.1 are shown below

	Hawk	Mouse
Hawk	−20	+60
Mouse	0	+20

Now consider a strategy which adopts the arbitrary rule 'owner wins, intruder loses' to settle a contest. The payoffs to this strategy against others of the same type is $\frac{1}{2}\ 60+\frac{1}{2}\ 0=30$ (assuming that it is owner and intruder with equal probability). If, in this population a 'mutant' strategy arises which simply plays hawk with probability p and D with probability $(1-p)$, ignoring the arbitrary asymmetry, hawk's payoff is $\frac{1}{2}\ [60p+20(1-p)]+\frac{1}{2}[-20p]+10p=10$ which is less than 30 for any value of p (p has to lie between 0 and 1). Hence adopting the arbitrary rule is an ESS.

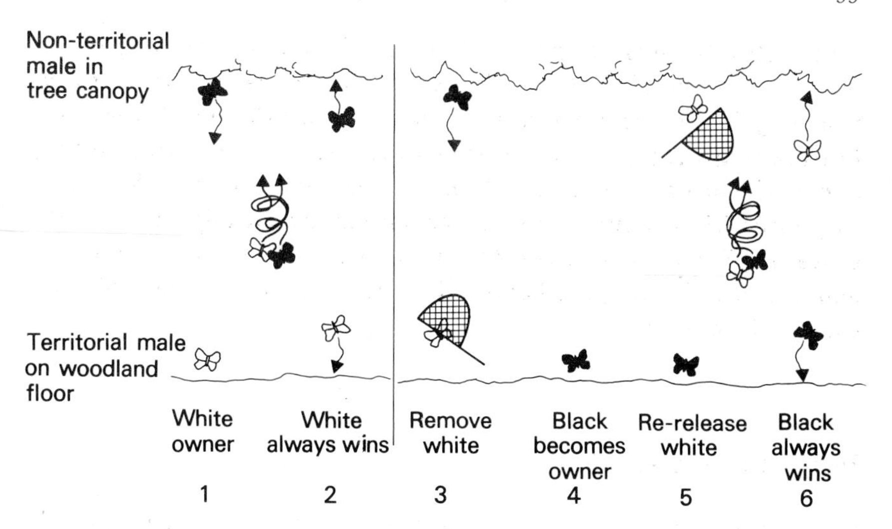

Fig. 10.2. An experiment which shows that the rule for settling contests for territories in the speckled wood butterfly, *Pararge aegeria*, is 'the owner always wins'. This experiment was done three times, each time in a different territory with a different pair of individually colour marked males. One male is represented as black and the other as white. Who wins the contest depends on who is the owner, even if the owner has only been in occupancy of the territory for a couple of seconds (from Davies 1978b).

ignoring the asymmetry. This is at first sight a rather startling conclusion but there is at least one well documented example in which an apparently arbitrary asymmetry is used to settle contests (Davies 1978b). Males of the speckled wood butterfly (*Pararge aegeria*) defend patches of sunlight on the forest floor in which they court females. Intruders invariably retreat in contests for a sunspot, but the outcome of a dispute between two individuals can be reversed by only a few seconds of prior residence (Fig. 10.2). This contrasts with the 'prior residence' effect in birds and fish which we discussed in the previous section, where the effect only works if the resident has been established long enough to gain information about the territory and hence expect a higher gain. In the butterflies, the gain to the two males from winning is equal but the totally arbitrary rule of 'resident wins' is used to settle disputes quickly. Sun patches are not in short enough supply to warrant a prolonged dispute.

If the rule 'resident wins' can be used as an arbitrary way of settling disputes, one might expect the rule 'intruder wins' to also crop up from time to time. One apparent example is the Mexican social spider (*Oecobius civitas*) which lives in aggregations on the underside of rocks. Each spider has a nest and if one is drawn out of its nest it may seek refuge in the hiding place of another spider. 'If the other spider is in residence when the intruder enters, it does not attack but darts out and seeks a new refuge of its own. Thus once the first spider is disturbed, the process of sequential displacement may continue for several seconds, often causing a majority of the spiders to shift from their home refuge to an alien one' (Burgess 1976). Another equally striking example is described in the following letter addressed to *The Times* on December 7th 1977 from a Mr. James Dawson: 'For some years I have noticed that a gull using a flag pole as a vantage point invariably makes way for another gull wishing to alight on the post and this irrespectively of the size of the two birds.'

10.3.3 *Graded signals*

Our discussion of ritualized combat has implied that the participants use either formalized displays or escalated fighting. If the contest is settled by displaying, neither contestant should signal until the last possible moment that it is going to give up. Morris (1957) noted that many displays are performed in a rather constant manner ('typical intensity') regardless of the strength of motivation of the performer. This is just what one would expect if these displays have evolved as a means of winning contests and not as a way of providing opponents with as much information as possible about the subtle variations in motivation of the signaller (Morton 1977). However, many animals, especially birds and mammals, have a whole series of graded threat displays which indicate (at least to the human observer) the exact balance of aggression and fear in the performer. For example the body posture and degree of flattening of a cat's ears give a good indication of how likely a cat is to attack or retreat during a contest (Hinde 1970). This seems to present a paradox: following Maynard Smith, we have argued that contestants should be selected to conceal their exact motivational state and display with typical intensity, and yet it seems that many animals do precisely the opposite. We must admit that the solution to the paradox is not clear, and we can only offer some guesses. Graded signals are, in effect, a form of gradual escalation; instead of an abrupt switch from ritualized to escalated fighting, there is a gradual transition involving closer and closer approach to escalation. If graded signals are used in assessment, an individual which uses a 'high intensity' threat display must either value the resource highly (and hence be willing to risk more in a fight) or have a high fighting ability. But what is to stop an individual which does not value the resource highly, or does not have a high fighting ability, from bluffing by means of a high intensity threat? As with other assessment signals we have discussed, the answer may be that high intensity signals are costly. Perhaps graded threat signals reflect a gradation of cost that an individual is willing to incur in order to win a contest. The animal is like a man at an auction sale: the best way to win the resource with as little cost as possible is to start with a low bid and go higher only if necessary. 'Cost' to the animal could mean either an energetic cost of performing the display, or, more likely, a risk of retaliation with escalated fighting by a rival. One can see intuitively why high intensity threat should have a higher risk of eliciting attack by the rival. Neither individual knows how far the other is willing to go, but if A plays the highest-cost move below escalated fighting and B is willing to go further, the only option for B is to escalate. If, however, A plays a lower-cost move, B can out-bid A without escalating. Although we suggest this as a possible explanation for graded signals, our general point is that such signals are actually something of a puzzle. It has usually been assumed that a signaller benefits by conveying its exact motivational state to others (Smith 1977), but the nature of the benefit is not obvious.

10.3.4 *Courtship and assessment*

The ideas of assessment and probing discussed in section 10.3.2 can also be applied to the analysis of courtship signals. The traditional view of courtship displays is that they allow females to select a mate of the right species, and serve to synchronise the sexual arousal of male and female, by overcoming male aggressiveness and female coyness (Bastock 1967). Synchronising sexual arousal is a proximate consequence of courtship, but the ultimate significance of overcoming female and male inhibitions, at least in species with pair bonds and parental care, may involve mutual assessment by both sexes. In long-lived monogamous birds such as the kittiwake *Rissa tridactyla* (Coulson 1971) and Manx shearwater *Puffinus puffinus* (Brooke 1978) pairs are normally constant from one year to the next, but they split up if breeding is unsuccessful in one season. Pairs also become more successful as a result of breeding together. The implication is that birds should carefully assess their mates before pairing, and use the displays of courtship to get an indication of the mate's quality as a parent. For example, in many birds, the male feeds the female during courtship. In a study of the arctic tern, Nisbet (1973) (1977) found that a male's ability to bring food to the female during courtship was a good indicator of his ability as a parent in feeding the chicks. Nisbet had no direct evidence that females assess their mates by courtship feeding ability but he notes that pairs often break up at the feeding stage. More direct evidence for assessment in courtship comes from the remarkable work of Erickson and Zenone (1976) on barbary dove (*Streptopelia risoria*) courtship. The traditional view that male courtship is a means of arousing the female (it certainly has this effect) would not predict that a male would reject an eager partner, but this is precisely what male barbary doves do. If the female goes into the bow posture (an advanced stage of courtship) too quickly during courtship, the male starts to attack her. In Erickson's and Zenone's experiment, the 'eager' females were produced by a pretreatment of stimulating them with the courtship of another male, so the reaction of the test males makes very good sense: they rejected females which showed signs of having philandered. The male barbary dove contributes considerably to the care of nestlings, so that assessment of mate fidelity and avoidance of cuckoldry are of great importance to him.

10.4 Inter and intraspecific deceit

Interspecific decit is so well known as to pass almost without comment. Batesian mimicry, twig-mimicking insects, and angler-fish lures are examples of successful deception in predator-prey relationships (Wickler 1968). Complex social signals are also mimicked in interspecific deception: beetles such as *Atemeles pubicollis* parasitise wood ants by faking the host's intraspecific signals, for example they induce ants to regurgitate food and groom (Hölldobler 1971). Deceit, or deliberate misleading, ought to be commonplace too in *intraspecific* signalling. Whenever there is any form of assessment, for example in combat, courtship or between parents and offspring, bluff, exaggeration and deceit might be profitable strategies. In spite of this, ethologists have failed to find many unequivocal cases of successful intraspecific deceit (Otte 1974). Some possible, but as yet untested, examples of intraspecific deceit are pseudo female behaviour by male sticklebacks and salamanders (see 7.10) and vocal mimicry in birds. Morris (1952) and Otte (1974) suggest that male ten-spined sticklebacks (*Pygosteus pungitius*) adopting cryptic female coloration may try to steal fertilisations or eggs from territorial males, using the female colour pattern to gain access to nests. [Rohwer (1978) suggests that egg stealing itself may be a form of deception; he argues that males could use stolen eggs as an advertisement of their previous courtship success to encourage new females to mate with them.] The function of vocal mimicry in birds probably varies from species to species (Jellis 1977) but one recent suggestion is that mimicry is used by territorial birds to deter rivals. Rechten (1978) points out that mimics often copy large aggressive, or predatory species. She suggests that this mimicry may dissuade rivals from trying to settle nearby by making the area appear to be inhabited by competitors and predators. A similar argument has also been proposed for intra specific copying of songs by birds (Krebs 1977b) (see 11.4.1). The large numbers of examples of interspecific deceit and relative lack of instances of intraspecific faking calls for an explanation. Is there something special about deceit in intraspecific communication? We have already suggested that both probing and assessment may limit the extent of intraspecific bluff although not eliminate it altogether, but perhaps there is also an additional more general reason for the apparent absence of more widespread intraspecific deceit. Successful deceit, whether between or within a species, depends largely on two factors: the deceit must be relatively rare, so that on average it pays the responder to react the way it does, and the responder must at least sometimes be unable to distinguish between fakes and the real thing. The responder's discrimination ability is limited by unpredictability: it cannot tune its selective response to a signal beyond a certain level of precision, because the true signal is bound to vary slightly from one time to the next. This variability allows the deceiver to get away with a fake signal. These two limiting factors apply equally to inter and intraspecific deception, but as Wallace (1973) first pointed out, intraspecific deceit is further limited by the fact that the deceiver and responder both belong to the same gene pool. Although Wallace did not put it in these terms, he essentially showed that intraspecific lying is not an ESS because its benefit is frequency dependent. The argument is as follows. Imagine a mutant 'liar' (A) in a population of non-liars (a). Assuming the mutant raises its relative fitness by deceiving conspecifics, the A gene will spread. However, as it spreads, the likelihood will increase that liars attempt to deceive others with the gene A. Wallace suggested that the A individuals might recognise one another and lie only to bearers of the gene *a*. If this is so then the spread of *A* genotypes automatically leads to the end of the habit of telling lies because eventually there will be no *a* genotypes left to be deceived. If, instead, there is no recognition of other *A* individuals by liars, and the lying habit has little cost, the *A* gene could still spread (albeit more slowly) to fixation, but once all individuals are liars, the lie does not confer any relative fitness advantage in *A* individuals, and the concept of deception as a means of gaining an advantage is no longer relevant. The deception becomes a convention adopted by all members of the population and is no more a lie than signing a letter 'Your obedient servant'. Wallace's argument is convincing, but he considers only two strategies, 'liar' and 'non-liar'. This is because he assumes (incorrectly, see Otte 1974) that only big lies will be successful, so that graded lying need not be considered. The picture is not so simple when one includes in the argument the possibility that bluff could escalate gradually through evolutionary time. In this case, we might expect the cycle described in Wallace's model in which successful deceivers sweep through the population, to be continually repeated. When one form of deceit has spread to fixation so that it no longer confers an advantage on its genotype, a further exaggeration of the deceit will start to spread. The general conclusion is that bluff and deceit are always advantageous, but they are limited by probing and assessment.

There can be no doubt that an informational view of animal communication is helped by a consideration of assessment and perhaps also of deception. But we prefer to avoid the very idea of information, whether true information or false. Wilson remarks that "If a zoologist were required to select just one word that characterizes animal communications systems, he might well settle on 'redundancy'. Animal displays as they occur in nature tend to be very repetitious, in extreme cases approaching the point of what seems like inanity to the human observer" (Wilson 1975). But it is only redundant and inane if you think the animals are trying to convey *information*. Substitute terms like manipulation, propaganda, persuasion, or advertising, and the 'redundancy' starts to make sense.

10.5 Persuasion and aesthetics

10.5.1 *Advertising*

Advertisements are among the most familiar communication devices in our world, and we should have learned by now that they have little to do with the conveying of information. Sometimes they tell the truth, more often they tell lies, but these terms are usually not even applicable. Advertisements are not there to inform, or to misinform, they are there to *persuade*. The advertiser uses his knowledge of human psychology, of the hopes, fears and secret motives of his targets, and he designs an advertisement which is effective in manipulating their behaviour. One of the favourite techniques of advertisers, which seems to work, however astonishing our rational selves may find the fact, is redundancy—'repetition to the point of what seems like inanity'. Many advertisements make no attempt to say anything about the product: they simply display its name over and over again.

Packard's (1957) exposé of the deep psychological techniques of commercial advertisers makes fascinating reading for the ethologist. A supermarket manager is quoted as saying 'People like to see a lot of merchandise. When there are only three or four cans of an item on a shelf, they just won't move'. The obvious analogy with lek birds does not lose its value merely because the physiological mechanism of the effect will probably prove to be different in the two cases. Hidden cine cameras recording the eye blinking rate of housewives in a supermarket indicated that in some cases the effect of the multiplicity of bright-coloured packages was to induce a mild hypnoidal trance. Again, there may be a functional, if not a causal, analogy in the field of animal signals, and we will return to hypnosis later.

10.5.2 *Social psychology*

Social psychologists have interesting findings relevant to the general subject of persuasion and 'attitude change' (Howland *et al.* 1953, Baron *et al.* 1974, Berkowitz 1975). There is a problem in the interpretation of these results. Social psychologists are especially interested in the effects of persuasion on verbally expressed 'attitude'. Behaviour, say purchasing or voting behaviour, is measured separately and found to be predictable, to a greater or lesser extent, from the previously measured attitude. The evolutionary biologist is interested in behavioural outcomes, and he finds the two-stage reasoning of the social psychologist superfluous: in this sense he may have more in common with the commercial advertiser who presumably is less interested in what his victims say they think about his product than in whether they go out and buy it.

The social psychological approach to persuasion can be considered under three headings: (1) what makes for an effective persuader? (2)

what makes for an effectively persuasive message? (3) what makes an individual vulnerable to being persuaded? Many of the answers to these questions turn out to accord with common sense. For instance attractive or admired individuals are especially effective persuaders, which is why footballers and film stars are paid large sums to pronounce on subjects about which they have no special expertise. Other findings are less obvious, and some are intriguing. The technique of inducing sales-resistance by 'inoculation' is one such. If an individual is in danger of being persuaded by a set of arguments or beliefs, he can be 'immunized' by prior exposure in the laboratory to a 'weakened dose' of the same arguments. He is then less likely to be persuaded when he meets the real thing. The technique has been shown to work, the parallel with vaccination extending even to details of timing (McGuire 1969). A variant of it is, perhaps, the political satirist's art of exposing his audience to a ridiculously parodied version of his opponent's arguments, often using vocal mimicry, like a mockingbird.

We propose no animal analogies for footballers and idols, nor are we suggesting that bird mimicry is an adaptation for boring females with satirical parodies of rivals' songs. What is valuable for ethology is not the findings of social psychologists, but their questions (1, 2, and 3 above). Ethologists interested in animal communication have borrowed from human psychology, as we have seen, concepts related to language and information. We are now suggesting that if we look to human psychology at all we would do better to concentrate on the psychology of salesmanship and persuasion.

10.5.3 *Monitoring and control*

The classical ethological view emphasized the motivational state of the actor, and treated signals as formalized readouts of the actor's internal state. Following our earlier analogy of the lie-detector machine, the reactor might be thought of as being provided with the equivalent of electrodes implanted in the actor's skull, by means of which he could monitor changes in the actor's internal state, and hence predict the actor's future behaviour. Natural selection is thought to favour actors who cooperate in having their intentions read—the recording electrodes are welcomed, perhaps even provided by the actor. In this chapter, we prefer to concentrate on the motivational state of the *reactor*, as being manipulated by the actor. We may continue to use the electrode analogy, but ours are stimulating, not recording electrodes, and they are implanted in the reactor's skull, not the actor's. What actually are these stimulating electrodes? Whatever they are, they must make use of the reactor's sense organs. It is reported that a flickering light tuned to the frequency of human EEG rhythms can have dramatic effects on behaviour, inducing epileptic seizures in susceptible people, and in one case a man felt 'an irresistible impulse to strangle the person next to him' (Grey Walter 1953). Who needs electrodes when the reactor has eyes?

Flickering lights are worth mentioning because it is easy to imagine them as the external equivalents of stimulating electrodes, pulsing away at the brain's own sensitive frequencies. But man is vulnerable to much more subtle influences than this. Flickering light is just one of the visual aids sometimes used by hypnotists whose primary weapon is verbal suggestion. A hypnotized subject can be persuaded to perform pointless actions in response to irrelevant stimuli, even long after he comes round from the trance and without his recalling anything about the original instructions. Human hypnotists use verbal suggestion, but there seems no obvious reason why some similar persuasive force should not be used by non-verbal animals. At the 1973 International Ethological Congress in Washington D.C., K. Nelson gave a memorable paper entitled: 'Is bird song music? Well, then, is it language? Well, then, what is it?' At least as plausible as either language or music is the possibility that bird song should be regarded as akin to hypnotic persuasion.

But it may be that these are not all that different from each other. There may be a continuum between hypnosis as it is commonly understood and ordinary verbal persuasion, with the 'spellbinding' oratory of a Hitler or a Billy Graham falling between. There may be little difference between regarding bird song as music and regarding it as hypnosis. 'Hypnotic' rhythm and 'haunting' melody are clichés in the description of human music. The drug-like effect of the nightingale's song on the poet's nervous system ('a drowsy numbness pains my sens, as though of hemlock I had drunk') might be at least as influential on the nervous system of another nightingale.

10.5.4 *Aesthetics*

Complex bird songs repay critical musical (Hall-Craggs 1969) and Gestalt-theoretic (Thorpe & Hall-Craggs 1976) analysis. The notion that bird song might have some aesthetic content has, oddly, been linked to the idea that this aspect of it is functionless in the Darwinian sense. Proponents of this view have spoken of the biological functions of song ('informing' others of the species and hormonal state of the singers, etc.), and have then gone on to speak of aesthetic reasons for singing as though these were extravagant luxuries superimposed on mundane, biological functions (e.g. Armstrong 1973). With Darwin, we prefer to think that the complex aesthetic beauty of bird song is there because natural selection has favoured it as such. We agree with those authors who say that the traditional views of the functions of communications—transmission of information as to species, sex, breeding condition, etc.—are pitifully inadequate to account for the musical elaboration of bird song. But these authors, having rightly rejected mundane information-purveying as the sole function of song, leapt too hastily to what they saw as the only alternative—music, performed for the enjoyment of the singer (Hartshorne 1973). They forgot oratory, persuasion, hypnosis. Oratory is unnecessary if the purpose is simply to convey information. Oratory comes into its own when the audience is resistant. In the case of singing to deter territorial rivals, the audience can obviously be regarded as resistant. The same might have been doubted in the case of singing to attract mates, but recent theoretical insights, already referred to, strongly suggest that even courtship should often be regarded as a battle of the sexes. As Williams (1966) has put it '. . . genic selection will foster a skilled salesmanship among the males and an equally well-developed sales resistance and discrimination among the females'. The fact that the same music is used both to repel (rivals) and to attract (females) need not surprise us. Martial music is 'strirring' even though it stirs one group of people to patriotic courage while simultaneously stirring another group into a panic.

If this Darwinian view of the aesthetics of bird song is accepted, it is still quite possible that individuals do in some sense 'enjoy' their own singing. The singer is, after all, a member of the same species as his audience, and his nervous system is presumably vulnerable to the same kinds of stirring stimuli. That his own singing is reinforcing for a bird is indicated by operant conditioning experiments (Stevenson 1967), although unfortunately the control sounds with which the birds' own songs were compared were limited. More interestingly, the reinforcing properties of bird song are strongly implied by the 'template' theory of song development (Konishi & Nottebohm 1969). Many young birds appear to teach themselves to sing by matching a wide spectrum of babblings against a stored template or mental image of what the song ought to sound like. The template may be a kind of tape recording of a conspecific heard earlier in life, as in the white-crowned sparrow

Zonotrichia leucophrys. Alternatively, for example in the song sparrow *Melospiza melodia*, the template seems to be provided even in individuals who have never heard a conspecific. In either case, during the period when the young bird learns the motor patterns of song, the template functions as a reinforcer, albeit a highly complex and elaborate one. We can rephrase the template theory in the language of aesthetics. Because of the way its nervous system is built, any individual song sparrow, of either sex is emotionally affected by the song of the species. Depending on the context, this influence shows itself either as sexual attraction, or as intimidation, or as self-reinforcement, just as a resplendent cavalry uniform may intimidate enemies, rouse the courage of self and comrades, and sexually attract female camp followers. The template strategy of development is economical, since it exploits a source of information which is already built into the species nervous system for other reasons.

10.6 Conclusions

We are contrasting two attitudes to the evolution of animal signals. One attitude, which we have here called classical, emphasises cooperation between individuals. Cooperation is facilitated if information is shared. Selection favours those actors who make it easy for reactors to 'read' their internal state, and hence to act as if in anticipation of the actor's behaviour. The other attitude, which we espouse, emphasises the struggle between individuals. If information is shared at all it is likely to be false information, but it is probably better to abandon the concept of information altogether. Natural selection favours individuals who successfully manipulate the behaviour of other individuals, whether or not this is to the advantage of the manipulated individuals. Of course, selection will also work on individuals to make them resist manipulation if this is to their disadvantage, just as natural selection works on prey animals to make them less likely to be caught by predators. In both these cases an evolutionary arms race will develop. Predators evolve adaptations so that they do sometimes catch prey in spite of anti-predator adaptations. In the same way, actors do sometimes succeed in subverting the nervous systems of reactors, and adaptations to do this are the phenomena which we see as animal signals.

Part 3
Strategies in Space and Time

Introduction

This final section of the book tackles the problem of how an individual should deploy its behavioural options in space and time.

The first two chapters consider the use of space by animals. Within areas of suitable habitat, individuals or groups are often spaced out more than would be expected by chance. This applies equally well to the regular spacing of several kilometres between buffalo herds as to the few millimetres that separate neighbouring barnacles on a rock. Spacing out implies that the animals are avoiding each other in some way, and the area that the individual or group occupies can be termed a territory. In Chapter 11 Davies considers various questions about territories. What are the benefits and costs associated with territoriality? Nectar feeding birds defend patches of flowers, squirrels defend nut-laden trees and limpets evict intruders from their algal gardens. The main benefit derived from defence in these species is to improve feeding efficiency.

When resources, such as food, are patchily distributed then those males that are able to defend the richest patches can often attract the most females for mating. Some territories seem to be solely concerned with mate attraction, for example the tiny bare patches of ground at leks where some antelopes and birds perform their displays. We have already seen that living in groups can decrease predation (Chapter 3) but in some circumstances spacing-out is the best way to combat predators. In addition to these benefits, there are also costs of territorial defence, and the few detailed studies that have been made support the intuitively reasonable hypothesis that animals defend territories only when the benefits accruing through defence outweigh the costs incurred. Chapter 11 also discusses the consequences of territorial behaviour for population regulation and the mechanisms by which territories are defended. Experiments have shown that song and scent can deter potential intruders but we also have to explain why intruders should obey such signals.

The environment consists of a mosaic of different habitat types; not all will be equally suitable for a particular species. In Chapter 12 Partridge shows that almost all animals have habitat preferences. She describes how these preferences can be measured and discusses which cues animals use to choose their preferred habitat: studies of the development of habitat preferences reveal inherited differences between closely related species. In some cases it has been shown that, as we would expect, individuals increase their fitness by having preferences. Experiments are described which demonstrate this adaptive basis for habitat selection in peppered moths, polychaete worms, fruit flies and titmice. Often the best place for any one individual to be depends on where all the other individuals are. Partridge considers the problem of how choosy an animal should be in its selection of a place to live. When searching imposes a cost it may do better to settle and claim a space in a less optimal habitat. Finally the problem of specialization and the role of habitat selection in speciation are considered.

The final two chapters discuss strategies in time. Previous chapters in the book have discussed optimal choices made by an animal which has already started to feed, court or defend a territory. But how does the animal decide in the first place whether to feed or court or defend its territory? This much more difficult problem is discussed by McCleery in Chapter 13. We can measure the efficiency of feeding behaviour in terms of food intake (Chapter 2), the efficiency of courtship in terms of mating success (Chapters 7 & 8) and the efficiency of territorial defence in terms of increased benefits accruing through the eviction of intruders (Chapter 11), but if we are to trade off the various behavioural options in an individual's repertoire, we need some common currency for the measurement of all activities. For example, consider Niko Tinbergen's classic experiment which demonstrated that when a parent gull removed the egg shells from its nest after the young had hatched, the nest became less susceptible to predation. Clearly this shows that egg-shell removal has survival value; but how does the parent gull decide when to remove the egg shells? How does it trade off the benefits from this behaviour against other behavioural options such as brooding the chick, preening or going off to collect food?

In Chapter 13 McCleery argues that the benefits of performing different activities could in theory be measured on a common scale. The relative advantages in the long term of performing different activities must be measured in terms of the effects of doing them on the survival and reproductive success of the individual and its kin. In the short term each behaviour can be thought of as having a cost; this is the chance of dying minus the chance of reproducing per unit time when engaged in the particular activity. An optimal decision-making animal should organize its behaviour in time to minimize the total cost incurred, or to put it the other way round, to maximize the total benefit. The costs incurred by an animal at any moment in time are likely to be associated either with actually doing a particular behaviour (e.g. when a heron eats a fish it stands a small chance of choking to death), or with the physiological deficits from which the animal is suffering (a hungry heron might die of starvation). Thus in choosing the optimal behaviour to perform next the animal has to take into account both the risks linked to its internal state and the environmental risks. It also has to assess how effective a particular behaviour would be in restoring physiological deficits (e.g. it might not pay a very hungry heron to hunt for fish in bad weather). McCleery describes two techniques for solving this type of optimal decision problem: a relatively cumbersome method (dynamic programming) and a more elegant analytical method (Pontryagin's maximum principle).

Can the costs associated with each activity and physiological deficit be actually measured? As yet this has not been done. Instead two less direct approaches have been made. One is to guess at the cost function and predict the decision rules which should follow from it. This has been done quite successfully with the choice between feeding and drinking made by doves in a Skinner box. The other method is to look at the decisions made by an animal and try and calculate what function the animal must have been minimizing in order to make the observed decisions. This technique has been tried with courtship in newts, in which the choice facing a male is to continue courting the female on the bottom of a pond or to surface and breathe. Clearly this area of research is still in its infancy, but it is one of the most important new developments in behavioural ecology.

Chapter 13 considered the minute to minute decisions an animal has to make to optimize its use of time. In Chapter 14 Horn looks at a set of problems that confront an animal over a longer period; how does an individual optimize its reproductive output during its lifetime? Under what sort of environmental conditions may we expect an animal (or a plant) to put all its energy into growth and competitive ability as opposed to reproducing? Should an animal have lots of young and not care for them at all, or lavish its attention on a favoured few? Why do atlantic salmon breed over several seasons, while pacific salmon spawn but once and then die? Should the offspring disperse into new habitats? Horn derives some rules for answering questions like these.

Chapter 11
Ecological Questions About Territorial Behaviour

NICHOLAS B. DAVIES

11.1 What is a territory?

Various criteria have been used to define the term territory, including 'defended area' (Noble 1939), 'exclusive area' (Schoener 1968) or a 'fixed, exclusive area with the presence of defence that keeps out rivals' (Brown & Orians 1970). In this chapter I prefer to adopt a less rigid definition and will recognise a territory whenever individual animals or groups are spaced out more than would be expected from a random occupation of suitable habitats (Fig. 11.1).

The spacing may vary from just a millimetre or two between adjacent barnacles on a rock to the several kilometres that separate neighbouring herds of buffalo. Spacing out may be maintained by overt aggression (as in many birds) or by avoidance of subtle signals such as scent (as in many mammals). The area occupied is sometimes fixed in space. For example breeding tawny owls (*Strix aluco*) are spaced out in woodland, each pair occupying a fixed, exclusive area for the whole of its adult lifetime (Southern 1970). On the other hand, although the hunting ranges of individual cats (*Felidae*) may overlap, at any one time the individuals are spaced out because they avoid each other temporally by smelling the deposits on scent posts of previous passers-by (Leyhausen 1965). In all these cases, wherever spacing out is due to interactions between individuals or groups, the occupied area will be referred to as a territory. Territories have been described for a wide variety of animals, both invertebrates and vertebrates (Brown & Orians 1970, Wilson 1975).

The following sections will consider various questions that can be asked about the phenomenon of spacing out. In the literature a lot of confusion has arisen because the various types of question have been mixed up. Firstly I will ask a functional question, what is the selective advantage of spacing out? I will describe how the various costs and

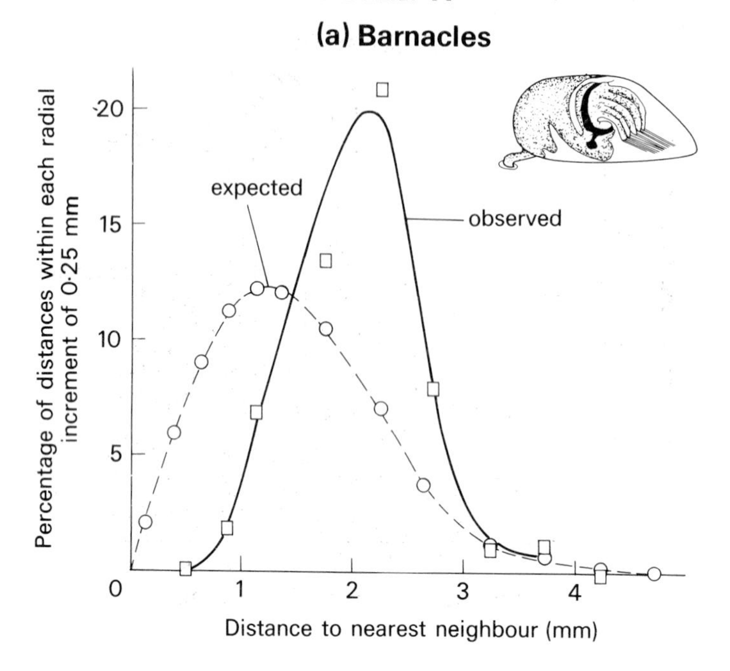

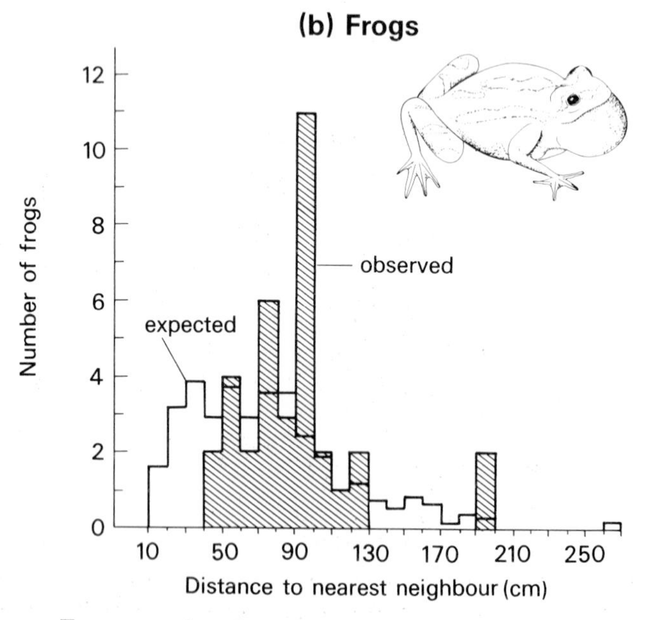

Fig. 11.1. Four examples of spacing out. In each case the animals are spaced out more than would be expected from a random distribution on the available suitable habitats. (a) Cyprid larvae of the barnacle *Balanus crenatus* settle on rocks so as to keep at least a distance of their own length away from nearest neighbours (Crisp 1961). (b) Calling male

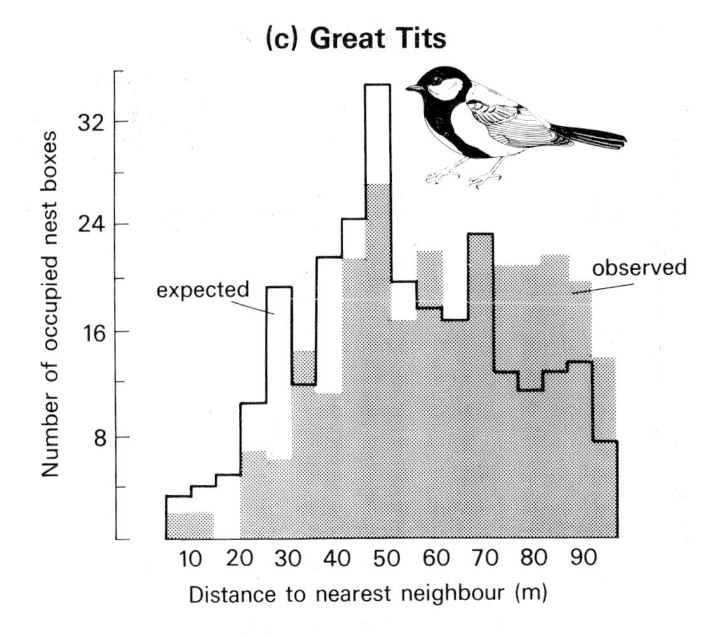

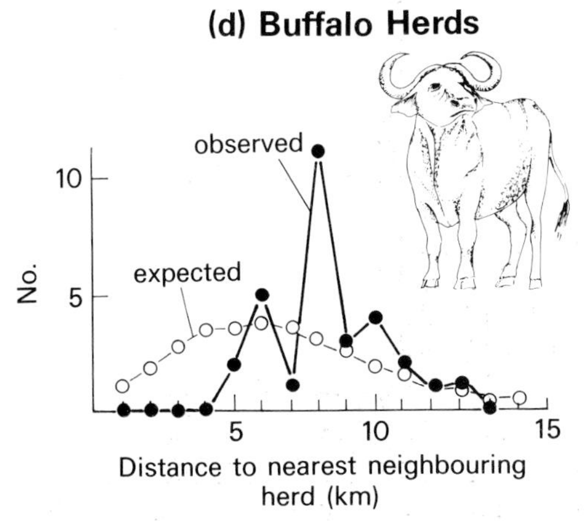

tree frogs *Hyla regilla* space out on the available calling sites at the breeding pond (Whitney and Krebs 1975b). (c) In Wytham woods, the great tit, *Parus major*, breeds exclusively in nest boxes. Nest boxes close together are occupied less often than would be expected on a random basis (Krebs 1971). (d) Buffalo (*Syncerus caffer*) herds in the Serengeti are spaced out more than would be expected on a random basis (Sinclair 1977).

benefits of spacing can be assessed in an attempt to understand why the spacing patterns we now observe in a species have been favoured during evolution. Secondly the consequences of spacing out will be examined, particularly in relation to animal populations. Finally I will ask the question, what is the mechanism by which animals achieve and maintain spacing out?

11.2 The benefits and costs of spacing out

Ever since the earliest description of territory (Howard 1920, Nice 1941) people have asked, what is the function of territorial behaviour? As more examples of territory were described it became clear that there was no single function because territories were used for a wide variety of different activities and so the question was rephrased as, what are the functions of territorial behaviour? (Hinde 1956). In this section I prefer to ask, how does spacing out influence fitness? This approach began with Brown (1964) who pointed out that territoriality will have both benefits and costs. He introduced the concept of economic defendability, which simply says that we would only expect an animal to defend a territory when there will be a net benefit in terms of fitness from doing so (Fig. 11.2).

There are two main methods for tackling the problem of how spacing behaviour influences fitness. Firstly we could look at changes in spacing behaviour with changes in ecological conditions. For example several studies have shown that spacing behaviour varies depending on the predictability of resources in time and space (birds, Crook 1965; primates, Crook 1970; antelopes, Jarman 1974). In general, where resources are unpredictable and patchily distributed, animals live in groups. Where they are predictable and dispersed the animals often exhibit spacing out (see also 1.3). Within species there may be great flexibility in social behaviour in relation to changes in resources. Spotted hyenas (*Crocuta crocuta*) live in clan territories in the Ngorongoro Crater, where their food supply is predictable and abundant, while in the Serengeti where food is very seasonal they wander over wide areas and do not defend any fixed areas (Kruuk 1972). In winter some small birds, such as pied wagtails (*Motacilla alba*), defend permanent territories where food is predictable while other conspecifics roam about in flocks and exploit transient, patchy food supplies (Davies 1976a). Male dragonflies (Odonata) localise their reproductive activity around mating or oviposition sites at ponds. When female arrival at the pond is predictable in time and space, males compete for territories at these optimum times and places (Campanella & Wolf 1974). When female arrival has low predictability, males do not defend fixed areas but wander all over the pond instead (Campanella 1975).

The second, more direct method is to measure differences in fitness between individuals who behave in different ways, for example fitness differences between territorial versus non-territorial individuals or between individuals with large versus small territories. In practice, however, fitness is rather difficult to measure; to do so, we would have

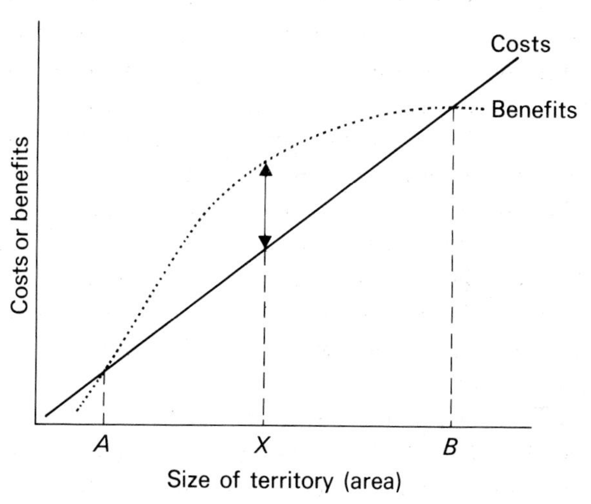

Fig. 11.2. A simple, graphical model which describes the idea of economic defendability of a territory. As the size of the territory increases, so do the costs of defence. The benefits (for example, in terms of food rewards) will increase at first and then will level off as the resource becomes superabundant in relation to the animal's needs. The animal should only defend the territory between A and B (where benefits > costs) and the maximum net benefit (benefit–cost) will be at a territory of size X.

to assess the relative contributions to the gene pool in successive generations for different individuals. Therefore, rather than attempt to measure the costs and benefits of spacing behaviour in terms of fitness, most studies have concentrated on certain parameters which are assumed to influence fitness directly. In the following sections I will describe cases where attempts have been made to measure the costs and benefits of territorial behaviour in terms of food, mates and predation.

11.2.1 *Territories and competition for food*

Indirect evidence

Many studies have indirectly suggested that one benefit from spacing out is concerned with food acquisition. For a wide variety of animals, home range size is positively correlated with body weight of the individual or group of individuals that inhabit it (lizards, Turner *et al.* 1969; birds, Schoener 1968; solitary, terrestrial mammals, McNab 1963; primates, Milton & May 1976). There is also some indication that type of diet influences home range size. For example, primates with more foliage in their diet have smaller home ranges than those that feed mainly on fruit and flowers, presumably because the latter is a more widely dispersed food resource (Milton & May 1976). However in all these studies it is difficult to distinguish cause from effect; we are still left with the question, is animal weight or diet a cause of home range size or a result of home range size which is determined by other factors?

Cost-benefit analysis

It has often been shown that territory size is smaller where food is more dense (birds, Stenger 1958, Cody & Cody 1972; fish, Slaney & Northcote 1974; limpets, Stimson 1973) or more nutritious (Moss 1969). Simon (1975) was able to induce short term changes in territory size in an inguanid lizard by experimental manipulation of the food supply; when food was added, the territories became smaller and when it was taken away again they returned back to their initial sizes. In some studies, more detailed assessment of the amount of food available on a territory has been made and this has been compared with the food requirements of the territory owner. In nectar feeding hummingbirds (Gass *et al.* 1976) and sunbirds (Gill & Wolf 1975), although the size of territories and their floral composition may vary enormously, the nectar supply that each territory contains is just enough to support an individual's daily energy requirements (Fig. 11.3). In tree squirrels (*Tamiasciuris*) the size of the territory also depends on the richness of the food supply and each territory contains approximately enough food to maintain a squirrel throughout the year (Smith 1968).

So far we have only been concerned with the benefits associated with a territory, but what about the costs? Whether defence of a food resource is economical or not will depend on whether the energy saved by gaining exclusive use of an area exceeds the energy expended in its defence. Presumably sunbird and squirrel territories contain just enough food to satisfy the individual's requirements, and no more, because a larger territory would involve greater costs for little extra benefit. In both the sunbird and squirrel studies there was some evidence that the animals were taking the costs into consideration in their territorial behaviour. Except for a brief period during the breeding season, male and female squirrels defend separate territories, caching

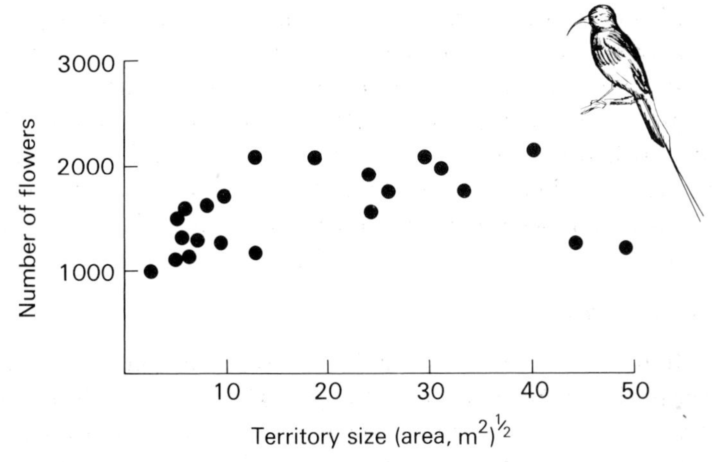

Fig. 11.3. Although the size of the territories of the golden winged sunbird (*Nectarinia reichenowi*) varies enormously, each territory contains approximately the same number of *Leonotis* flowers (from Gill & Wolf 1975). The energy expended by a sunbird during a day can be determined by time budget analysis (Wolf & Hainsworth 1971, Wolf *et al.* 1975) and this can then be compared with the energy content of the flowers in the territory. The result found is that the number of flowers defended yields just enough nectar to support the sunbird's daily energy requirements.

stores of conifer seeds in a central location within the territory. Smith (1968) showed that by living on a separate territory in this way, a squirrel would expend less energy collecting food and taking it to its cache than it would if it shared a territory with its mate. Horn (1968) also showed that, theoretically, when food is stable and dispersed in space (as it is in the squirrel case), the costs incurred in foraging trips to feeding sites are smaller if individuals are spaced out in separate territories. However when food is unpredictably concentrated in patches, a clumped dispersion, with all individuals in a central colony, is more efficient in terms of minimising travelling costs. In fact Brewer's blackbirds (*Euphagus cyanocephalus*), which exploit such patchy food supplies, do nest colonially (Horn 1968). It is important to realise that the simple models of Horn and Smith only consider one cost associated with the use of a territory, namely the travelling costs incurred in foraging. Obviously there are other costs involved such as defence of the territory against intruders.

In the sunbird study, Gill and Wolf (1975) were able to measure the time that territory owners spent in various activities such as defence (chasing off intruders), feeding and sitting. They showed that the nectar level per flower inside a territory was higher than that in undefended flowers. This was because defence of the territory prevented other sunbirds feeding at the flowers and thus allowed resource renewal over a longer period of time between feeding visits. (Stimson 1970 showed a similar effect for territorial limpets, *Lottia*, grazing films of algae on rocks.) An additional benefit from exclusive use of a patch of flowers may have been that the owner could regulate its visits to the renewing nectar supplies in a more efficient manner. The presence of intruders would presumably interfere with evaluation of efficient return times (Charnov *et al.* 1976a, see also 2.4.3). Gill and Wolf found that because territory owners fed at higher nectar levels, they were able to obtain their daily energy requirements in a shorter time and thus could spend more time in less expensive activities such as sitting. The most important result was that the energetic costs of territorial defence were easily offset by the benefits of the energy saved from the shortened daily foraging time on the territory.

Sunbirds and squirrels are not always territorial. When food levels are very high intruder pressure may increase and then the owners give up defence of their territories presumably because it becomes uneconomical. However these are only qualitative observations; Carpenter and MacMillen (1976a) have gone a stage further. By directly quantifying the costs and benefits associated with territoriality they could successfully predict the changes from territorial to non-territorial behaviour from calculations of when the costs exceeded the benefits. They studied the Hawaiian honeycreeper (*Vestiaria coccinea*), another nectar feeding bird that defends flowers. The benefits of territoriality were measured in terms of the raised level of energy available as a result of preventing others using the food supply. The costs were measured in terms of the energy expended in territory defence. For territoriality to be of net benefit,

(Basic cost of living) + (Added cost of being territorial) < (Yield to individual if not territorial) + (Extra yield gained by territorial defence of the resource)

i.e. $$E + T < aP + e(1-a)P$$

where the yield without territorial defence is a fraction a of the total productivity P. When territorial defence is perfect ($e=1$) the bird can gain all of the remainder of the potential productivity, namely $(1-a)P$. Territorial defence should occur between two thresholds. At one extreme, food may be so abundant that the bird can obtain all its requirements in the absence of any defence. Territory defence should begin as (aP) becomes too small to meet the cost of living (E); that is when $aP < E$. Thus the upper threshold for territoriality is when

$$P = \frac{E}{a}$$

At the other extreme the productivity may become so low that even with the extra yield gained through territorial defence, the bird is still unable to meet its daily energy requirements and it would do better by going off to find another patch of flowers. This threshold will occur when

$$P = E + T$$

These parameters were measured by time budget studies and then conversion factors were used to change time spent into energy expended (Carpenter & MacMillen 1976b). The values obtained were 13·4 k.cal. for E and 2·3 k.cal. for T per 24 hours. They also measured the energy lost to thieves per day for non-territorial birds on a foraging area and found that intruders accounted for 75 per cent of the nectar produced, thus $a=0·25$. Knowing these values the thresholds for economical territorial defence could be calculated. When defence is perfect ($e=1$), the lower threshold is 15·7 k.cal. per 24 hours, which is equivalent to the nectar produced by a territory containing 60 flowers. The upper threshold is 53·6 k.cal, equivalent to 207 flowers. The model's predictions for territoriality were tested for ten birds, seven of which were observed to be territorial and three non-territorial. The behaviour of nine of the ten birds agreed with the model's predictions (Fig. 11.4).

This simple model has several limitations. Firstly it assumes that defence costs are constant, whereas it seems likely that intruder pressure will increase with the number of flowers on a territory. Secondly, Carpenter and MacMillen did not consider the profitability of alternative foraging strategies. For example, the lower threshold must be the absolute limit because the bird would die if it went below it! The exact point at which it becomes profitable to give up the territory and move elsewhere will depend on the profitability of other areas and other strategies. Nevertheless this study has been discussed in some detail because it illustrates one of the few attempts that have been made to quantify the costs and benefits of feeding territoriality, rather

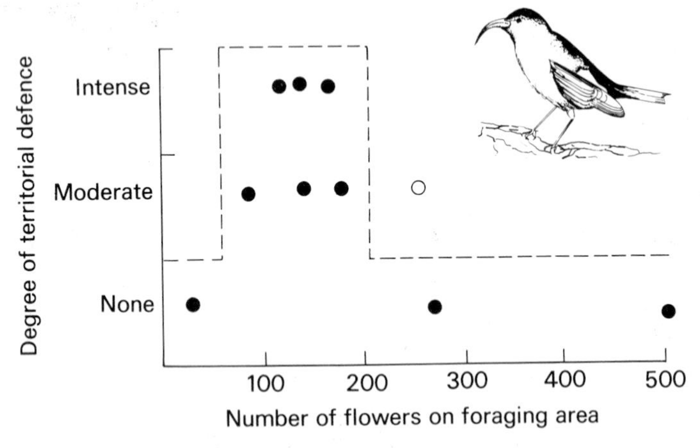

Fig. 11.4. Cost-benefit analysis of territoriality in a nectar-feeding bird, the Hawaiian honeycreeper, *Vestiaria coccinea*. A model, described in the text, predicts that the birds should defend a territory between two threshold levels of resource abundance (dotted lines). Ten birds were observed and their behaviour was classed as intensely-, moderately-, or non-territorial, depending on their degree of defence of the food resources they were exploiting. The behaviour of nine of the ten birds (solid circles) fitted the predictions of the model while one bird (open circle) did not (Carpenter & MacMillen 1976a).

than merely present qualitative observations. This cost-benefit approach adopted in the studies of the nectar feeding birds could profitably be applied to other situations. One of the problems of this method is evaluating the time period over which the animal is optimising the trade-off between the costs and benefits. In nectar feeding birds these appear to be assessed on a daily basis and changes in territory size and from territorial to non-territorial behaviour are rapid, matching the rapid flux in the resources exploited. In other cases optimisation may be over a much longer time span and we may not always expect an animal to cease being territorial just because on a daily basis the costs of defence exceed the benefits obtained (Fig. 11.5).

Superterritories—can territoriality be spiteful?

These studies described above support the hypothesis that an animal will defend a territory of a size that is just sufficient to satisfy its own

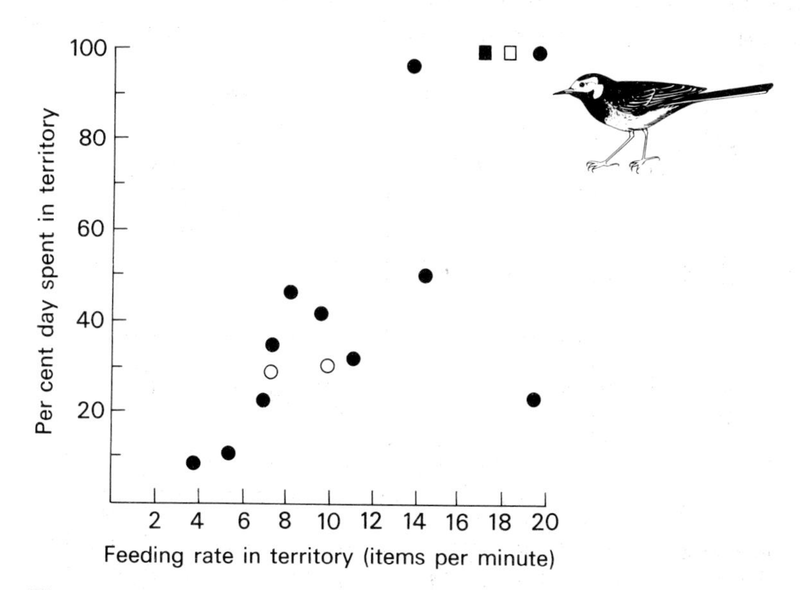

Fig. 11.5. Flexibility in the social behaviour of pied wagtails *Motacilla alba*. Each point is the result of a dawn-to-dusk continuous watch of an individual bird during a winter's day. Different symbols represent different individuals. When feeding conditions in their territories are good, territory owners spend the whole day there. On days when feeding is not so good, they leave their territories and join flock birds nearby to feed. However, even when feeding conditions on their territories are lower than those in the flock, the owners continue to spend some time on territory and chase off intruders. This is an example of long term optimisation, where costs incurred in defence on days of poor feeding may bring benefits later on in the winter when the territories provide the best feeding areas (Davies 1976a).

requirements. Verner (1977), however, has suggested that individuals should defend territories much larger than is necessary for their own survival and reproduction. He points out that selection acts on an individual's performance relative to that of all others. Thus an individual could increase its fitness either by improving its own absolute performance or by inhibiting that of other conspecifics. Verner suggests that one way of achieving such spiteful behaviour would be to defend a 'super territory' that prevented other conspecifics from using vital resources.

There are no data to support this idea; even when territories are apparently larger than is necessary for short term benefits, the animal may be securing a territory that will support it at times in the future when resources are scarcer (e.g. Hirons 1976). There are also theoretical objections to Verner's model. Even if the spiteful behaviour had no cost, it is difficult to see how it could evolve. The fitness of those who defend 'super territories' is increased only to the degree to which they reduce the average fitness of the non-spiteful members of the population. Thus the selective advantage of a spiteful strategy is very low when it first appears in a population (Rothstein in press). Even if holders of super territories did become common then a purely spiteful strategy would begin to suffer because it would be played against other spiteful individuals. For example, if a wood was filled with individuals holding super territories so that just enough space was left for one normal sized territory, then an individual that accepted this would settle and reproduce while an individual that only accepted a super territory would not. Thus a purely spiteful strategy is not an ESS. However it is possible that more complicated strategies (such as play spite against normal individuals but not against other spiteful individuals) could be stable.

11.2.2 *Territories and competition for mates*

Many animals only defend territories during the breeding season and here spacing behaviour seems to be concerned with the acquisition of mates. Emlen and Oring (1977) have applied the idea of economic defendability to mating systems. Individuals of one sex (usually the males) will maximise their inclusive fitness by attempting to control access to mates (females) (see 7.1). Whether this is economical will be influenced by the spatial and temporal availability of mates, which will determine the costs and benefits associated with the control. Sometimes males defend the females themselves, for example harems (see 7.5.2). Often, however, males defend areas of ground to which females come for mating. These territories can broadly be divided into those that contain some resource essential to females and those that do not. This may not be a sharp distinction and in practice there is probably a continuum between the two extremes.

Territories with resources

Males can control their access to mates indirectly by defending resources that are critical to the females such as food, nests or oviposition sites. Whenever these resources are patchily distributed, a situation may arise where there is a mosaic of male territories of different qualities. We may expect that it will be those males which can defend the best quality territories who will achieve the greatest reproductive success.

Orians (1969) has presented a graphical model which proposes that differences in quality between male territories may result in polygyny, even though the sex ratio may be unity (Fig. 11.6). This model makes three predictions. First we would expect to encounter polygyny most

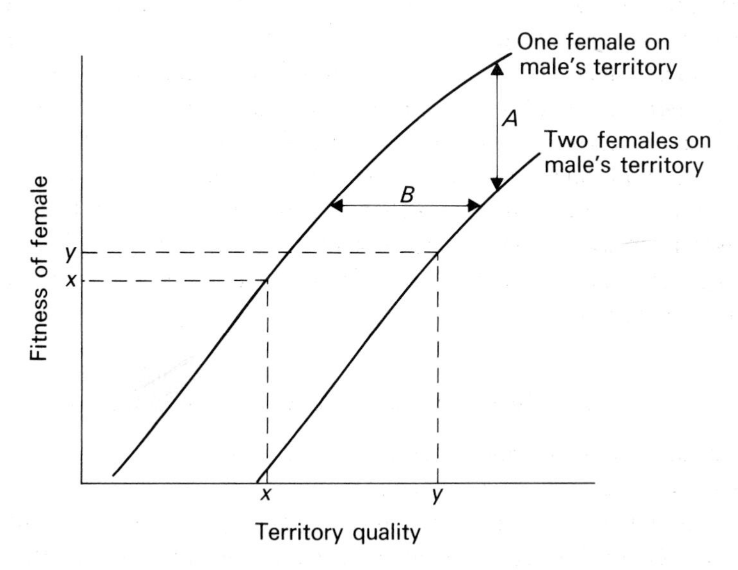

Fig. 11.6. The polygyny threshold model (from Orians 1969). It is assumed that reproductive success of the female is correlated with environmental factors, such as the quality of the male's territory in which she breeds, and that females choose mates from the available males. In the model, a female suffers a decrease in fitness A by going to an already mated male compared with the fitness she could expect if she had a male all to herself. Despite this drop in fitness, provided the difference in quality between the territories is sufficient (B = the polygyny threshold), a female may expect greater reproductive success by breeding with an already mated male. For example, a female who shared a male on territory y would do better than a female who had a male all to her self on territory x.

often in those habitats where environmental resources are the most patchy, and thus where there is a wide spectrum in the quality of male territories. In accordance with this prediction, thirteen out of the fourteen species of North American passerine birds that are regularly polygynous breed in marshes, prairies or savannah-like habitats where there is a wide range in productivity, even between small adjacent areas of habitat (Verner & Willson 1966). One problem with this analysis is that most of the species considered belong to one family (Icteridae); for European passerines there does not seem to be such a correlation between food dispersion and polygyny (von Haartman 1969). In the sandpipers (Calidridinae), when resources are evenly dispersed on the tundra where they breed, both sexes are resident on large territories and monogamy prevails. Where resources, such as food, are patchily distributed so that there are differences in quality between male territories, there are various degrees of polygamy (Pitelka *et al.* 1974).

Secondly, the model predicts that females will select males on the basis of territory quality. There is some evidence that male hummingbirds with the best feeding territories (Wolf & Stiles 1970, Stiles 1973) and male dragonflies whose territories contain the best oviposition sites (Campanella & Wolf 1974) do in fact obtain the greatest number of matings. In the long-billed marsh wren (*Telmatodytes palustris*), up to 50 per cent of the males may be polygamous. Territories are established before pairing. Bigamous associations form even when bachelor males are defending territories. Apparently bachelor males fail to attract mates because they have inferior territories. Those males who have the largest territories obtain the most females; thus territory size of bigamous males is larger than that of monogamous males which is in turn larger than that of bachelor males (Verner 1964). In this species food availability on the territory may also influence the number of females attracted (Verner & Engelsen 1970). Similarly, in the three spined stickleback (*Gasterosteus aculeatus*) males with larger territories are more successful at luring gravid females to their nests for egg laying (van den Assem 1967). By preferring males on large territories females increase their reproductive success. This is because one of the major causes of egg loss is predation by other males who try to interfere with the nest; males with large territories are less susceptible to such interference (Black 1971).

In other cases the resource differences between male territories that influences the number of females that a male obtains is not the size of the territory but the quality of vegetation it contains (Fig. 11.7). In the red-winged blackbird (*Agelaius phoeniceus*) certain types of vegetation provide safer nesting sites from predators and hence result in greater fledging success. Those males whose territories contain the most of this suitable vegetation attract the most females (Holm 1973). Similarly, in the frogs *Rana clamitans* (Wells 1977a) and *Rana catesbeiana* (Howard

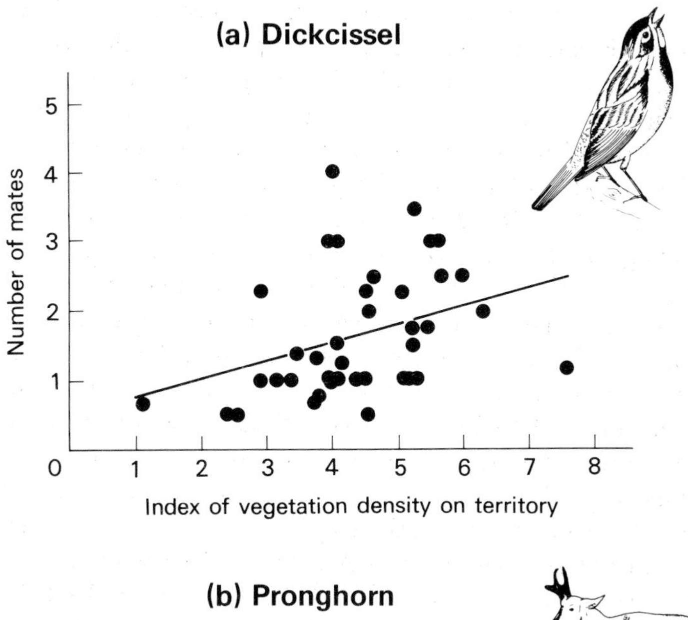

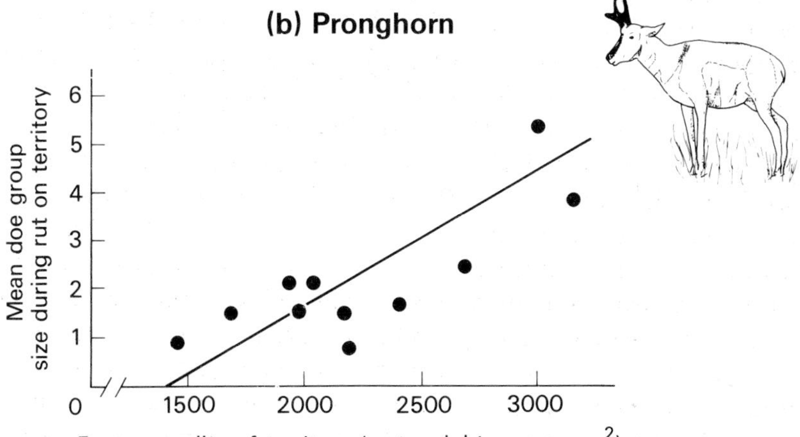

Fig. 11.7. Two examples where a male's reproductive success depends on the quality of his territory. (a) In the dickcissel (*Spiza americana*) males with denser vegetation on their territories attract more mates, probably because this provides better nest sites (Zimmerman 1971). (b) In the pronghorn (*Antilocapra americana*), males whose territories provide the best foraging areas attract the largest number of females (Kitchen 1974).

in press a), males whose territories contain the best oviposition sites are able to attract the largest number of females for mating.

All these examples are of males apparently competing for resources that directly influence female reproductive success. However, theoretically this need not always be the case. It is quite possible that females could choose certain territory qualities (such as large size or type of vegetation) simply because they are attractive. Males whose territories contained such attractive features would obtain the most females and females who mated with these males would increase their fitness because their sons would also choose attractive territories and thus, in turn, attract many mates (O'Donald 1963). This is exactly the same argument as that used by Fisher (1958) in relation to male appearance and sexual selection (see also 7.2).

Finally, Orians' model predicts that those females who decide to mate with already mated males will have at least as high reproductive success as those who choose unmated males. In fact Holm (1973) found that, in the red-winged blackbird, those females who mated with already mated males did better than those who mated monogamously. The model does not necessarily predict this result. The first females to arrive would presumably go to those males with the best quality territories. Eventually a point may be reached where the next female to arrive would do better by going to an unmated male, because territory suitability declines with the number of females already there. The model could describe a settlement pattern phenomenon where females always choose the best available territory. This will simply depend on where the last female went and so if we measured reproductive success at any one time, we could find that those females who shared a male were doing slightly better, just as well as or even slightly worse than females who were on a territory by themselves.

We are still left with the problem of whether females are choosing males on the basis of the quality of their territory or whether they are assessing male quality itself. For example if we find that males with larger territories have higher reproductive success, is this because females prefer the larger territories or is it because females prefer certain types of males, who just happen to be the ones who defend the large territories? (Davis & O'Donald 1976b, Trivers 1976).

Territories without resources: Leks

In some cases neither the females themselves nor the resources that they require are economically defendable. This may arise because the resources are superabundant, unpredictable in time and space or very costly to defend (Emlen & Oring 1977). In these situations the males often aggregate at traditional, communal display grounds, or leks, where they occupy small patches of ground and compete for dominance status. Although the territories on the lek may be just tiny, bare patches of ground no more than a few centimetres (manakins, Lill 1974; frogs, Emlen 1976b) or at most a few metres in diameter (grouse, Kruijt & Hogan 1967; antelopes, Buechner & Schloeth 1965), they are vigorously defended against other males. Females visit the leks in order to mate. In all the leks that have been studied, whether they are birds, antelopes, dragonflies or frogs, almost all of the copulations are performed by a few of the males, even though all the males on the lek may be sexually responsive (Fig. 11.8). The successful males are often those that occupy certain spatial positions, often the central territories on the lek. In dragonflies, the most successful males are those that are able to command territories at the pond during the time of the day that most of the females arrive (Campanella & Wolf 1974). Thus in many lek species, like the grouse and the antelopes, the most successful males are those defending the best place, while in others, like the dragonflies, they are those defending at the best time.

As with the cases of mating territories with resources, we have the problem of whether the females are selecting male quality or territory quality. In a study of manakin leks, Lill (1974) failed to find any aspect of male morphology or display that correlated with mating success. In an experiment, he removed two of the successful males from a lek and within minutes their vacant territories were reoccupied by other males. Even though the new males displayed in a different manner from the previous occupants, females continued to visit these same territories for mating. This experiment may suggest that it is some characteristic of the territory that influences female choice, rather than some characteristic of the male himself. However Lill was unable to find any territory characteristic that correlated with mating success. Alternatively, it is possible that the new males who occupied the vacated territories were the 'next best' males and females were choosing the best males available. Wiley (1973) also failed to find any correlation between courtship displays of male sage grouse (*Centrocercus urophasianus*) and mating success and similarly Buechner and Schloeth (1965) found that in antelope leks, although the occupancy of the territories changed frequently, the females still preferred the same areas in the lek for mating. So maybe females are choosing some arbitrary, traditional cue, such as the central territories, thereby forcing males to

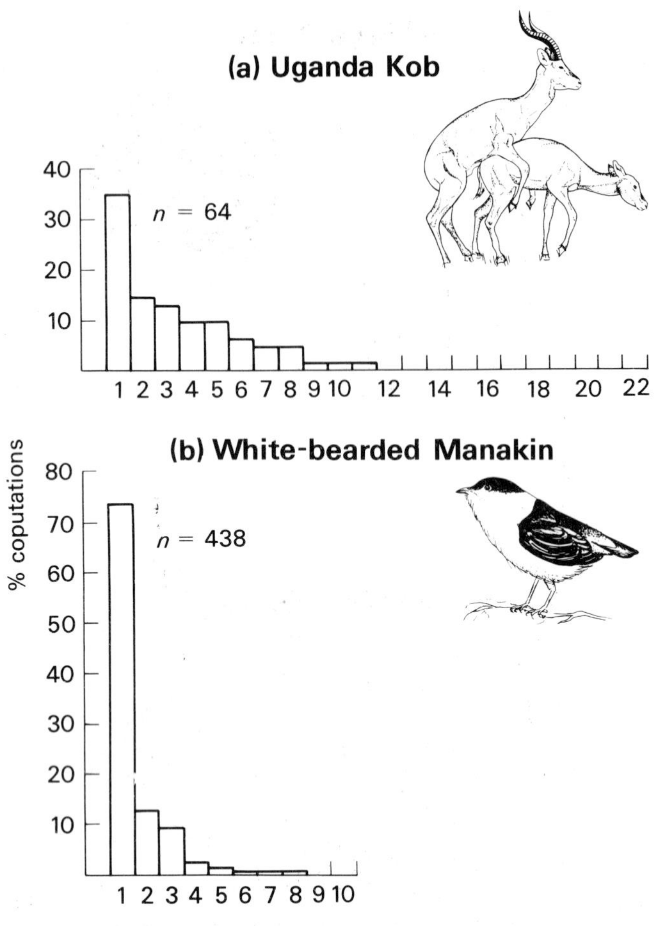

compete for these locations and then selecting to mate with the winners of the competition. In effect they let the male-male competition do the sorting for them (Emlen 1976b, Cox & Le Boeuf 1977, see also 7.4).

On the other hand, two studies suggest that it is male display characteristics that influence female choice. In the black grouse, *Lyrurus tetrix* (Kruijt & Hogan 1967) it appears that those males with the most efficient courtship tactics are the most successful, while in the ruff, *Philomachus pugnax* (Hogan-Warburg 1966) the choice of the female may be determined by differences in individual behaviour and plumage patterns of the males. Thus the problem of what exactly it is that the females are choosing remains unresolved.

Although the potential benefits from defence of a central territory

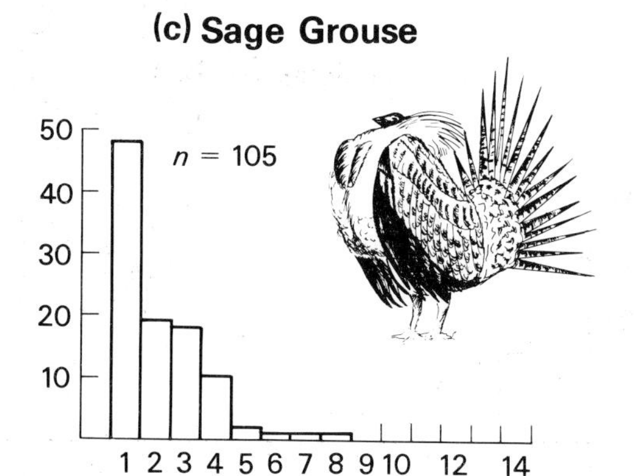

% copulations

(d) Black Grouse

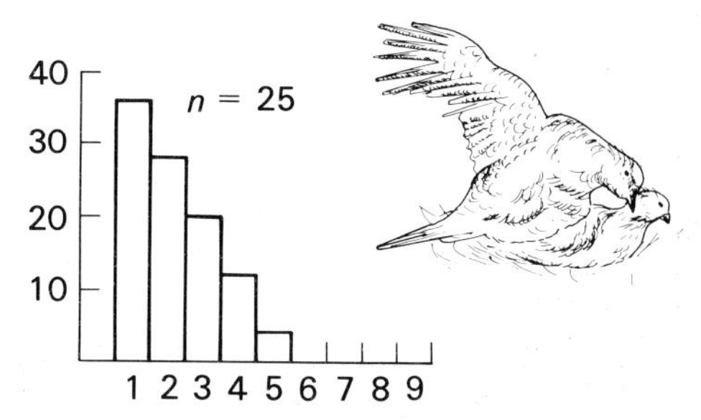

Territorial males present on lek, in rank order of mating success

Fig. 11.8. In each of these four lek systems the males are ranked in descending order of the percentage of the total copulations they performed in a breeding season. In all four cases, almost all of the copulations were performed by a few of the males. The successful males were those that defended the central territories on the lek. (a) Uganda kob, *Adenota kob thomasi* (from Floody & Arnold 1975.) (b) White-bearded manakin, *Manacus m. trinitatis* (from Lill 1974). (c) Sage grouse, *Centrocercus urophasianus* (from Wiley 1973). (d) Black grouse, *Lyrurus tetrix* (from Kruijt & Hogan 1967).

on a lek are enormous, the costs must also be considered. One of the costs may be an increased risk of predation on displaying males on the lek. Wiley (1974) reviewed grouse social systems and found that the type of habitat influenced the dispersion of the displaying males. Only in species living in open country do males congregate on leks. In forest species, where visibility is restricted, the risks of predation on aggregations of grouse may have militated against the evolution of leks, and in these species the males remain dispersed. Defence of territories on the lek must also impose high energetic costs and perhaps also a risk of injury through fighting off other males. In grouse, males start breeding at a later age than females. Wiley (1974) suggests that such deferred reproduction in males may be advantageous if breeding imposes a higher risk of mortality than not breeding (e.g. through predation and fighting), and if reproductive success increases with age. In other words, a male grouse must trade off the costs of attempting to breed with the potential benefits he will gain. Only when he is old enough to have a good chance of successfully competing for a territory will it be worth-while risking the increased fitness costs that this involves.

Not all of the males on a lek adopt conspicuous, territorial strategies. There are often many males who employ silent, sneaky mate-searching behaviour (e.g. birds, von Rhijn 1973; frogs, Howard in press b; fish, Constantz 1975; insects, Alexander 1975). It is sometimes assumed that the central, territorial males have the highest reproductive success. This may well be true at any one moment in time, but to test this hypothesis we really need to know the lifetime reproductive success of males adopting different strategies at a lek.

In some cases the male strategy seems to be age dependent; it is the young males who are non-territorial or who defend peripheral territories and they move centripetally as vacancies arise (Kruijt & Hogan 1967, Wiley 1973). Here, presumably, males who obtain territories do indeed have greater lifetime fitness.

In other cases, however, the male strategy may be genetically determined. In the ruff (Hogan-Warburg 1966, van Rhijn 1973), 'resident' males defend territories on the lek, while 'satellite' males adopt the sneaky strategy of attempting to steal copulations, for example while a resident is busy chasing off an intruder. There is a distinct plumage difference between these two sorts of male; the residents have dark head tufts and usually dark neck ruffs as well, while the satellite males are adorned with white tufts and ruffs. This difference in status between the types of males is at least in part genetic.

The best strategy for any one male at a lek must depend on what all the other males are doing; for example if all male ruffs were residents then it may pay a new arrival to adopt the satellite strategy. The proportion of males adopting the various strategies at a lek may be an

ESS (see 1.2 and 10.3 for discussion of the ESS concept). Although some territorial males may have a very high reproductive success, the variance for this strategy is high and it is quite possible that, over a lifetime, the average reproductive success of males employing non-territorial strategies is just as great (see also Gadgil 1972).

In the topminnow (*Poeciliopsis occidentalis*) territorial males are black and aggressive and attract females with elaborate courtship displays. Non-territorial males are brown-coloured like the females and, ignoring all the formalities of courtship, simply dash up to a female and attempt a quick copulation (Constantz 1975). The rate at which males performed mating attempts was about the same for each strategy, though Constantz could not be sure if each mating attempt resulted in sperm transfer. If a territorial male was removed then a non-territorial male took its place and within a matter of minutes it became black and aggressive. This may be an example where the payoff for a mate searching strategy is frequency dependent and where different strategies result in the same reproductive success.

11.2.3 *Territories and predation*

Depending on the ecological circumstances, predation may favour clumping or spacing-out. When predators search for camouflaged prey, selection often favours spacing-out. Tinbergen *et al.* (1967) showed this in an experiment where they laid out artificially-camouflaged hen's eggs in plots of different densities. Wild crows came down to prey upon the eggs and crowded plots, where eggs were close together, suffered a much higher mortality than scattered plots. The selective pressure favouring the spacing out was area restricted searching by the predators. Whenever the crows found an egg they increased their search in the immediate vicinity so that there was an increased risk for any other eggs nearby (Croze 1970). Predation has also been shown to select for spacing out in some passerine bird nests (Horn 1968, Krebs 1971 see, Fig. 11.9a).

In other situations, where mass attacks on predators by a number of individuals at once are more effective in reducing predation than individual attacks, then clumping in space may be favoured (see also 1.3.3). If the owners of a nest attack a predator over a distance from their nest that is greater than the average distance between nests, then there will be more potential attackers within a colony than on the outside and thus coloniality will be selected for (Kruuk 1964, Patterson 1965, Birkhead 1977, see Fig. 11.9b). Against this is the potential

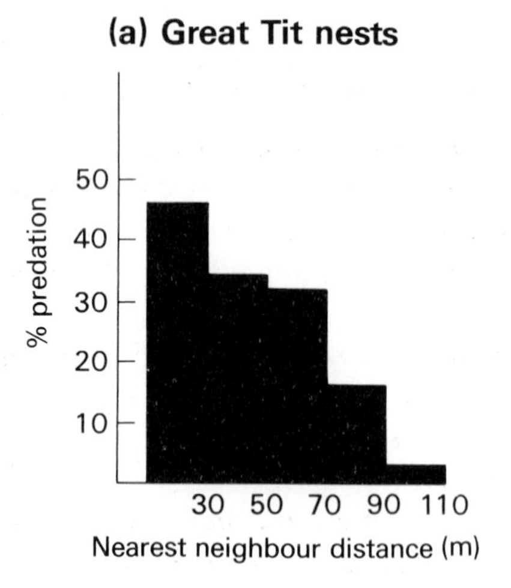

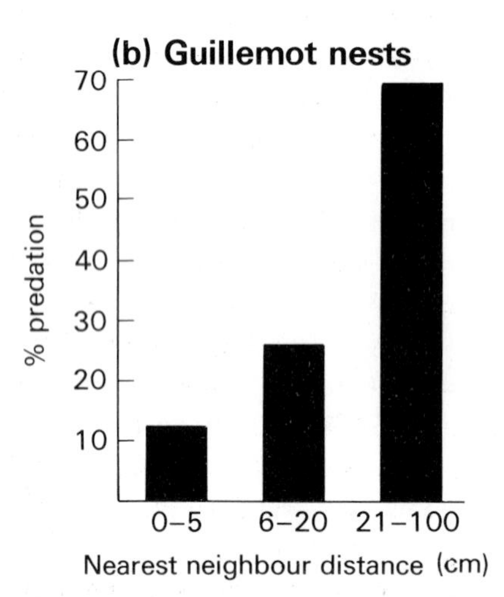

Fig. 11.9. Territory size and predation. (a) In Wytham Woods, the main predators of nestling great tits, *Parus major*, are weasels, *Mustela nivalis*, which may rob up to 50 per cent of the nests in some years (Dunn 1977). The closer a nest is to another nest, the greater the chance of predation. Thus selection favours spacing out of nests (from Krebs 1971). Unlike the tits, Guillemots, *Uria aalge*, (b) defend their eggs and chicks against predators and in this case selection favours clumping because group defence is more effective (from Birkhead 1977).

advantage of spacing out to combat area restricted searching of predators, so the spacing pattern adopted may be a compromise and will depend on such factors as the main predators involved (Kruuk 1964).

11.2.4 *Spacing as a compromise*

Although in the previous sections spacing has been described in relation to single parameters such as food, mates and predation, these will all act together to determine fitness. Spacing behaviour will probably always be a compromise moulded by conflicting selective pressures.

A good example of such a compromise is the dispersion of wading birds on their intertidal feeding grounds; this varies enormously, both within and between species, from densely packed to widely scattered flocks. Predation pressure will select for closely knit groups (Page & Whitacre 1975) (see also Chapter 3) but this may not always be compatible with efficient foraging. In those waders that hunt by sight, by searching for visual cues on the mud surface, the smaller the average nearest-neighbour distance (i.e. the more compact the flock) the lower the feeding rate (Goss-Custard 1976). Thus interference between foraging birds selects for spacing out and waders feeding in this way disperse in loose groups. However, when waders are hunting by touch, probing for prey beneath the mud surface, feeding rate does not decline with decreases in nearest-neighbour distance. In this case, where efficient foraging is not incompatible with compact flocking, the birds occur in dense groups. Therefore the spacing patterns within a flock seem to be determined by a compromise between predation and foraging selective pressures (Goss-Custard 1970).

Similarly, Horn (1968) showed that concentration of blackbird nests into a central colony will be favoured by selection for efficient exploitation of a variable, patchy food supply. However, within the colony, predation pressure selects for spacing out. Once again the spacing pattern adopted must be a compromise.

11.2.5 *Interspecific territoriality*

Territorial defence versus individuals that exploit different resources will inflict costs for very little resulting benefit. Although it is usually conspecifics that exploit the same resources, not all conspecifics may pose the same threat. For example in the lizard (*Anolis aeneus*) different sized individuals (different ages) take different sized prey. The degree of aggression between individuals depends on their relative sizes; maximum aggression occurs between residents and intruders of the same size, where food overlap is also a maximum (Stamps 1977).

There is usually even less overlap in resource exploitation between members of different species and, as expected, most animals tolerate overlapping territories with other species. In birds, Orians and Willson (1964) suggest that interspecific territoriality may occur most often between species which have similar ways of exploiting food resources. This may arise because insufficient time has passed for divergence in their ecology or in situations where two species live in structurally simple environments where overlap in ecology is most likely to occur (e.g. Catchpole 1972). For example, in marshland habitats yellow-headed blackbirds (*Xanthocephalus xanthocephalus*) and red-winged blackbirds (*Agelaius phoeniceus*) defend mutually exclusive territories. Although there is some difference in habitat selection, both species feed their nestlings on similar types of insect food (Orians & Willson 1964).

Studies of pomacentrid fish provide the best evidence that interspecific territoriality is related to overlap in food resources. Over 90 per cent of the intruders that *Pomacentrus flavicauda* evicted from its territory were of other species. Low (1971) found that 35 of the 38 species that were chased off were food competitors, while all of the 16 species that were allowed to trespass unmolested exploited different food supplies. Ebersole (1977) was able to relate the degree of aggression to other species by *Eupomacentrus leucostictus* to the degree of overlap in diet. Similarly, territorial limpets defend their gardens of algae not only against conspecifics but also against other species of grazing limpets (Stimson 1970, 1973, Branch 1975).

On the other hand, Murray (1971) has argued that interspecific territoriality is not adaptive and merely occurs through mistaken identity. This is a very unlikely view, especially considering the great variability in appearance and behaviour of the species that elicit aggression in the pomacentrid fish, and the very distinctive songs that elicit interspecific aggression in some song birds (Catchpole 1977). In fact the evidence suggests that interspecific territoriality is beautifully adapted to the amount of overlap in resource requirements.

Interspecific aggression may also often occur in species that feed on mobile prey that is liable to disturbance from other individuals, irrespective of whether they are feeding on the same type of food. Reed warblers (*Acrocephalus scirpaceus*) do not defend any permanent feeding territories but they vigorously defend the bushes in which they happen to be foraging at the time, both against conspecifics and any other species of small bird. This is because their feeding method, which involves capture of flies (Diptera) just as they prey take off, is very susceptible to interference (Davies & Green 1976).

It also seems likely that predation pressure may select for inter-specific spacing, especially where two species nest in similar places and thus where a predator may be searching for both at the same time. In the great tit (*Parus major*), not only does another great tit nest nearby increase the probability of predation (Krebs 1971), the presence of a nearby blue tit (*Parus caeruleus*) nest has the same effect (Dunn 1977).

11.3 Consequences of spacing out for populations

11.3.1 *Removal experiments*

Does territorial behaviour limit population density in a given area or does it merely result in spacing out of residents (Fig. 11.10a)? Many studies have shown that when territory owners are experimentally removed, or die naturally, their places are rapidly taken over by new-comers (birds, Orians 1961, Watson 1967, Krebs 1971; mammals, Healey 1967, Carl 1971; fish, Clarke 1970; dragonflies, Moore 1964; butterflies, Davies 1978b; limpets, Stimson 1973). The deduction from this observation is that, prior to removal, potential settlers had been prevented from occupying the territories by the presence of residents. These experiments show that territorial behaviour can limit population

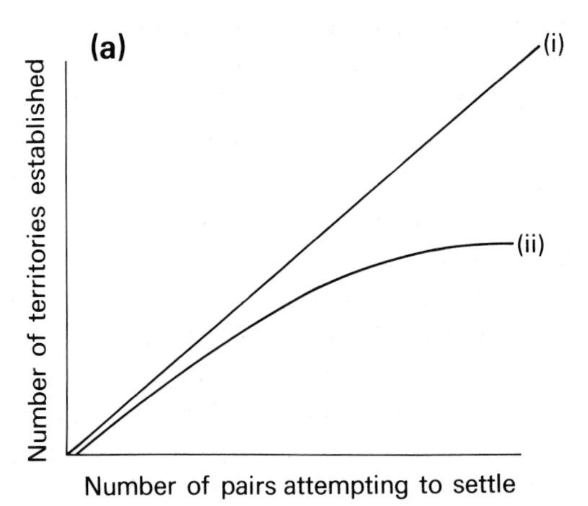

Fig. 11.10(a). Two possible consequences of territorial behaviour. In (i) there is no limit to territory size, so all pairs attempting to settle get territories and territorial behaviour just results in spacing out. In (ii) there is a lower limit to territory size, so that some pairs fail to get territories and have to leave the area. In this case territorial behaviour does have consequences for population regulation. The way to test for situation (ii) is by removal experiments (see text).

density on a local scale (Fig. 11.10a, case ii). It is important to realise that this does not mean that a 'function' of territorial behaviour is to limit population density, as Wynne Edwards (1962) suggested. Such reasoning implies selection acting at the level of the group or population and there are strong theoretical reasons for supposing that this does not usually occur (Maynard Smith 1976a, see 1.4). What it does mean is that limitation of population density can be a consequence of territorial behaviour that has been selected for in some other context by means of individual selection. For example, removal experiments have shown that territorial behaviour in the great tit limits population density in woodland and this arises as a consequence of the fact that predation pressure has selected for spacing out of nests (Krebs 1971).

In the removal experiments it is important to determine where the replacements came from and how they would have fared anyway if they had not occupied a vacant territory. In the great tits, the replacements came from hedgerow territories outside the wood, which were suboptimal in terms of reproductive success (Krebs 1971). In red grouse (*Lagopus l. scoticus*), replacements were non-territorial individuals that lived in flocks, which would not have bred and would probably have died in the absence of a territory (Watson 1967). In both these examples, therefore, territorial individuals have a selective advantage and territoriality has a density limiting effect (Fig. 11.10b).

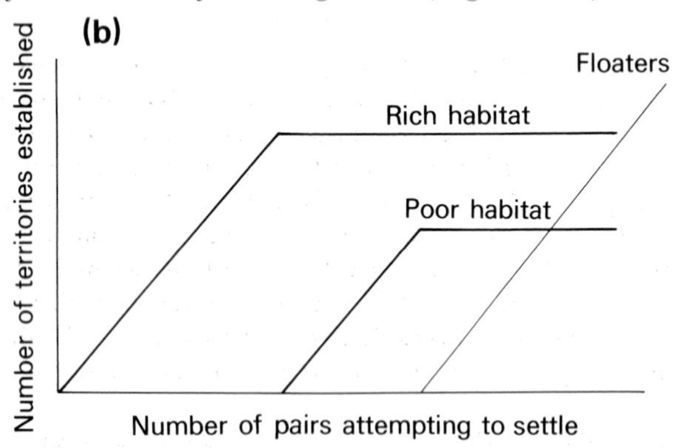

Fig. 11.10(b). A model of territoriality as density limiting—an extension of case (ii) in Fig. 11.10(a). Hypothetical effects of territorial behaviour on breeding densities in two habitats, assuming a fixed, minimum territory size. In this model, birds occupy the rich habitat first until it is full, then other arrivals are forced to fill up the poor habitat. Finally, when all the available habitat is occupied, remaining birds are 'floaters' and fail to get territories (from Brown 1969). The data from Krebs' (1971) study of titmice and Watson's (1967) study of red grouse fit this model.

In other cases individual reproductive success may be the same in different habitats and territorial behaviour may simply provide a gauge whereby potential settlers judge the suitability of occupied areas (Fig. 11.10c). In a study of titmice in Holland, Kluijver and Tinbergen (1953) found that in years of low population density most of the birds bred in mixed woodland while in years of higher density some individuals bred in pine woodland as well. Over the years the population density in the

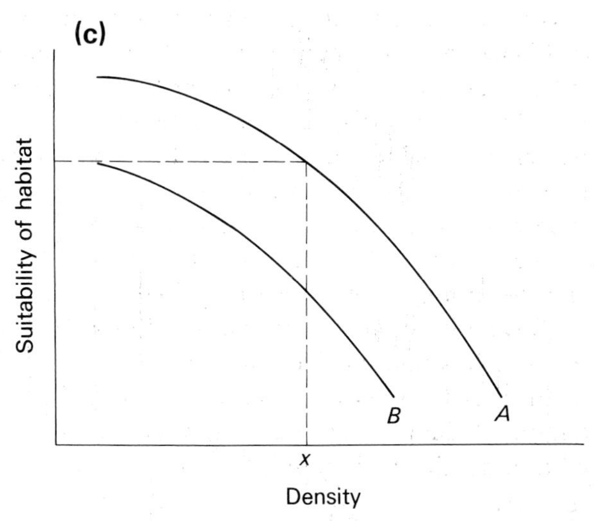

Fig. 11.10(c). A model of territoriality as a means of density assessment —an extension of case (i) in Fig. 11.10(a). Territorial behaviour does not limit density in either habitat. It is assumed that each individual chooses to settle in the best available habitat and that suitability of the habitat decreases with the number of occupants. The first arrivals will settle in habitat A. When the density reaches x, habitat B will be equally attractive so the next arrivals will go to either habitat A or B. Thereafter, the habitats will be filled so that the suitability of each remains the same (from Fretwell 1972). The data from Kluijver and Tinbergen's (1953) study of titmice seem to fit this model.

mixed woods varied much less than that in the pinewoods and Kluijer and Tinbergen suggested a 'buffer-effect', whereby territorial defence set an upper limit on the density in the mixed woods, such that in years of high population some pairs were forced to find territories in the less dense pine woods. However, in titmice reproductive success declines with density (Perrins 1965) and furthermore, in the Dutch study there was no difference in the reproductive success per pair in the two habitats (Brown 1969). So one explanation is that each individual was settling where it could expect the greatest reproductive success and territoriality had no density limiting effect at all (Fig. 11.10c) (see also 12.6.3).

Therefore the data of Krebs (1971) support the density limiting model while those of Kluijver and Tinbergen (1953) support the density assessment model. The difference is probably just one of degree and may depend on the difference in quality between the habitats studied. Where the difference is small then birds breeding at low density in a 'poor' habitat may do just as well as those breeding at a higher density in the 'optimal' habitat. Where the difference between habitats is greater, then territorial behaviour may result in some individuals suffering a lower reproductive success.

In some cases, such as the tawny owl, territorial behaviour has a strong density dependent effect on mortality and the population in a wood may remain very stable from year to year (Southern 1970). However in the great tit, although the removal experiments showed that territorial behaviour does limit density, it does not limit density to the same level each year. This is because of great variability in territory size from year to year which means that the effect of territoriality on the number of birds breeding in the wood is only weakly density dependent (Krebs 1970).

11.3.2 *Proximate and ultimate determinants of territory size*

What causes these fluctuations in territory size? Unlike some cases described above (section 11.2.1) variations in territory size in the great tit may not be related to changes in environmental resources such as food supply (Krebs 1971). A possible proximate cause of the variation is simply the way in which the birds settle on the territories each spring. In a model, Maynard Smith (1974a) showed that synchronous settlement of birds on their territories could result in twice the final density as asynchronous settlement. This was also shown experimentally by Knapton and Krebs (1974) who compared the effects of settlement patterns after simultaneous versus successive removals of territorial song sparrows (*Melospiza melodia*). Where settlement of new territorial owners was asynchronous, the first pairs to arrive spread out and left gaps between their territories that were too small to accommodate other pairs that arrived later. In synchronous settlement the territories were arranged so that each pair had the smallest acceptable territory and the final density was higher. Therefore one reason that tawny owl densities in woodland are so much more stable than those of great tits may be that there is much less variability in settlement patterns due to the fact that the owls are longer lived birds, so that territorial boundaries remain almost exactly the same from year to year.

Thus, it is important to distinguish the proximate and ultimate determinants of territory size. Suppose, for example, that we find a relationship between food supply and territory size (see 11.2.1). This could arise by several different proximate mechanisms. Food itself could be the proximate factor; the animal could decide on its territory size by monitoring the food supply directly. Alternatively the ultimate link between food supply and territory size could be caused by the fact that more birds attempt to settle where food is the most abundant and this settlement pattern determines territory size; for example the animal could select a certain territory size by monitoring intruder pressure.

The cost-benefit argument discussed in section 11.2 is concerned with the ultimate factors, or selective advantage, that determine the choice of a particular sized territory and it implies that there will be an optimal territory size (see Fig. 11.2). Although there has not been much work in this field, Dill (in press) has produced a detailed model and discussion of this problem.

11.4 Proximate mechanisms for the maintenance of spacing

So far functions and consequences of spacing behaviour have been discussed. Now I want to turn to a different question and ask, how is spacing out maintained? For example, both in pied wagtails (Davies 1976a) and speckled wood butterflies (Davies 1978b), the rate of intrusions onto occupied territories is much less than that onto empty territories. This implies that intruders can perceive that a territory is occupied and refrain from trespassing.

11.4.1 *Keep-out signals*

What are the keep out signals involved in territorial defence? Territory owners appear to have a variety of displays which they use in defence of their territories. Peek (1972) proposed a three tier defence system in song birds, which may also apply to other animals such as frogs (Whitney & Krebs 1975b). He suggested that song is used as a long range warning signal to repel potential traspassers at a distance; visual displays are used at an intermediate range to repel actual trespassers and finally, if an intruder persists, it is chased and attacked.

Peek (1972) tested this idea of graded keep out signals by means of two experiments with red-winged blackbirds. This species has a patch of bright red and yellow feathers on the upper wing which it displays to intruders. Peek found that when these feathers were painted black, the owners were much less successful in maintaining their territory than normal birds. In another experiment Peek muted territorial owners by cutting the hypoglossal nerves which supply the syrinx. These muted birds were also much less successful at keeping out intruders than were sham operated territory owners. This last experiment clearly shows that vocalizations are involved in territory defence but it does not demonstrate whether the keep out signals is the song or the calls of the bird. (But see D. G. Smith 1976, who repeated this experiment and found that bilateral sectioning of the hypoglossal nerves had no effect on territorial defence.)

The idea that song is a means of territorial defence seems so eminently reasonable that it has often been accepted as a fact, even in the absence of any hard evidence. It is only recently that the first experimental test of this hypothesis has been attempted. In a previous study, Krebs (1971) showed that when he removed male great tits from their territories, they were replaced within a few hours by other males (section 11.3). These new birds were either owners of poor quality territories outside the wood or were non-territorial floaters. The rapidity with which they took over the new vacancies suggested that they must have regularly monitored the wood to look for empty spaces. The experiment summarised in Fig. 11.11 shows that song was one cue which told intruders whether a territory was occupied. Empty territories that were occupied by loudspeakers broadcasting great tit song were much more effective at keeping out replacements than territories that were silent or broadcast control sound (see also, Krebs 1977a).

Great tit males have a repertoire of up to about six different song types and Fig. 11.11 shows that repertoires were more effective as keep out signals than were single songs. Why does a repertoire of songs enhance the effectiveness of the keep out signal? In the great tit, reproductive success declines with density (Perrins 1965) so when a newcomer decides where to settle it must presumably assess both habitat quality and the density of birds already present to determine the suitability of the area (Fig. 11.10c). Newcomers may assess the density of residents in an area by listening to songs and Krebs (1976) suggests that habituation may be the mechanism for achieving this density assessment; by habituating quicker to fewer songs, a bird

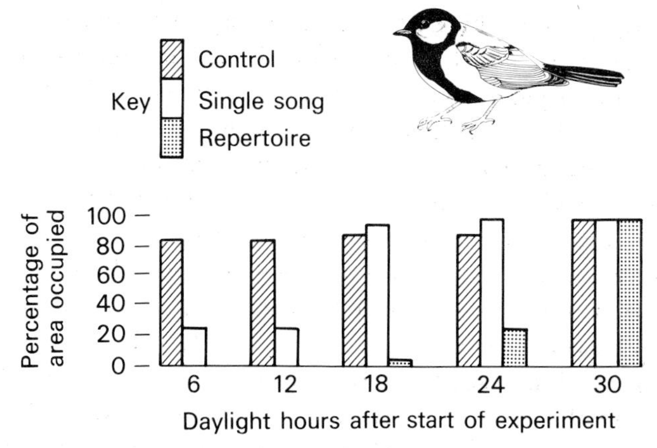

Fig. 11.11. An experiment which shows that bird song can act as a keep-out signal in territory defence. Territorial great tit males were removed from their territories and replaced with loudspeakers that either played a control sound (a note on a tin whistle) or played recorded tit song. New birds, which came to replace the removed males, took much longer to settle in those territories that broadcast great tit song than in control territories. Song repertoires were a more effective keep out signal than single songs (from Krebs *et al.* 1978).

would be more likely to settle in a less dense area. Song repertoires may be a way by which resident birds cheat; by singing a variety of songs a resident may create the impression that there are more birds present and decrease the apparent suitability of the area (see also 10.4). The repertoire of songs thus tricks the newcomer into making an inappropriate density assessment and increases his likelihood of moving on somewhere else. Krebs (1977b) has called this the 'Beau Geste Hypothesis', and in accordance with his idea is the fact that a territorial male is more likely to change its song type at the same time as it changes song perches than if it stays on the same perch, a behaviour which would enhance the effectiveness of the illusion that there was more than just one male present (Krebs *et al.* 1978).

Many passerine birds show a great deal of individual variation in song. Several studies have shown that territorial males respond more strongly to playbacks of songs of strange males than they do to songs of neighbours (Weeden & Falls 1959, Emlen 1971, Krebs 1971, Brooks & Falls 1975). Individual recognition of neighbours is both by their song characteristics and location (Falls & Brooks 1975). This differential response is presumably adaptive in that it saves a territory owner wasting time on defence versus neighbours that present less of a threat than strangers.

Cody (1974b) has pointed out that interspecific territoriality is sometimes associated with similarity between the species concerned in those characteristics, such as voice or appearance, that are assumed to be concerned with territory defence. He has termed this phenomenon character convergence, but the problem with this hypothesis is that because interspecific territoriality often occurs between congeners it is difficult to distinguish it from similarity through common genetic descent. Experiments are needed to determine for example whether the song of species A acts as an effective keep out signal to individuals of species B.

Often territorial defence is by means of more subtle signals such as scent. Many species of mammals mark their territories with urine and faeces deposits or with scent produced by specialized glands and it has been assumed that these signals may somehow be involved in territorial defence (Ewer 1968). However, as with bird song, there have only been a few experimental tests of this hypothesis. Scent deposits in the territories of rabbits (*Oryctolagus cuniculus*) cause intruding conspecifics to exhibit increased alertness and readiness to flee (Mykytowycz 1968). In wolves (*Canis lupus*), each pack's territory is marked with urine and when neighbouring packs encounter the alien scent, they increase their own rate of scent marking so that concentrations of marks accumulate along the borders of each territory (Peters & Mech 1975). The critical evidence needed to show that scent deposits act as a keep out signal is that they cause aversion by intruders. In the wolf study, there was some anecdotal evidence that this was the case and in mice (*Mus musculus*, Jones & Nowell 1974) and foxes (*Vulpes vulpes*, Macdonald 1977) males can be inhibited from trespassing onto a territory by the smell of the owner's urine scent alone.

In some recent, exciting work Hölldobler has discovered a complex system of pheromones involved in the territorial behaviour of ants. Both in the honeypot ant (*Myrmecocystus mimicus*) and in the African weaver ant (*Oecophylla longinoda*), workers assemble nest mates to help in territory defence against intruders, by laying down recruitment pheromones (Hölldobler 1976a, Hölldobler & Wilson 1977a). *Oecophylla* lays down odour trails of pheromone to recruit nest mates to newly discovered areas within the territory (Hölldobler & Wilson 1977a). Workers of *Oecophylla* also use a pheromone to advertise their territories and deter invasion by alien workers. Hölldobler and Wilson (1977b) showed that this territorial pheromone served as a keep out signal even in the absence of the ants themselves. In an experiment, workers were introduced onto an arena, half of which was marked with their own colony pheromone and half marked with scent droplets from an alien colony of the same species. The exploring ants inspected the droplets with their antennae, where chemoreceptors are located, and showed much more caution and initial aversion to the area marked by alien scent.

11.4.2 *Obeying keep-out signals*

Once we have been able to identify the mechanism by which spacing out is maintained, we are still left with a difficult problem. Why do intruders obey the keep out signals? On perceiving, through scent or song, that a territory is occupied why doesn't an intruder persist and fight in an attempt to oust the owner, instead of beating a cowardly retreat?

In nature most contests for resources, such as territories, are ritualised and almost always one contestant retreats before there is a serious fight (see also 10.3). Maynard Smith and Price (1973) showed that we need not resort to 'good for the species' arguments (Lorenz 1966) to explain why individuals refrain from serious fighting. Conventional settlement of contests can evolve through individual selection.

In effect, when an individual competes for a territory it has to ask, do I gain more, in terms of fitness, by fighting or retreating? Where its only chance of securing a vital resource, such as a mate or a territory, depends on its winning one contest, then it should fight and risk the costs that this may entail. There are in fact very few examples where animals habitually go in for such a strategy of all out war. This is presumably because there is a high risk of injury in a serious fight (e.g. Geist 1974a), whereas the disadvantage of an immediate retreat is offset by the chances of future contests for the resource.

What cues should be used for a conventional settlement of a contest for a territory? Some contests may provide the individuals with a measure of the resource holding potential or absolute fighting ability of the combatants (Parker 1974c). In such cases we may expect the stronger animal to win and the weaker animal to retreat without escalation (see 10.3). For example, the honeypot ants (*Myrmecocystus mimicus*) only rarely engage in actual physical combat; when they do the result is usually fatal for both opponents. Instead, when territorial borders are challenged, neighbouring colonies recruit their worker force and engage in ritualised tournaments at the territory boundary, involving stereotyped displays in which few ants get injured. When one colony can summon many more displaying ants than the other to the tournament arena, then the weaker colony gives way without a serious contest (Hölldobler 1976a). In the sheet-web spider *Linyphia triangularis*, contests for territories (a female's web) are usually won by the males with the larger chelicerae (Rovner 1968).

However, in theory, an arbitrary rule that is totally irrelevant to the fighting ability of the contestants, can be used to settle a contest quickly (Maynard Smith & Parker 1976, see 10.3.2). An example where contests for territories are settled by an arbitrary rule is in the speckled wood butterfly (*Pararge aegeria*); the winner of the contest is just simply whoever happens to be the resident on the territory at the time (Davies 1978b, see 10.3.2).

11.5 Conclusion: on the art of asking precise questions

I want to end this chapter by emphasising the importance of asking clear and precise questions about territorial behaviour. As an example I will briefly summarise the questions and answers involved in the great tit studies, frequently referred to in this chapter.

(1) How is a territory recognised? In the breeding season great tit nests are spaced out more than would be expected from random occupation of available sites. The area occupied by individuals or pairs can be usefully described as a territory.

(2) What are the functions of the territory? Territorial defence involves both costs and benefits. One benefit of spacing out in the great tit is a decreased chance of nest predation. Defence costs presumably impose an upper limit on the territory size that can be economically defended.

(3) What are the consequences of territorial behaviour? Removal experiments show that, as a consequence of spacing out, population density is limited on a local scale. However territoriality does not limit density to the same level each year because of great year to year fluctuations in territory size.

(4) What determines territory size? One factor causing variations in territory size in the great tit may simply be the variations in settlement patterns on the territories each spring. Therefore a proximate factor determining territory size (settlement pattern) can be different from a selective pressure that promotes it in the first place (predation).

(5) How are territories defended? Song acts as a keep out signal. The reason intruders obey the signal is probably that a serious fight for a territory is costly and an intruder can benefit more by retreating and waiting for a vacancy or by breeding elsewhere.

Chapter 12
Habitat Selection

LINDA PARTRIDGE

12.1 Introduction

Habitat selection is the choice of a type of place in which to live. It is obvious that all animals are found in a more or less restricted range of environments. It is usually much less obvious how these restricted distributions are achieved in practice, but in some instances choice has been shown to be involved.

The word 'habitat' has been used in a variety of senses. One usage is interchangeable with that of the word 'biotope', meaning a particular community. For example, certain biotopes have been described as 'woodland habitat' or 'savanna habitat'. In a second sense habitat has been used to mean the distribution of animals in relation to physical and chemical variables. (Whittaker *et al.* 1973, but see Kulesza 1975 and Whittaker *et al.* 1975). I will use the word to describe the conglomerate of physical and biotic factors which together make up the sort of place in which an animal lives. For example, these factors could include shade, humidity, prey items and potential nesting sites.

Habitats may be described with different degrees of precision. For example, an animal species may be restricted to a particular plant, and also to a particular part of that plant. In the latter case, the term microhabitat may be used to cover the finer distribution. The habitat of an animal may vary with time, for example seasonally or with the age of the animal.

Habitat selection has profound consequences. The choice of a particular habitat has the effect of placing the animal in a particular environment. Life in that environment may modify the phenotype of the animal in a variety of ways. A habitat imposes a particular selective regime on the animals within it. Hence differences in habitat choice within a species may lead to local differences in gene frequency. Differences in habitat selection between species may increase the genetic differences between them by placing members of the different species under different selection pressures. In addition, life in a particular environment provides an animal with particular kinds of experience from which it may learn. Such learning may increase genetic differences between species, or give rise to phenotypic differences between members of the same species. The altered phenotype resulting from learning may be better adapted to its environment, and this may reduce the effects of selection on its genotype. Thus choice of habitat is a prime mover in determining the way in which an animal's heredity will interact with its environment to produce its eventual phenotype, and a gene affecting habitat preference will inevitably affect the phenotype of its bearer in other respects.

It must be evident from the foregoing remarks that habitat selection has ramifications in many directions. There have been many studies of both the processes involved in habitat selection, and its ecological and evolutionary consequences. This chapter will describe some cases where habitat preferences have been demonstrated (12.2), some cues used in habitat choice (12.3), ways in which habitat preferences interact with other factors to produce distributions in nature (12.4), the development of habitat preferences (12.5), their adaptive significance (12.6), habitat specialists and generalists (12.7) and the role of habitat selection in speciation (12.8).

12.2 The demonstration of habitat preferences

Geophysical events and the powers of dispersal of a species will set the ultimate limits to its geographical range. These limits will be modified by the species' competitive interactions with other species, which will also affect local distribution within the geographical range. It is only within these constraints that choice of habitat can operate. It clearly follows that habitat *distribution* may not follow directly from habitat *preference* or *choice*; inter- or intra-specific competition can exclude animals from preferred habitats and force them into less suitable areas (see 12.4). An animal species might be found in a particular habitat because individuals which happened by chance to arrive in that habitat survived; others which went elsewhere did not. Such a process certainly occurs in plants, but it is not known to occur in animals, because all animals can move and hence select a habitat at some stage in their life history. It is however possible in some groups, for example marine animals with a sessile adult stage, that the powers of locomotion of the larvae are sometimes not sufficient to bring them into a suitable habitat for settling.

12.2.1 *Quantitative descriptions*

Ecologists are often interested in the extent to which habitat preferences are involved in bringing about the habitat distributions which are observed in nature. The starting point for such investigations is usually an attempt to correlate animal densities with environmental variables. Often the investigation is concerned with the contribution of habitat differences to ecological segregation between potentially competing species. Ecologists accept that there must be selection for divergence between the niches of individuals of competing species which inhabit the same area. Some form of ecological segregation between potentially competing species is therefore expected, and one way in which this can be achieved is by habitat differences.

There have been an enormous number of studies of this problem, many of them on birds. For example, James (1971), Kikkawa (1968) and MacArthur *et al.* (1962) have all used multivariate statistical approaches to describe the distributions of bird species in relation to the nature of the vegetation. Often in such analyses some of the habitat variables which are measured are correlated with one another (Green 1971). Green advocated the use of multiple discriminant analysis to obtain a more economical description of habitat differences, and applied this technique to habitat description for ten species of bivalve molluscs. In all the above examples there were habitat differences between the species, although the nature and extent of these sometimes varied seasonally. It is, of course, logically impossible to show that two species occur in identical habitats; the consideration of one more habitat variable which had not been included could reveal some segregation.

Different combinations of a small number of independent habitat variables may be sufficient to segregate a much larger number of species. For example, seven species of *Anolis* lizards in Puerto Rico can be split into three groups on the basis of their occurrence on perches of a certain diameter and height above the ground. Within each of these three groups the structural habitat is very similar, but there are large differences within groups in the degree of shade in the habitat (see Fig. 12.1). These latter differences appear to reflect temperature requirements, which result in some species living in shade at higher altitudes and others in less shady areas at lower altitudes. At each altitude, the different species occur on different sorts of perches (Rand 1964). Choice does seem to be involved in producing these habitat differences since Kiester *et al.* (1975) have experimentally demonstrated choice in the field, but there may also be exclusion between the species.

It is debatable whether the time of day at which an animal is active should be regarded as part of its habitat, but there are instances

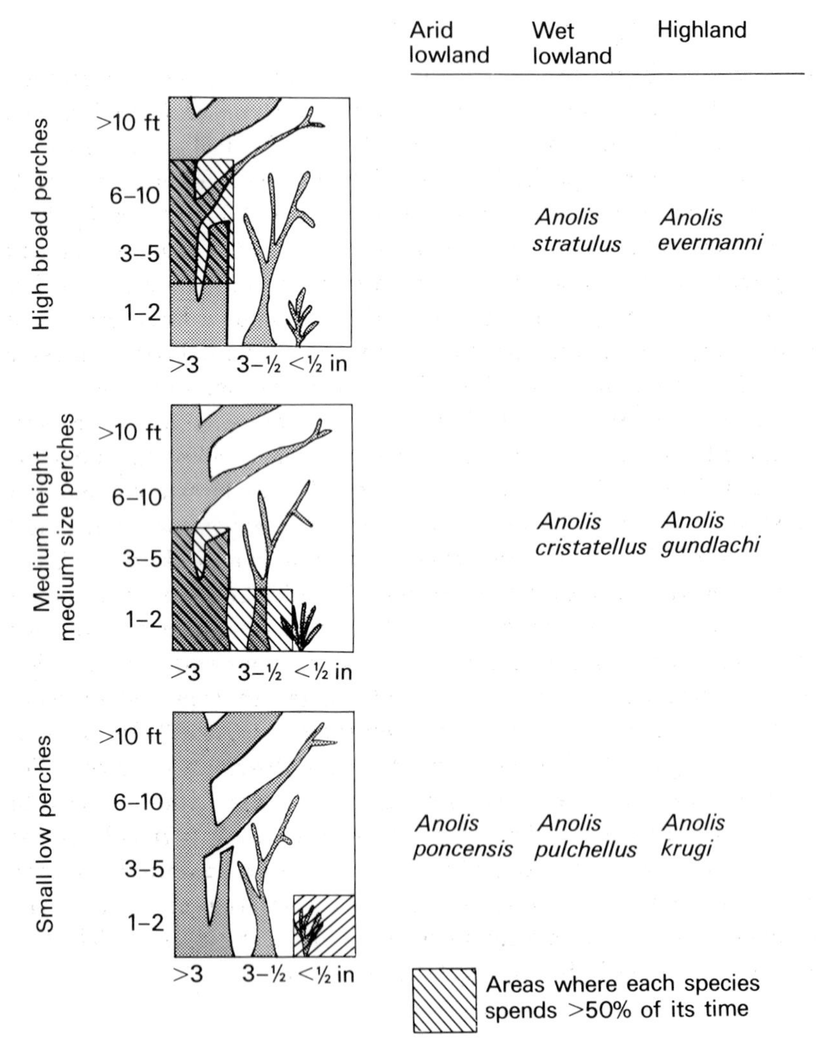

Fig. 12.1. Ecological segregation between seven species of *Anolis* lizards by a combination of altitude and the nature of the perches used. These differences are probably mediated by the shade requirements of the seven species. Modified after Rand (1964).

where differences in time of activity may contribute to ecological segregation. For example, many habitats support both nocturnal and diurnal predatory birds which overlap considerably in diet (Cody 1974b). Different individual lizards (*Sceloporus jarrovi*) are active at different times. As a result defended territories overlap spatially because the animals concerned do not interfere with one another (Simon & Middendorf 1976).

12.2.2 *Field experiments*

In a few cases it has been possible to demonstrate habitat preferences in the wild, and to show that these correlate with natural distributions. There are practical problems involved in such studies. In order to demonstrate habitat preferences *per se* in the wild, it is necessary to remove potential competitors of the same and different species from the area where the tests are to be done, and then to introduce animals singly and note where they spend most of their time. It is practical to do this only for a few species, and such experiments have been done mainly using small mammals. Most of the experiments have involved a study of the differences in habitat between more than one species.

Douglass (1976) has studied the habitat preferences of two species of voles of the genus *Microtus* in the wild. He used enclosures in natural vegetation from which other small mammals had been removed. The two species, *Microtus montanus* and *M. pennsylvanicus* spent most time in vegetation of different species composition, and the different types of vegetation were similar to those in which the same two species were found in completely natural conditions (Hodgson 1972). Douglass also found evidence for competitive exclusion between the two species, when they were together. His experiments suggest that the preferences of the two species and competition between them are both likely to be involved in bringing about their different habitat distributions in the wild.

Differences in habitat preferences can also occur within species. In the United States the prairie deer mouse *Peromyscus maniculatus* has segregated into two races, one of which is restricted to prairie habitats, the other to forest. In enclosures in natural vegetation members of the two races are found more often in the type of vegetation in which they occur more frequently in nature (Wecker 1963).

12.2.3 *Laboratory experiments*

Where experiments in the wild are not feasible, the 'next best' approach is to provide the animals with a choice between samples of semi-natural habitat in the laboratory. This approach can be misleading because it may be difficult to be certain that the habitat samples contain the cues to which the animal responds in the wild. Thus an absence of preference in the laboratory does not necessarily indicate that there is no preference in the wild situation.

A laboratory experiment has been used to investigate foliage preference in birds. In the wild, blue tits (*Parus caeruleus*) occur mainly in broad-leaved trees while coal tits (*P. ater*) are found mainly in conifers (Snow 1954, Gibb 1954, 1960). Wild blue and coal tits kept in an aviary and presented with a choice between branches from conifer and broad-leaved trees showed a difference in preferences: the blue tits spent more time in broad-leaved branches than did coal tits, while coal tits spent more time in conifer branches than did blue tits (Gibb 1957) (see Fig. 12.2a). Thus the difference between the species was similar to that found in the wild. Gibb's experiments were done with mixed-species groups of birds, so that the effects of competition cannot

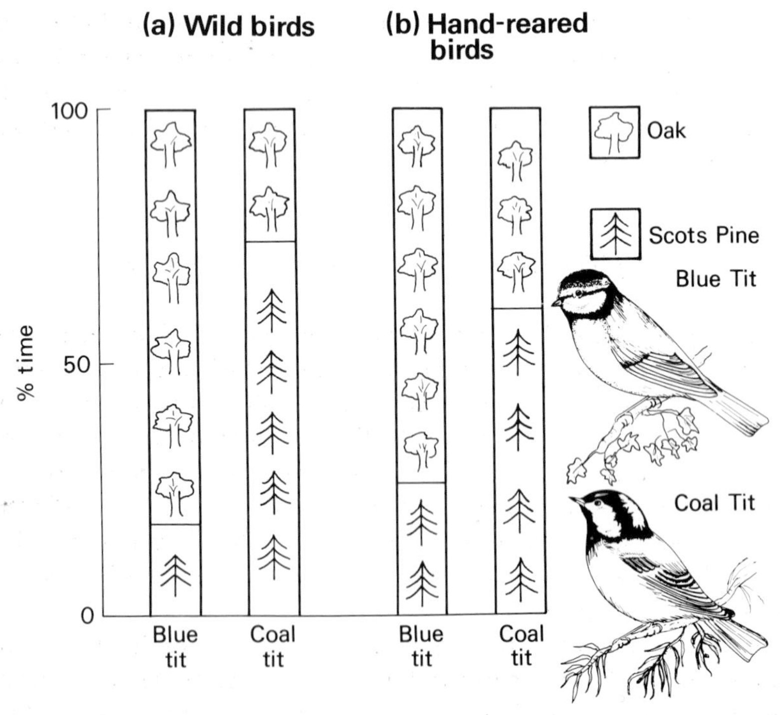

Fig. 12.2. Choice of oak and Scots pine branches by (a) wild (modified after Gibb 1957) and (b) hand-reared (modified after Partridge 1974) blue and coal tits. Blue tits show a higher preference for oak than do coal tits.

be ruled out, but small numbers of birds were involved and abundant vegetation was provided.

Similar experiments have been done to investigate intraspecific differences in habitat preference. For example, investigating substrate preference in the polychaete worm *Spirorbis borealis*, MacKay and Doyle (1978) have shown that populations in sheltered areas settle on the alga *Ascophyllum nodosum*, while in more exposed areas they occur more often on *Fucus serratus*. Larvae taken from the two populations and tested for the probability of settling on the algae in the laboratory show similar differences to those in the wild: members of each group are more likely to settle than animals from the other population on the alga on which they occur more frequently in the wild.

If preferences in the laboratory do not correlate with distribution in the wild, the discrepancy may point to different mortality rates in the different environments in the wild. For example in the laboratory the fingernail clam (*Sphaerium transversum*) prefers mud to sandy mud or sand. However, in the wild the animals do not differ in density between the three environments (Gale 1971). Gale concluded that parasitism, predation or competition was over-riding the effects of preference in regulating clam distribution in nature, but it is also possible that the clams use some cue which was absent in the laboratory experiments when chosing a habitat.

12.3 The cues used in choosing a habitat

Laboratory studies of the ways in which animals distribute themselves in relation to environmental variables tell us which variables could be important in nature, but not which cues the animal actually uses. It is, unfortunately, difficult to control all other variables while manipulating the one under study in the field. Studies of this problem therefore tend to be done in the laboratory, and rely on a correlation between the animals' behaviour in the laboratory and their distribution in the field to provide evidence that the variable under study is actually used by the animal in nature.

Studies of temperature and humidity preferences in insects are common, although preferences in the laboratory often do not correlate with distributions in the field (Wigglesworth 1965). One instance where such a correlation has been found concerns the temperature preferences of the larvae of three species of dung-breeding flies. In the laboratory the larvae of the three species prefer the temperature range which accords with that of their normal breeding places (Thomsen & Thomsen 1937).

There are instances where environmental variables have been shown to interact with each other in producing animal distributions. For example, the temperature preferences of three species of *Drosophila* are affected by humidity (Prince & Parsons 1977) and the depth preference of the reef fish *Acanthurus triostegus* is affected by the cover availability (Sale 1968). The functional significance of such interactions will be discussed in section 12.6.

The proximate environmental cues to which animals respond when selecting a habitat may not be the same as the ultimate factors which have brought about the evolution of the response. For example, blue tits prefer to live in oak woodland where they are best able to feed. However they cannot be assessing the food itself when deciding where to breed because territories are established before the leaves and caterpillars have appeared. The proximate cue could be something like the shape of the oak trees, or some other cue that distinguishes them from other sorts of tree.

The complexity of cues used in selecting a habitat may be related to the way in which habitat preference develops in an individual. Differences between congeneric species in the cues used in selecting habitats are often simple. For example, woodlarks require a few standing trees in their habitat while skylarks will only live in treeless grassland. Lack (1971) has suggested that such simple differences are probably controlled genetically. It may be difficult to control preferences for complex stimuli genetically. This problem is discussed further in section 12.6.

12.4 Factors which interact with habitat preferences to produce natural distributions

Competition within or between species, and the action of predators or parasites could shift the distribution of animals away from that expected on the basis of their habitat preferences.

The present habitat preferences of a species may have resulted partly from previous competition with other species. The arrival of a new species with competitive superiority in part of the habitat space of the established species would lead to selection in favour of those individuals of the established species who preferred that part of the habitat space not occupied by the new species. There would therefore be net selection for restriction of the habitat preference of the established species, bringing about less overlap in habitat preference with the new species.

12.4.1 *Population density and intraspecific competition*

The effects of competition within species can be complicated. Depending on the availability of preferred and less-preferred habitats in relation to population density a number of situations can occur (Brown 1975b, von Haartman 1971). If population density is low, the preferred habitat for a species may not be fully utilised, so that other habitats will not be occupied. It is difficult to demonstrate this situation experimentally, because a species may be absent from certain habitats because these are not acceptable at all, rather than because population density is too low to necessitate their use. If population density varies from year to year, a narrower range of habitats may be occupied in years of low population density, providing some indication that the habitats abandoned in years of lower density are less preferred. Such a situation is documented for the brown lemming *Lemmus trimucronatus* in Alaska. This species shows great cyclic fluctuations in numbers. In years of peak numbers the lemmings occur in all terrestrial habitats. In years between peaks, they occur chiefly or solely in areas of optimal habitat (Pitelka 1973). Chaffinches and great tits show a similar pattern (Glas 1960, Kluyver & Tinbergen 1953). The yearly fluctuations in numbers for both species are greater in pine woods than in mixed woods, suggesting that pine woods may be occupied to different extents depending on population density, whereas the preferred mixed woods are always fully occupied. Chaffinches tend to occupy mixed woods before conifer woods in the spring which again suggests that mixed woods are preferred. As population density increases further, there may come into existence a surplus population which is prevented from breeding because all suitable breeding sites are occupied (see 11.3).

If the population density is high, and many less-preferred habitats are available, then a large proportion of the population may occupy these less-preferred areas. In general, if preferred habitats are also optimal, the density of animals will be higher in more suitable habitats (see 12.6.7)

12.4.2 *Interspecific competition*

Indirect evidence

Competitive interactions with other species can also affect habitat distributions. Contiguous allopatry, where two species have abutting but non-overlapping geographical ranges, is usually assumed to indicate that the species involved are too similar in ecology for coexistence in the same area to be possible. In a few cases there are some differences in habitat between contiguously allopatric species, usually associated with differences between the environments in the two geographic ranges. Such a situation is common in frugivorous birds (Gilliard 1969).

Habitat differences are often found in the area of overlap between closely-related species. In a few cases these differences have been demonstrated to be brought about by competitive exclusion. This is suggested by habitat displacement, where the habitat of one or both species is modified in the area of sympatry in such a way as to reduce the overlap in habitat which would occur if they retained the habitats used in allopatry. In section 12.8 I will describe such a situation in the Canary Island chaffinches. A similar pattern is found in voles of the genus *Microtus* (Murie 1971).

Indirect evidence for the effect of competition on habitat distribution comes from the phenomenon of competitive release on islands. In general island faunas are impoverished in comparison with those on nearby mainlands, and several studies have shown that animals on islands are broader in their habitat use (Cody 1974b, Van Valen 1965). However, caution must be used in the interpretation of such data because the nature of the habitats on islands may differ from that on the mainland (Grant 1972).

Werner (1977) has also found indirect evidence of the action of interspecific competition in producing habitat segregation between three species of sunfish which co-exist in lakes. These three species take prey items of different average size, and by knowing the distribution of prey sizes in the lakes where the fish live Werner calculated the extent to which the three species compete for food. Using the species packing theory developed by May and MacArthur (1972) and Roughgarden (1974), Werner could show that two species, the largemouth bass (*Micropterus salmoides*) and the bluegill (*Lepomis macrochirus*) would be able to co-exist, while the third species, the green sunfish (*L. cyanellus*) which takes food intermediate in size to the other two species could not. Werner then quantified habitat use in the three species, and could show that the bass and bluegill completely overlapped in the habitats which they utilised, while the green sunfish fed in a different habitat in very shallow water near the shoreline.

Direct evidence

Experiments involving removal of one species provide more direct evidence for the action of competitive exclusion. There have been studies using this technique including a particularly elegant series of investigations of the habitat distributions of small mammals. In general, voles of the genera *Microtus* and *Clethrionomys* tend to occupy different habitats; the former live in grassland, the latter in woodland. This habitat segregation is maintained on islands where both genera are represented by at least one species, but on islands where only a single vole species is present it often also occupies the habitat of the absent genus (e.g. Cameron 1964). A field experiment has shown that both *C. gapperi* (Grant 1969) and the prairie deermouse (*Peromyscus maniculatus*) (Grant 1971) are inhibited from making use of grassland by *M. pennsylvanicus*. In enclosures from which *M. pennsylvanicus* has been removed, *C. gapperi* or the deermouse will move into grassland. Using the same technique, Morris (1969) has shown that *C. gapperi* inhibits *M. pennsylvanicus* from entering woodland. Grant (1970) has shown in laboratory experiments that agonistic encounters between individuals are associated with the competitive exclusion. This kind of interspecific territoriality between two species, as opposed to overlap with microhabitat differences, has been suggested to occur more often (a) in structurally simple environments, (b) if two species feed in very similar strata in complex vegetation, and (c) if the presence of many other species prevents the two species from diverging ecologically (Orians & Willson 1964, see 11.2.5).

Habitat segregation via competitive exclusion has also been demonstrated in barnacles. In general the barnacle *Chthamalus stellatus* is found higher up rocky shores than is *Balanus balanoides*. By removing *Balanus* Connell (1961) was able to show that the *Chthamalus* is excluded from the lower shore, where it will survive in abundance if *Balanus* is absent. *Balanus* individuals smother, crush or undercut *Chthamalus* as they grow in size. The intensity of this interspecific competition is reduced by the action of the whelk *Thais lapillus* which preys on both barnacle species, thereby lowering the population size of each.

Competitive exclusion may cause animals to use different parts of a three-dimensional environment. Morse (1970) has shown that in mixed-species flocks of birds the foraging space of a species which happens to be absent from a flock is taken over by the species which normally forage adjacent to it. Jenssen (1973) has shown that the lizard *Anolis opalinus* perches higher in the vegetation in the presence of *A. linatopus*, again suggesting competitive exclusion. The effects of competition on habitat used are discussed further in sections 12.7 and 12.8.

Factors other than competition may alter animal distributions away from those expected on the basis of habitat preferences. Natural selection of any kind which acts differently within two habitats could have such an effect. However, if habitat selection is adaptive (see 12.6), selective loss should be higher in less-preferred habitats, so that its effects on distributions should enhance those of habitat selection.

12.5 The development of habitat preferences

Studies of the development of habitat preferences have taken the form of demonstrations either that differences in preferences are partly genetic in origin or that preferences are modifiable by early experience of particular environments.

If two animals are reared in identical environments and differ in their habitat preferences, then the difference must be hereditary. If individual blue tits (*Parus caeruleus*) and coal tits (*P. ater*) are reared in aviaries with no experience of vegetation and then presented with a choice between oak and scots pine branches, coal tits perch more on the pine branches than do blue tits, who perch more than the coal tits on oak branches (Partridge 1974) (see Fig. 12.2b). This difference corresponds to the habitat difference between these two species in nature, and to the difference between wild birds in aviaries (see 12.2). Such hereditary differences in habitat preference can occur within species (see 12.6.2).

In a study of habitat preferences in two races of the prairie deermouse, Wecker (1963) has shown that preference for grassland or woodland is heritable. In a further study of the effects of early experience, he showed that the preference can be reinforced by experience of the preferred habitat, but cannot be reversed by experience of the less preferred habitat.

Positive effects of early experience of a habitat on preference for that habitat have been shown in insects (Thorpe 1939, Cushing 1941), fish (Gilbert 1918), amphibians (Wiens 1970), birds (Klopfer 1963) and mammals (Anderson 1973). The functional significance of flexible habitat preference is discussed in 12.6.

Changes in habitat preference brought about by learning and cultural transmission may be responsible for changes in both habitat distribution and geographical range. For example, in Germany the mistle thrush, *Turdus viscivorus*, was originally found in conifer forests. In 1925 the birds began to breed in small groups of deciduous trees in cultivated areas and eventually in parks in cities. This population spread and increased very rapidly, almost certainly by the cultural transmission of habitat preferences to the young birds (Peitzmeier 1949, quoted in Hilden 1965).

12.6 The adaptive significance of habitat selection

To show that habitat selection is adaptive, it is necessary to demonstrate that animals choose to live in those sorts of places where they will have maximum chance of survival or reproductive success. An adaptive trait must also be in some sense inherited. As with any other phenotypic character, habitat preference could be transmitted in a number of different ways. For example, young blue tits could inherit a preference for oak trees. On the other hand they might inherit a tendency to prefer the sorts of trees to which they are exposed early in life, perhaps because they are taken there by their parents. Or they could acquire a preference for trees where they find most food or are least subject to predatory attack.

In practice it is difficult to assemble all the relevant information for any one example of habitat selection, and perhaps this is why there are so few examples where this has been done. Estimates of survival and reproductive success for one species in more than one habitat can be difficult to obtain because animals are unlikely to occur naturally in less preferred habitats if preferred ones are available. Without some basis for comparison of fitness in different habitats it is impossible to demonstrate that fitness is being increased by habitat choice. There is also a problem with comparisons within a species. If there is competition for preferred habitats, and some animals are forced into less-preferred habitats, it may be that those animals which succeed in obtaining access to preferred habitats are fitter. Any comparison between their success and that of the animals in less-preferred habitats will be confounded by this original difference in fitness. However these difficulties have been overcome in a few cases.

12.6.1 *Some examples*

Peppered moths

Studies of the peppered moth, *Biston betularia* are a striking example. This species has a melanic mutant which has become common in polluted areas, while the non-melanic form is found in unpolluted areas (Kettlewell 1958). The colour difference between the two forms is controlled by a single major locus. The two forms were placed on backgrounds of different colours in aviaries, and insectivorous birds were more likely to take moths which did not resemble the colour of the background on which they were resting (Kettlewell 1955a, 1956). Kettlewell also showed that if marked moths were released in polluted woods then the pale form was recaptured at lower frequency than expected on the basis of the frequency at which it was released. On the other hand, the melanic form disappeared more rapidly than expected in unpolluted woods. Kettlewell attributed these differences to different rates of predation on the two forms, the pale form being taken more often in polluted woods where lichens are absent and the tree bark therefore darker in colour than in unpolluted woods where the melanic form is more vulnerable to predation.

In most woods both pale and dark areas of bark are available for the resting moths. It would clearly be adaptive for the moths to prefer backgrounds on which they were less conspicuous. When confronted with a choice between light and dark backgrounds, the pale form chooses light backgrounds significantly more frequently than does the melanic form, although the difference is not a large one (Kettlewell 1955b, Kettlewell & Conn 1977). It is not known whether the melanic moths have a preference for dark backgrounds and the pale form for lighter backgrounds, or whether both forms are matching the background to their own colour. By painting the scales around the eyes, Sargent (1968) has ruled out the latter possibility for two different species of moths, but this information is not available for the peppered moth. If the two forms of *Biston* do have different background colour preferences, then the genes controlling the preference must be closely linked to the melanism locus: if the two characters became dissociated, lower fitness would result.

Polychaete larvae

MacKay and Doyle (1978) have examined substrate selection by larvae of the polychaete *Spirorbis borealis*. They examined the probability with which larvae settled on the algae *Ascophyllum* and *Fucus* when they were presented singly in the laboratory. Larvae produced by adults from sheltered areas were more likely to settle on *Ascophyllum* than those from exposed areas, while larvae from exposed areas were more likely to settle on *Fucus*. *S. borealis* adheres more strongly to *Fucus* than to *Ascophyllum* and especially under turbulent conditions removal from the substrate is a major source of adult mortality (Doyle 1975). *Ascophyllum* is often the only alga available in sheltered areas, while both algae occur in exposed areas. Hence in exposed areas animals are selected to prefer *Fucus* to which they are better able to adhere, while in sheltered areas they are selected to be prepared to settle on *Ascophyllum* which is the only substrate available to them. The differences between the populations in substrate preference are in part genetic, and there is genetic modification of the dominance of the genes affecting habitat preference so that in each habitat the genes for the more advantageous preference are dominant (MacKay & Doyle in prep.).

Hummingbirds

Where variation in habitat preference within a species in nature is inadequate for comparison to be made, it may be possible to manipulate the situation. One example where this has been done concerns nest site selection by two species of hummingbird. Both species tend to nest under branches and by measuring conditions in artificial styrofoam nests placed in different sites Calder (1973) has shown that a branch above the nest reduces radiative heat loss to the open sky. Convective heat loss is probably also reduced. Prevention of heat loss is particularly important to hummingbirds because of their high surface area to body volume ratio.

Drosophila

Other workers have manipulated environmental variables which interact in a biologically significant way. For example, in a study of three *Drosophila* species Prince and Parsons (1977) have shown that flies preferred lower temperatures when the air was dry than when it was saturated with water vapour. The flies also preferred lower temperatures if they had been kept in a dry atmosphere before the experiment was started. High temperature and low humidity increase the desiccating power of air, so the flies preference for lower temperatures if they were

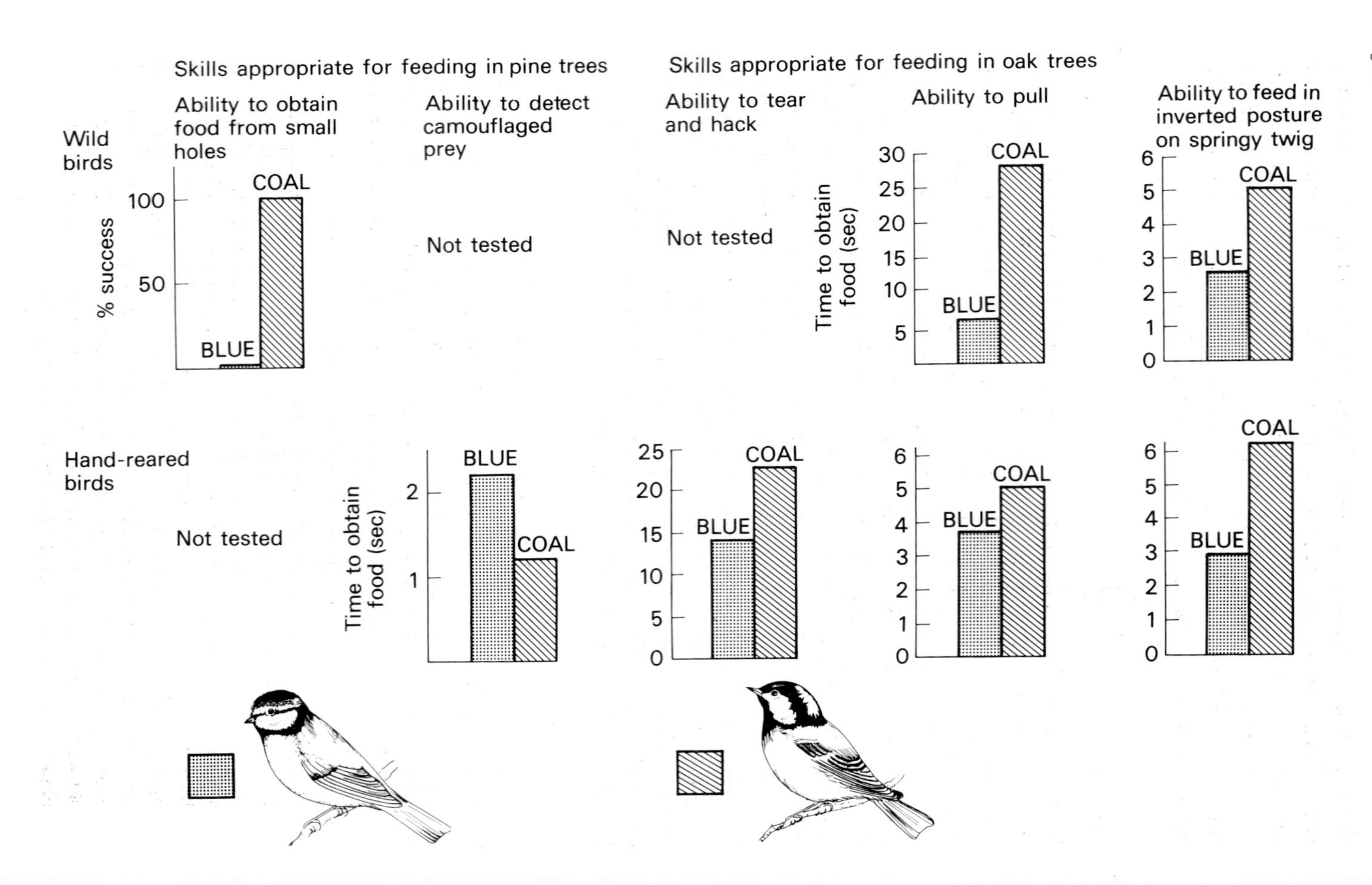
Skills appropriate for feeding in pine trees
Skills appropriate for feeding in oak trees
Ability to obtain food from small holes
Ability to detect camouflaged prey
Ability to tear and hack
Ability to pull
Ability to feed in inverted posture on springy twig
Wild birds
% success
100
50
BLUE
COAL
Not tested
Not tested
Time to obtain food (sec)
30
25
20
15
10
5
BLUE
COAL
6
5
4
3
2
1
0
BLUE
COAL
Hand-reared birds
Not tested
Time to obtain food (sec)
2
1
BLUE
COAL
25
20
15
10
5
0
BLUE
COAL
6
5
4
3
2
1
0
BLUE
COAL
6
5
4
3
2
1
0
BLUE
COAL

dried before or during the experiment was adaptive in preventing water loss. Prince and Parsons were not able to tie in the different temperature and humidity preferences of the three species with their natural environments, or with their survival and reproductive success under the different conditions, perhaps in part because of inadequate information about the microhabitats of wild *Drosophila*.

Titmice

It is also possible to use interspecific comparisons to gain some insight into the functional significance of habitat selection. Blue tits show a stronger preference than coal tits for broad-leaved as opposed to conifer branches (see 12.2, 12.5). By testing members of both species on artificial feeding tasks in the laboratory I have been able to show that the species differ in their feeding skills, and that each species is better than the other at exploiting the type of food supply which it encounters in its own habitat (Partridge 1976a) (see also 2.7). These differences appear to be hereditary (Partridge in press) (see Fig. 12.3). If blue and coal tits are potential food competitors, then such a finding indicates that habitat selection in these two species is likely to be adaptive. The question has become: 'Are blue tits better than coal tits at exploiting the kind of food supply found in broad-leaved woods'?; rather than 'Are blue tits better at exploiting the sort of food found in broad-leaved woods than in pine woods?'.

12.6.2 *How choosy should an animal be?*

It must be adaptive for an animal to choose a particular habitat in which to live. However, as pointed out, preferred habitats may be limited in availability because of competition from other animals and it may be hard for an animal to find suitable habitats when these are

Fig. 12.3. Feeding skills of wild and hand-reared blue and coal tits. Coal tits live mainly in pine trees, where the insect prey consists of small items requiring probing and acute eyesight for their exploitation. Blue tits live mainly in oak trees, where the insect prey requires forceful hacking, tearing and pulling feeding techniques, and the ability to hang upside down on small twigs and leaves. A series of artificial feeding sites were designed to test the skills of captive birds under controlled conditions in the laboratory. Note that the top left-hand histogram shows per cent success rate, while the others all show the mean time to obtain food. Based on Partridge (1976a and in press).

rare. These problems raise the question of how 'choosy' animals should be when selecting a habitat. There may be a cost associated with waiting a long time for the most preferred habitat. This may be why the planktonic larvae of *S. borealis* from sheltered areas settle more rapidly than those from exposed areas. If predation while in plankton is a source of mortality then rapid settling may be an advantage in sheltered areas where only one algal substrate is available. However, the advantage of such behaviour may be outweighed by the advantage of waiting until a preferred substrate is found in the more exposed areas. In a theoretical analysis, Doyle (1975) has shown that it would be adaptive for marine larvae to have an exploratory phase before selecting a habitat provided that the average increase in fitness because of information about potential habitats gathered during the exploratory period outweighs the chance of death in the plankton. Marine plankton often become less choosy as the exploratory phase proceeds (e.g. Knight-Jones 1953). Doyle has also shown that this behaviour would be adaptive in an environment where a mixture of habitats is available, and the animal runs the risk of failing to settle at all if its acceptance of sub-optimal habitats is too low, but also runs the risk of failing to settle on the best available habitat if its acceptance of suboptimal ones is too high. Levins (1968) considered a case where an animal has limited time in which to find a suitable habitat, after which time it dies. There are two usable habitats which occur at different densities, with different probabilities of survival connected with them. The conclusions from the analysis were that it becomes worth opting for a less good site if searching ability is low, or if the difference between more and less suitable sites is small.

12.6.3 *Flexibility in habitat preference*

Habitat selection may remain flexible after initial settling. Migratory birds often select different habitats in their winter and summer ranges (Lack 1971), and there are instances of individual animals selecting different habitats in different years (e.g. von Haartman 1949). To what extent such differences are controlled by choice and by competitive exclusion is unknown.

In a real population, difference between fitnesses in the two habitats will often depend partly on population density, and sometimes also on the frequency of genes affecting habitat preference. As a good environment becomes more crowded, a new arrival might fare better by opting to settle in the less good, but also less crowded environment (Fretwell

& Lucas 1970) (see Fig. 12.4). As mentioned earlier in this section, to demonstrate that habitat selection is adaptive, it may be necessary to compare fitness in more and less preferred habitats. In fact, the effects of competition may render such a comparison invalid, because the presence of other animals may lower fitness in the preferred habitat to the level of fitness in the less crowded less preferred one (see also 11.3).

In section 12.5 several examples were described where an animal increases its preference for a habitat to which it is exposed early in life. There are a variety of reasons why such behaviour might be

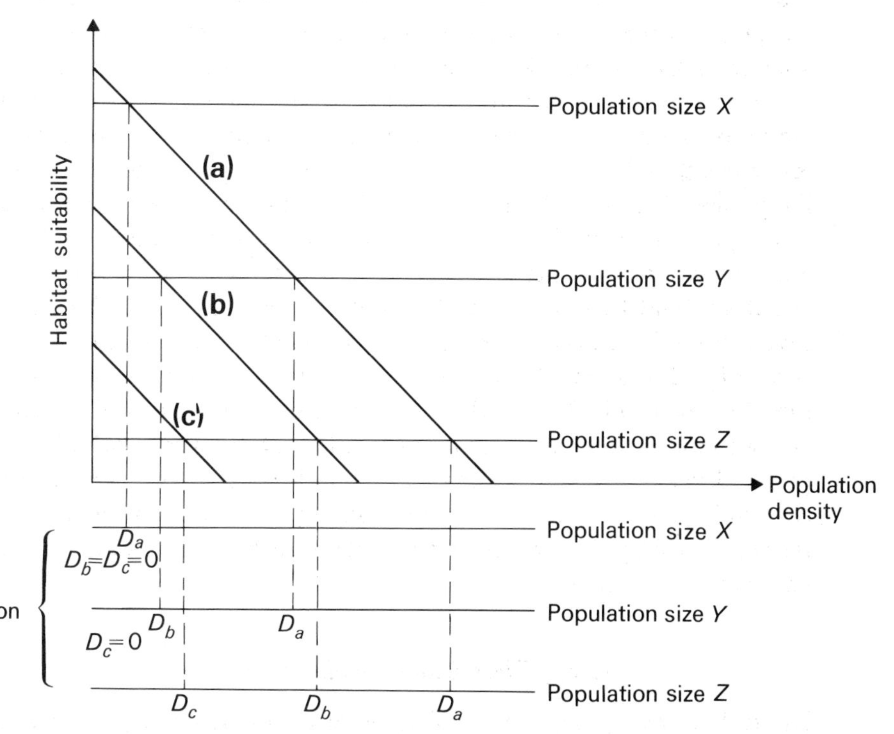

Fig. 12.4. The effect of population density on habitat suitability and hence on habitat distribution. Three habitats are shown, (a) with highest suitability, (b) intermediate and (c) with lowest suitability. Horizontal lines representing three total population sizes, $x<y<z$, are drawn, each intersecting the suitability/density lines at equal suitabilities. Every animal in the population has an equally suitable home area, low habitat suitability being compensated for by an increase in the area per animal. At the three population sizes considered, the density of animals in the three habitats are shown on the habitat distribution axes, below the graph. Da = density in habitat (a), Db = density in habitat (b) and Dc = density in habitat (c). Modified after Fretwell and Lucas (1970).

adaptive. In species which only occupy one habitat the young are automatically exposed to the appropriate habitat. If it is easier for preferences for complex stimuli to be learnt than to be genetically determined, then the young of such species should learn to prefer the habitat in which they are raised.

However, in species which live in a variety of habitats which differ in their suitability, a young animal may grow up in a sub-optimal habitat. The animal should not acquire a preference for such a habitat, but rather should live in and tolerate a suboptimal habitat while retaining preference for more appropriate habitats. However, there is evidence that one animal at least does not behave in this way. Great tits nesting in hedges have lower reproductive success than those in woodland. Despite this, adult hedgerow birds rarely move into woodland territories which are vacated, while juveniles will readily do so (Krebs 1971). It appears that the experience of breeding is associated with an increase in preference for the habitat where breeding occurred. It may be that successful breeding *per se* is the important cue. This may be why many young birds from species occupying a wide range of habitats tend to nest in similar habitats to their parents (Hilden 1965); the fact that they themselves were successfully raised may indicate that the area was suitable, although other areas might have been even more suitable.

However, there is another reason why animals might learn to prefer an originally sub-optimal habitat, although it does not seem to apply to great tits. During its early experience a young animal may learn the location of resources and resting sites, how to catch prey and other useful skills. If it starts life in a sub-optimal habitat the animal may acquire these skills, and therefore increase its fitness in the sub-optimal habitat. As a result, this habitat may eventually cease to be sub-optimal for that particular animal, because the animal is now fitter than it would be if it moved to a habitat that was optimal before learning of skills started. Whether or not it would be adaptive for the animal to remain indefinitely would depend on the level of skill acquired, the speed with which new skills could be acquired, and the difference in fitness of the animal in the two habitats.

If habitat quality varies from time to time and place to place, it may be adaptive for a young animal to sample the available habitats and form a preference for the one where it 'does best', perhaps in terms of acquiring food or some other factor. Hence it might be adaptive for animals living in a wide variety of habitats to learn their habitat preference. This phenomenon could explain results such as Wecker's

(1963) where the animal's habitat preference could be increased only by experience with its usual habitat (see 12.5). It could be that the animal 'did well' in that habitat and less well in the one for which it did not acquire a preference. It is hard to see why an animal should have a predisposition to acquire a habitat preference for a certain sort of environment, since the same effect would be achieved by having a genetically determined preference. However, it may be that heredity dictates the broad habitat preferences, but the details are filled in by learning after the animal has sampled the available habitats. Such a mechanism might be valuable where habitat quality varied with time.

12.7 Specialists and generalists

Several of the examples described above provide examples where animals are better suited to the habitats which they prefer. In the absence of competitors a more generalist habitat use might have been possible for some of these animals. For example, blue tits can and do survive in conifer woods, and they might be more numerous in such woods were it not for competition from the better adapted coal tit. In other cases competition may be less important in bringing about restriction to particular environments. For example, in the case of the peppered moth the typical form may be completely prevented from surviving in very heavily polluted areas because of the action of predators. In the extreme case the animals' physiology may bring about the restriction: most fish species cannot obtain oxygen from air.

The breadth of habitat used by a species is of course determined by the behaviour of its constituent members. Van Valen and Grant (1970) and Roughgarden (1972) have pointed out that a species may occupy a wide range of environments either because individual animals are generalist in their habitat use, or because the species is made up of a number of divergent individual animals, each specialised in its habitat use. In the latter case Van Valen (1965) has suggested that if different sections of the population or species are each adapted to their specialised habitat then polymorphism will be promoted in populations or species with broad habitat use. This idea has become known as the niche variation hypothesis, which states that species with broader ecological niches should be more variable than those with narrow niches because of the action of disruptive selection. Some studies support this contention (Rothstein 1973, Grant *et al.* 1976) while others apparently do not

(Willson 1969, Soulé and Stewart 1970). There are of course formidable problems in making an adequate test of the hypothesis. It would be necessary to compare members of the same species from different areas or of different species, to demonstrate that habitat use was wider in some area or species, and then to show that some character of importance in exploiting the habitat was more variable in the areas or species where the habitat use was broader.

There is one beautiful example where these rigorous demands have been met. Grant *et al.* (1976) have compared the medium billed ground finch (*Geospiza fortis*) on the Galapagos islands of Daphne and Santa Cruz. Bill size is more variable on Santa Cruz than on Daphne, and associated with this the finches occupy a much wider range of habitats on Santa Cruz. Grant *et al.* were able to show that on Santa Cruz shallow billed birds tended to inhabit parkland while deep billed forms lived in woodland. The deep billed birds were better at cracking large seeds than the small billed birds, and the latter cracked small seeds more efficiently. There was a higher proportion of large seeds in the woodland habitat than in the parkland (see also 2.7).

Many ecologists believe that environments with high temporal heterogeneity will select for inhabitants with broad ecological niches (e.g. Levins 1968). Some evidence consistent with this view comes from the work of McDonald and Ayala (1974) who have shown that *Drosophila* populations kept in environments with high temporal heterogeneity in such factors as light, temperature and food show higher average heterozygosity than animals in less variable environments.

It has often been suggested that animals in the equable tropics face more constant environments than temperate animals. If this is true it follows that tropical animals should have narrower niches than their temperate counterparts. The same idea has been suggested for mainland as compared with island animals. These possibilities have been tested with respect to habitat preferences by Sheppard *et al.* (1968) and Klopfer (1967). Their results were inconclusive, and this aspect of the subject needs further study.

Soulé and Stewart (1970) believe that the time for which a community has existed will be an important determinant of the niche width of the inhabitants, and that the inhabitants of old communities will have narrower niches than animals in younger communities. This is, they suggest, because time is needed for speciation and as new species are produced and species diversity increases, the niches of individual species will become narrower.

12.8 Habitat selection and speciation

The involvement of habitat selection in speciation is a controversial subject. There is a widespread belief that most speciation is initiated in allopatry, and that in such a process habitat selection plays a fairly minor role, only after the speciating populations have come into contact. There has been argument about whether genetic divergence and speciation can occur in sympatry, and most of this argument has centred around the nature of the barrier to gene flow between the speciating populations. Differences in habitat preference within a species would provide such a barrier. Social behaviour can also act as a barrier to gene flow (Anderson 1964).

In general, the facts concerning the distribution and ecology of closely related species are consistent with both allopatric and sympatric speciation. In most cases where sympatric speciation is suspected it is impossible to prove that geographical isolation has not occurred, so that allopatric speciation can rarely be ruled out. However, sympatric speciation via habitat differences has been implicated in theoretical studies and in a few cases in nature. The theoretical possibility of sympatric speciation has been supported by Fisher (1930), Thorpe (1939, 1940, 1945), Mather (1955) and Thoday (1972). In all cases some form of disruptive selection is invoked.

Several genetic models have been made of disruptive selection caused by the distribution of a population between different habitats (often called niches by the authors). Such models show that if the fitness of different genotypes differs between habitats, then polymorphisms can be established with different equilibrium gene frequencies in the different habitats (Levene 1953, Deakin 1966, Levins & MacArthur 1966, Prout 1968, Bulmer 1972). Such differing gene frequencies can be maintained even in the presence of high gene flow between the different habitats. These studies are clearly related to the niche variation hypothesis of Van Valen (see 12.7).

Results consistent with these models have been produced both in laboratory experiments and in studies of wild populations. A polymorphism can be established in laboratory populations of *Drosophila melanogaster* subjected to disruptive selection for sternopleural chaeta number (Thoday & Boam 1959). There are many examples of allele frequencies differing between interbreeding groups in the wild. For example there are local differences in the frequency of allozymes in fish populations, associated with climatic variables (Mitton & Koehn 1975).

Local populations of the American eel *Anguilla rostrata* also have different allele frequencies. Eels from these populations all migrate to the same breeding area, which has led Williams *et al.* (1973) to suggest that differences between the local populations can be attributed to the selection pressures in the different areas. Care is needed in the interpretation of information of this kind; Rockwell and Cooke (1977) have shown that different local populations of snow geese (*Anser caerulescens*) have different frequencies of alleles affecting feather colour, and geese from different areas mix at the time of pair formation. However, the local differences in allele frequency are in fact maintained by assortative mating for the colour locus.

Maynard Smith (1966) and Van Valen (1965) pointed out that genes which differ in fitness from habitat to habitat will tend to become associated with genes affecting habitat loyalty, because an animal which selects a habitat in which it is fitter will leave more offspring than one which does not. A direct demonstration of this phenomenon is lacking. However, it has been shown that disruptive selection for habitat preferences can produce a response. One such study on the polychaete worm *Spirorbis borealis* has been described in section 12.6.1.

A similar result has been obtained with houseflies in the laboratory, although the genetic basis is not known. Pimentel *et al.* (1967) kept houseflies in communicating boxes in one of which there were 9 vials of fish food and 1 of banana food, and only eggs laid on banana food were allowed to develop. In the other box, there were 9 vials of banana food and 1 of fish food, and only eggs laid on fish food were allowed to develop. The result was the divergence of two sub-populations, one of which preferred to lay on fish food and the other on banana food despite 15% migration between the boxes in the latter stages of the experiment.

A new habitat may become available to part of a population through changes in the environment or through spontaneous changes in habitat preference. In such cases habitat preference may play a primary role in producing reproductive isolation between sympatric populations, as appears to have occurred in the speciation of phytophagous Diptera of the genus *Rhagoletis*. Species in one section of this genus coexist on walnut trees and show highly contrasting colour patterns which are important in reproductive isolation. In contrast, the species in another section of the genus each live on a different variety of native fruit tree and are very similar to one another in appearance. Speciation in this latter group appears to have been accompanied by a shift to a new plant family in every case. In corroboration of this, a new race (defined by its habitat preference) has originated from one of these species and inhabits apple trees which have been introduced to the United States by man; a second race attacks plums. Bush has suggested that changes in habitat (plant species) preference have caused sympatric speciation giving rise to morphologically similar species. In this group of species plant selection leads automatically to mate selection because mating occurs on the fruits and only one fly species is present so there has not been evolution of the colour characteristics usually associated with reproductive isolation between other species in the genus (Bush 1969a, b). The fact that races of *Rhagoletis* specialising on fruit trees introduced by man have definitely been produced in the absence of geographical isolation from the parent population makes Bush's hypothesis particularly plausible.

By crossing two species of monophagous flies Huettel and Bush (1972) have shown that host plant selection in these two species is controlled by a single locus. It may be, therefore, that changes in host preference in these insects involves the modification of at most a few key loci concerned with host selection. Bush (1974) has suggested that changes in host preferences have led to sympatric speciation in many obligate insect parasites of both plants and animals.

Disruptive selection is likely to be particularly effective in bringing about sympatric speciation where different host plants or animals are involved. Such habitats provide discontinuous selection pressures which are likely to affect many loci simultaneously. Most habitats differ from one another along a continuum, and are hence likely to produce less strong disruptive selection on populations which occur in more than one of them.

Habitat selection probably also plays a role during conventional geographic speciation. Hinde (1959) has pointed out that if differences in habitat preference between populations develop while they are allopatric, then when the populations come together the habitat differences may enhance reproductive isolation. Such a suggestion is consistent with the fact that hybridisation sometimes occurs in parts of species ranges where the usual habitat distinctions break down (Mayr & Gilliard 1952, Sibly 1957, Cory & Manion 1955).

Habitat differences could also be important in determining the nature of the coexistence between new species. It is generally agreed that coexistence within the same geographical range is accompanied by some ecological difference between the species. One of the ways in which this could be achieved is via differences in habitat. Some closely related species have abutting but non-overlapping geographical ranges, a situation known as contiguous allopatry. Pocket gophers in the United States often show this type of distribution (Miller 1964) and it may indicate that the ecological requirements of the species are so similar that geographical overlap is impossible because of competition between the species. Where species do overlap, a difference in habitat is often found. Lack (1971) pointed out that in European birds there is every gradation from species separated mainly by differences in range with a small area of overlap where they differ in habitat, to species separated mainly by habitat but having partly separate ranges. Where habitat differences between species are more marked in the area of overlap than outside it, there is some indication that the habitat difference may be important in allowing the geographical overlap. For example, on the Canary Islands the blue chaffinch *Fringilla teydea* occurs on the islands of Gran Canaria and Tenerife where it is found in coniferous woods. The European chaffinch *F. coelebs* also occurs on these two islands, but in broad-leaved woods. On two other Canary Islands, Gomera and La Palma, only the European chaffinch occurs and it is found in both types of woodland, suggesting that on the islands where both species are found *F. teydea* excludes *F. coelebs* from conifers (Lack & Southern 1949, pers. obs.). In some cases the habitat requirements of sympatric species may be so similar that they defend mutually exclusive territories (see 11.2.5).

In summary, speciation occurs when one part of a population becomes reproductively isolated from another. Reproductive isolation can result from geographical isolation, but it seems that habitat segregation can also be an effective mechanism. Populations distributed between more than one habitat and in consequence differing genetically may therefore represent an early stage in speciation via habitat differences.

Chapter 13
Optimal Behaviour Sequences and Decision Making

ROBIN H. McCLEERY

13.1 Introduction

The studies described so far in this book have mainly considered how a particular activity can be carried out by an animal in the best way, given the constraints of the environment and the animal's capabilities. The best is normally taken to mean the most successful in terms of the ostensible function of the activity; the best feeding strategy is that which maximises the net rate of energy uptake, of courtship that which maximises the number of successful matings and so on. This chapter is a review of some attempts to study the more general question of how an animal's choices between the different activities of which it is capable should be made to produce the overall sequence of behaviour best adapted to the environment. It is not an exhaustive survey, partly because there is an explosive growth of interest in this field at the time of writing, and partly because it seems more useful to put across the fundamental ideas with the aid of a few rather fully described examples than to skate over the surface of a large number of different studies.

The problem is most easily understood by considering situations where conflicting demands are being made on an animal. Consider a herring gull that is incubating its eggs one sunny spring morning. At any given moment its problem is to decide whether to go on sitting on its eggs or whether to do something else, e.g. go off to feed or drink, or to get up and chase an intruder off the territory. The relative merits of these options depend on all kinds of factors, such as the length of the time since the sitting bird fed, the likelihood of the mate returning soon or the risk of predation to unguarded eggs. Two types of question raised by this example are of interest here; firstly, how can we account for the decision processes, and secondly how good are animals at choosing the most advantageous options?

13.2 Decision making

13.2.1 *Rational choice and utility*

The crucial thing about a decision maker (assuming that it does not behave entirely at random) is that it must be 'rational'. In decision theory the term rational is used in a special sense and means two things: that options can be ordered, and that choices are made so as to maximise something. To be able to order the available options two criteria must be satisfied. Firstly given any two options A and B, it must always be possible to tell that either B is preferred to A, A is preferred to B, or that they are equally desirable. Secondly, all preferences must be transitive; i.e. if A is preferred to B, B to C then A is preferred to C (Edwards 1954). The thing that is maximised by a rational animal is called 'utility', which can be thought of as equivalent to the benefit derived from making a particular choice, and must be measurable in a currency common to the different options. It is argued that a decision making process which gives consistent results must be rational in the technical sense; thus if animals are able to make consistent choices between options it follows that their decisions must maximise benefit, or utility.

This line of argument has not gone completely unchallenged. Simon (1956) claimed that a simple theoretical choice making animal can be constructed to deal with multiple goals in an adequate way without such a common currency. The key concept in his formulation is that organisms perform satisfactorily, or 'satisfice', as distinct from optimising; that is, they perform so as to stay alive, rather than maximising something. Simon shows that the animal's time can be allocated to activities related to individual needs without creating any problem of overall allocation or coordination, or need for a 'utility function'. Simon's ideas have not generated much experimental work, and little attempt has been made to evaluate 'adequacy' explanations against 'optimality' explanations of the same phenomena; the work described in this chapter is partly based on the claim that any consistent choice making mechanism embodies a maximisation principle.

13.2.2 *Do animals make rational decisions?*

It follows from the previous section that to prove rational decision making in animals it is necessary to demonstrate consistent preferences between alternatives and transitive ordering of options. To find out how an animal weighs up different factors in making a decision we need to ask it to choose between alternatives of different combinations of the two factors; for example, how would it choose between an easily accessible small amount of food and a larger amount which could only be obtained with difficulty, or by exposing itself to danger? The weighing up can be represented by means of an indifference function (Fig. 13.1); this is a graph whose axes represent (strictly) the increasing utility of the two factors, in this case size of food reward and risk run in getting it. The lines in the graph join combinations of danger and

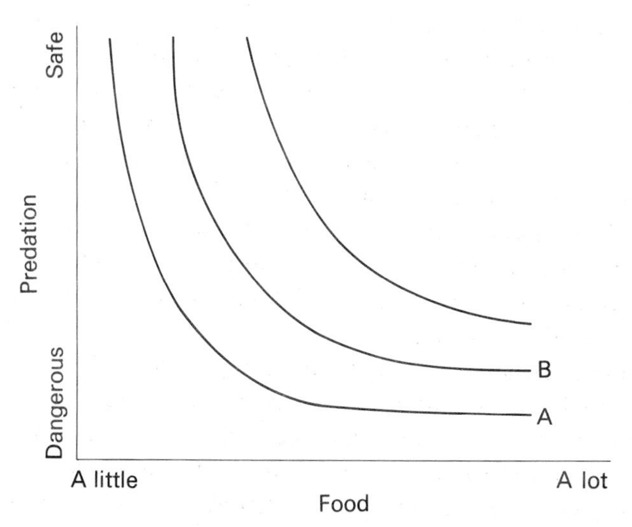

Fig. 13.1. Hypothetical indifference curves for an animal choosing food patches of different sizes in more or less dangerous places.

food amount which the animal is equally likely to choose; thus a patch of food in a very dangerous place is only likely to be taken if it is very productive. The curves are in fact contours of utility and the further out from the origin of Fig. 13.1 one goes the more advantageous the final choice will be; in other words points on curve A are all equally advantageous, but all points on curve B are preferable to all points on curve A. Studies using indifference curves to demonstrate rational choice in animals can be found in the psychological literature. For example, Logan (1965) asked rats to choose between food rewards of different sizes delayed for different times in the arms of a runway; his rats showed sufficiently consistent choices for him to be able to plot the indifference curves shown in Fig. 13.2. In a similar study Lea (1976) found that pigeons showed consistent choices between rates of food reward in a Skinner box (see also McFarland 1976). Attempts to show that animals will make a choice on the highest accessible contour of Fig. 13.1, i.e. that they do maximise utility, have been less successful, although Hodos and Trumbule (1967) claim that chimpanzees can do so.

13.2.3 *The common currency*

What is the 'common currency' in which the relative advantages (utilities) can be expressed? In several studies on functional design it has been taken to be energy. For example Wolf and Hainsworth (1971) state explicitly in their analysis of time and energy budgets in territorial humming birds that the time budget is subsequently to be translated into energetic costs (see 11.2.1). Huey and Slatkin (1976) also chose to do their accounting in energy terms, although they remarked that this could be translated into fitness. While Alexander (1967) argues that in morphology a simple design criterion such as 'minimise energy

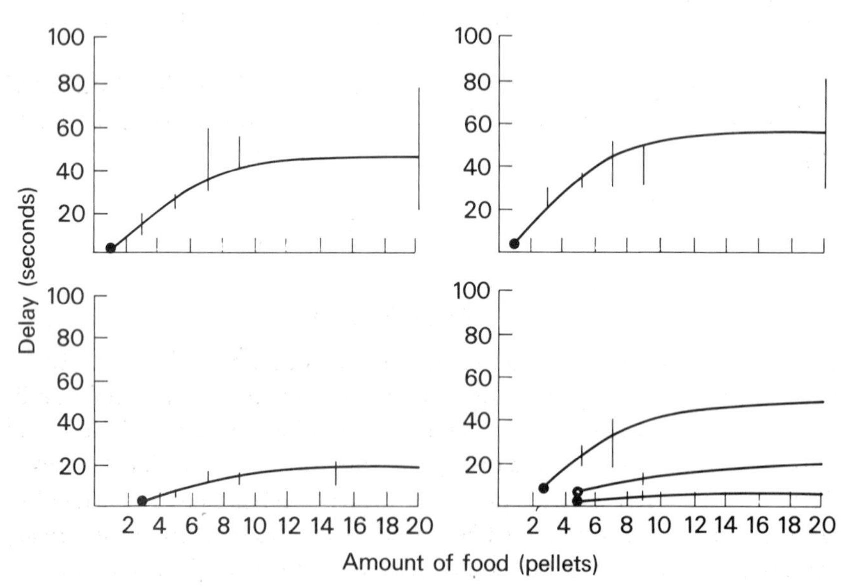

Fig. 13.2. Indifference curves for rats offered food rewards of different sizes subject to different delays in a maze. The vertical bars represent the range of delays within which the indifference value is estimated to fall for a particular reward size; smooth curves are fitted indifference equations. Note that 'utility' decreases on the y-axis as delay increases, i.e. this axis is reversed compared with Fig. 13.1. (From Logan 1965.)

costs' can be justified in terms of increased reproductive potential from so doing, it is far from clear that a similar argument will always work for behaviour. An appropriate common currency must be one related to the ultimate evolutionary advantage or disadvantage of indulging in certain behaviours, even though this is much harder to measure from the zoologist's point of view. The assessment of the relative advantages of different activities ought therefore to be made with reference to the effects of performing these activities on the reproductive potential of the individual and his close kin. Each activity performed by an individual can be thought of as incurring a certain probability of death and a certain probability of successful reproduction; this can be expressed as a single quantity called the 'cost' of a behaviour (in effect cost is the negative of utility) so that the cost per unit time of a behaviour is the chances of reproducing subtracted from the chances of dying while performing that activity, including the costs incurred from not performing other activities with which it is incompatible.

$$\text{Cost per unit time} = \frac{\text{chances of dying}}{\text{per unit time}} - \frac{\text{chances of reproducing}}{\text{per unit time}}$$

This is really just a formalisation of the way in which functional arguments are always justified in biology (cf. Nagel 1961, Williams 1966, Pulliam 1976).

13.2.4 *The objective function*

The relationship between the common currency and different options, i.e. the cost (or utility) of adopting each option under specified conditions, is called an objective function (function in the mathematical sense). The argument of section 13.2.1 amounts to saying that whatever decisions an animal makes, provided they are consistent (or rational in the sense defined above), must be made so as to maximise (or minimise) something; that something is the objective function. This leads on to the idea pursued further in section 13.7 that we can explain behaviour by working back from the decisions that were actually made to find the objective function which has been maximised, hoping that it makes some kind of biological sense; such a move does not of course test whether an animal is behaving optimally—it assumes it.

13.3 Optimality theory

13.3.1 *Why use optimality theory?*

One obvious reason for using optimality theory is the central role played by maximisation principles in decision making (see above). I shall sometimes refer to optimisation as maximisation of something and sometimes minimisation; this is because which process has to be done may depend in a more or less arbitrary way on the details of the mathematical formulation. Put simply, optimisation might either consist of minimising net cost or maximising net benefit; which it was would depend whether one had the equation written out in terms of cost or benefit.

Rosen (1967) points out that although optimality or minimisation principles such as the Principle of Least Action are widespread in Physics, the only justification for using them is the fact they they work; in Biology however there is an excellent rationale for such principles, namely the theory of evolution. Natural selection is a process which maximises 'inclusive fitness' or some other suitable quantity.

Many biologists feel that the rule of the evolutionary game is only to do better than your competitors rather than to maximise something; those alive now are those which have succeeded in this. But by surviving, animals presently in existence have optimised something since they could not have done any better than survive; 'trying harder' would not yield greater rewards. The problem arises from the confusion of the term 'optimal' in the sense 'maximises something (fitness)' with 'optimal' in the more commonplace sense of 'the best design'. The process of natural selection can be thought of as the comparison of variant structures in terms of their survival value and the choosing of the best. Comparison and choosing the best is a maximisation or optimisation principle, so the notion of optimality does seem to be inherent in evolution. But merely to remark that if animals are constructed in a certain way and they survive then such construction must lead to maximum fitness is to avoid the interesting issue, namely how does this construction contribute to fitness in the population geneticist's sense. The only sensible way to tackle this problem is to try to use optimal control theory to 'design' the best possible theoretical animal so as to have a yardstick against which to measure the performance of real animals (cf. Schoener 1971). It is important to remember that evolution can only have maximised fitness over the phenotypes which have in fact occurred. Hence we are confronted by the classic problem, what constrains evolutionary innovation? On the one hand it is clear that we should avoid facile explanations of apparently sub-optimal design in terms of evolutionary constraints, if only because this discourages the search for other factors which show the sub-optimality to be apparent rather than real; but to assert that any improvement in design that is physically possible must necessarily have occurred is equally unwise, since it leads eventually back to the empty assertion that what exists must be optimal.

In practice, as pointed out by Williams (1966) following Pittendrigh (1958), the principle of natural selection has not been used in a sufficiently disciplined fashion and most of the problems to which it has been applied make easy demands on the theory. The application of the mathematical apparatus of optimal control theory does go some way to take up the challenge offered to carry out more rigorous quantitative studies.

13.3.2 *Cost functions*

A rigorous approach to the problem of optimal design in animals consists of evaluating the actual costs which will be incurred by the animal choosing to perform different behaviours in a specified environment, and with a specified internal state. Such measurements can be made in real animals; for example, when an incubating herring gull leaves its nest unattended there is a substantial risk that the eggs will be lost through predation (Parsons 1971) or exposure to lethal temperatures (Drent 1970). The cost in the sense defined in section 13.2.3 increases exponentially with time, but the gull may be offsetting this cost against the chances of dying of hunger or thirst or being taken by a predator. The quantitative relationship between these risks and the causal variables is called a 'cost function'. It will be seen that it bears a resemblance to what was called an 'objective function' in section 13.2.4, for it is a component of the quantity which must be minimised by the optimum behaviour sequence, but there is a subtle distinction. Costs can, in principle, be measured directly. In the herring gull example above we might discover the risks to unguarded eggs by removing the parents and seeing how long the eggs survive; similarly the relative advantage of feeding can be partly assessed by detailed knowledge of fluctuations in the availability of food and by knowing the food reserves carried by a particular bird, which might be indicated by its weight. The objective function can only be determined by inference. In the herring gull case we would need a very detailed model of how all the internal and external factors affecting instantaneous fitness depend on the activities the animal performs; such a model would have to be evaluated by comparison with real sequences of behaviour which could be obtained by extended tracking and observation of individual birds. It might then be possible to work backwards from the model explaining these sequences to find what function was being minimised. This objective function would then have to be given a biological interpretation. In principle it might be possible to compare cost functions and objective functions in order to determine how well the animal is adapted. Since the objective function is minimised by definition it follows that if the cost function evaluated directly for an animal in a given environment is equivalent to the objective function then the animal must be optimal with respect to that environment. To put it another way, the cost function is the thing that the animal ought to minimise in order to behave optimally in a particular environment; it is thus a property of the environment. The objective function on the other hand is the function which actually is minimised by the animal, and is therefore a property of that particular animal. If the two things are the same it follows that the animal is doing what it ought to, i.e. it is perfectly adapted, and no rearrangement of its behaviour patterns could result in a better performance. In practice this is complicated by the fact that we cannot be sure that an objective function is unique; hence if they do not match up it is hard to know what this means, though if they do match we could be fairly confident that we understood the behavioural adaptations of our animal.

13.3.3 *What do optimality theories test?*

Perhaps the most remarkable thing about the relatively straightforward optimality theories such as those reviewed in Chapters 2 and 8, is that real animals seem to fit their behaviour quite well to the models which are proposed on largely logical grounds. A cynic might remark that this is because we only hear about those models which do work, but it might very reasonably be asked, how do we know when something is optimal? Could we distinguish the case where design is not optimal from that where it is but we have not evaluated the performance criterion correctly? It is sometimes asserted, wrongly, that optimality is an hypothesis for testing, but a moment's reflection will show that this cannot be so. All that can be tested in any instance is whether a particular optimality model is correct; failure of a model does not mean that optimisation as such is not occurring, since there may be another model which does fit the facts but which has not been thought of yet. Optimisation is a weak postulate, which is assumed to be true most of the time. In the 'direct' approach the hypothesis under test consists of a postulated set of cost functions which it is hoped the animal might be minimising by its behaviour. If it is not doing so this could mean one of at least three things:

a. the evaluation of the cost functions might be wrong;

b. a relevant cost of some option may not have been included;

c. the animal may not be perfectly adapted to its environment; in fact this is likely to be the case more often than not, except in environments that have been stable for a long time.

In the case of the other approach the animal is assumed by definition to be minimising the objective function; the only way of evaluating such a theory is by trying to interpret the objective function. The validity of this approach is considered further in section 13.7.4.

13.3.4 *Finding the cost functions*

Assuming that we are prepared to adopt the postulate that an animal optimises its behaviour with respect to some cost function, how can we proceed to show that this is consistent with the animal's behaviour? In principle it could be done by working out the costs associated with possible (actual or invented) sequences of activities which the animal might adopt and hope to find that the one actually chosen is the best; this means that we must be able to calculate the costs associated with any particular behaviour under specified conditions, a process involving two steps. Firstly the consequences to the animal of performing an activity must be known. These consequences are partly changes in the internal state (food reserves for example) resulting from actions; generally it will be necessary to have a great deal of background physiological knowledge about the animal so that a detailed model of the internal state and its changes can be formulated. Secondly the costs incurred by an animal from being in a given internal state must be measured; for example what are the chances of death for a herring gull after three days' food deprivation? These must be added to the risks directly incurred from indulging in certain activities; for a herring gull, what are the chances of being run over by the tractor while following the plough? As remarked above (section 13.3.2) it is possible in principle to measure these cost functions directly by measuring the changes in fitness resulting from being in a certain state. Ideally this should be done directly in the field; an attempt is currently under way with herring gulls during the incubation period (Sibly & McCleery in prep.. see McFarland 1977, for a preview).

In practice the direct approach requires an empirical base of gargantuan proportions which does not exist for any species at the present. Most studies of costs and benefits have involved guesses at likely forms for the cost function. For example, in their theoretical model of the costs and benefits of thermoregulation Huey and Slatkin (1976) use extensive knowledge of lizard physiology, but their benefit curve is a guess, based on the observation that many lizards appear to regulate their temperature to a value near the upper physiological limit. The studies detailed below similarly rely on informed guesswork for their cost functions.

13.4 Feeding and drinking in doves – a worked example

An attempt to characterise the optimal behaviour sequence for an animal has been made by Sibly and McFarland (1976) who investigated the choice between feeding and drinking behaviour in a Barbary dove which had been deprived of both food and water and was allowed to obtain these by pecking at keys in a Skinner box. This is obviously a highly artificial arrangement; doves normally forage by pecking up grain from the ground and, like other Columbidae, they can normally satisfy their thirst by pumping up water using their muscular oesophagus. The advantage of using this experimental paradigm is that a good deal is understood about the motivational factors thought to underly the behaviour (McFarland 1970, McFarland & Sibly 1975), and the experimenter is able to control the environment so there is less uncertainty about relevant causal factors than there would be in a more natural experiment.

13.4.1 *Motivation and causal factors—state spaces*

A prerequisite for investigating optimal behaviour sequences in any animal is an understanding of the relationship between the behaviour of the animal and its internal state. The internal state is a complex notion, involving the instantaneous values of all causal factors relevant to behaviour. Many of these causal factors are physiological variables such as level of food reserves, or blood osmolarity, for example; others may represent such things as the animal's estimate of food availability, or the danger of going to a particular place which could be the result of the animal's experiences with its environment. The internal state is strictly a set, in the mathematical sense, of values of all such variables. Such a set of numbers can be represented graphically by means of an n-dimensional Euclidean space (McFarland 1971, section 6.13) called a vector space, which is linear and has Cartesian coordinates. Any set of variables can be represented in this way but it is perhaps easier to think of a physiological example to begin with.

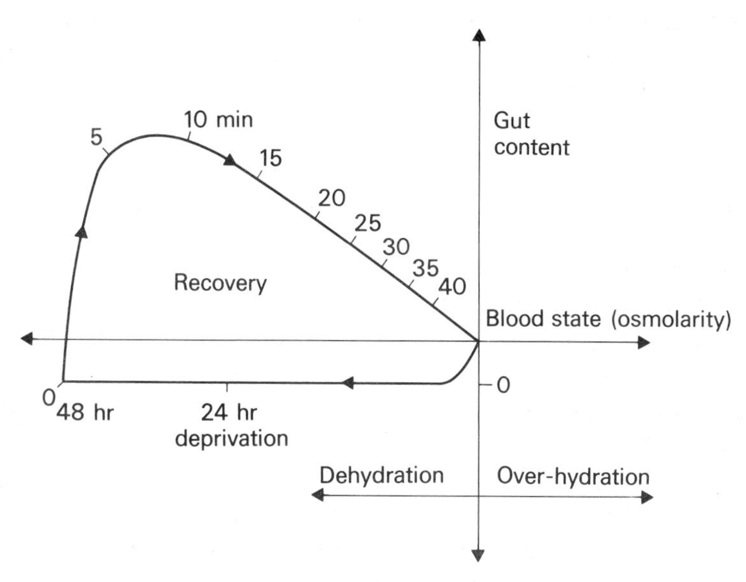

Fig. 13.3. Example of a physiological state space with axes representing blood state and gut content in relation to water balance. The trajectory traces changes in state caused by 48 hours deprivation followed by access to water. (From McFarland 1970.)

Consider two physiological variables such as blood osmolarity and the water content of the gut. These can be represented as two orthogonal axes defining a two dimensional space. The physiological state of the animal with respect to these two variables is then defined by a point in this space whose y-coordinate represents the gut content and whose x-coordinate represents the blood osmolarity at any given instant (Fig. 13.3). This space is an example of a physiological state space. As the state of the animal changes the point will appear to describe a trajectory in the space; notice that time is not explicitly represented in this diagram. The way in which behaviour affects this internal state can now be seen; not drinking (i.e. doing anything incompatible with drinking) will allow the blood to become dehydrated, and hence the state will change along the trajectory labelled '24 hrs deprivation' in Fig. 13.3, but when the animal is allowed (or chooses) to drink his state will change as shown by the 'recovery' trajectory. This representation of internal state can be generalised to include all causal factors by employing an n-dimensional space each of whose axes represents a causal factor.

The following terminology is used throughout this chapter. The *state* of an animal at any instant with respect to a set of variables is defined as the values of all the variables at that instant; which variables are involved is indicated by qualifying the word 'state'. Thus *physiological state* is the instantaneous value of all the physiological variables, *causal factor state* the instantaneous value of all the internal and external causal factors for behaviour. The instantaneous values of such sets of variables can be represented as a point in a vector space; this space is called the *state space*, again qualified by the nature of the variables involved. Thus physiological state space is a vector space in which a point represents the physiological state of the animal at an instant (see above, and Fig. 13.3); we usually write *physiological space* for short. It follows from these definitions that a 'physiological state' and a 'point in physiological (state) space' are the same thing, and that trajectory in physiological space represents the history of the physiological state over some specified time period.

Conventionally it has been considered that the behaviour is 'driven' in some way by its causal factors. In simple terms the different causal factors combine in various ways to produce tendencies for the various activities of which the animal is capable; the behaviour actually expressed is the one with largest tendency (see McFarland & Sibly 1975). Notice that the set of all tendencies can be represented in an n-dimensional tendency space (misleadingly called 'candidate space' in the literature) in the same way as the causal factors. Similarly behaviour can be described in terms of a behaviour state space whose axes represent rates of performance of each independent activity. Some parts of the behaviour space may be inaccessible, meaning that certain combinations of activities cannot be performed because of physical constraints; in particular behaviour cannot be performed at a negative rate. It is usual to represent the behaviour state by means of a vector $\mathbf{u}$ and the causal factor state by means of a vector $\mathbf{x}$ ($\mathbf{x}$ means all the values of $x_1, x_2, \ldots x_n$). As has been shown, the behaviour of the animal itself influences the internal state, so the whole arrangement is a feedback system, schematically described in Fig. 13.4. One result of this is that it is just as logical, mathematically speaking, to think of the animal controlling its internal state by means of its behaviour as it is to think of the behaviour being controlled by the internal state. In other words, the animal is to be thought of as a homeostat in which internal state is kept within some specified limits by means of both physiological and behavioural controls.

In the experimental arrangement of Sibly and McFarland (1976), the causal factors to be considered are rather few. The internal variables are simply the subject's need for food and water and his estimate of their availability in the experimental apparatus. Since the doves were maintained for long periods at a constant body weight when they were only fed and watered in the Skinner box it was assumed that the amount

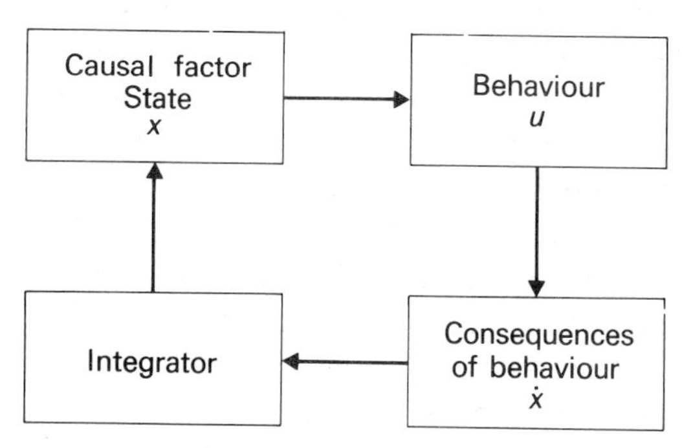

Fig. 13.4. Relationship between causal factor state and behaviour. (After Houston *et al.* 1977.)

taken during the experimental session was a reliable guide to the deficit at the start of the session; it was further assumed that the birds' estimate of availability was correct (i.e. that it corresponded to the maximum rate at which rewards could in fact be obtained) since they were thoroughly familiar with the experimental procedure before testing began. Other factors such as seasonal variations due to the breeding cycle were controlled by keeping the birds in constant winter conditions (short days and ambient temperature of 10°C). It is easy to see that the change in internal state resulting from behaviour depends on the rate at which the dove pecks the relevant key and the rate at which food or water rewards are dispensed. However, as Sibly (1975) has shown, the physiological deficit combines multiplicatively with the animal's estimate of the reward rate; hence the rate at which deficit is reduced depends doubly on the actual reward rate, partly because the physiological consequences depend on the rate of delivery by the system, and also because the rate at which the animal will choose to work depends on the 'incentive value' defined as the maximum rate at which rewards can be obtained.

13.4.2 *Costs and benefits*

In section 13.3.4 it was claimed that the costs could and should be measured directly as the cost per unit time in fitness terms which is incurred by being in any given state. However, Sibly and McFarland decided that it was not practicable to do this in the case of the Barbary dove under laboratory conditions. They decided on *a priori* grounds that the cost function was likely to be convex, that is that an increase in deficit when deficit was already large would be more costly than the same increase when deficit was small (Fig. 13.5). Obvious examples of such cost functions would be a quadratic $C \propto x^2$ or an exponential of the form $C \propto e^x$. For illustration of the method they chose the quadratic on the grounds that the mathematics of this problem is well known in mathematical control theory (cf. Elgerd 1967); few if any attempts have been made to solve dynamic optimality problems with more complex cost functions. In this example therefore the estimation of the cost functions is a guess. In addition to cost depending on the

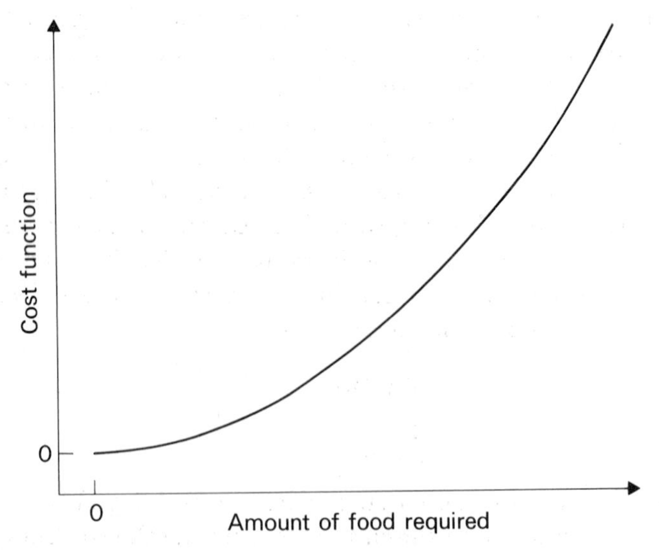

Fig. 13.5. Hypothetical dependence of cost on the amount of food required. This is an example of a 'convex' cost function. (From Sibly & McFarland 1976.)

square of the deficit, they also supposed that cost is an increasing function of the rate at which behaviour is performed.

13.4.3 *Separability of the cost function*

An important assumption of the method used by Sibly and McFarland is that the total cost of being in a certain state $\mathbf{x}$ and choosing the combination of rates of behaviour $\mathbf{u}$ can be separated into components related to internal state and those related to rate of behaviour; mathematically, $C=f(\mathbf{x}, \mathbf{u})$ and there are no terms containing both $\mathbf{x}$ and $\mathbf{u}$. This corresponds to a probabilistic independence between risks incurred by selecting a behaviour at a certain time and the internal state at that time; the risk of crossing the road to the pub is taken to be the same regardless of how thirsty you are. While this assumption seems reasonable in cases such as the Barbary dove example it could well not be valid elsewhere. Houston (1977) gives the example of newt courtship (see also section 13.7) where the benefit to the male of depositing a spermatophore will depend on how ready (and therefore how likely) the female is to pick it up.

13.4.4 *Finding the optimum solution*

The first thing to be said is that this type of dynamic optimum problem pushes the science of control theory to its limits. The procedure is to choose a certain combination of rates of behaviour, corresponding to a position in the behaviour space defined in the previous section, so as to minimise the total cost incurred while moving the internal state, represented by a vector $\mathbf{x}$ as defined above, to its final value. Cost was defined in section 13.2.3 as the instantaneous risk of being in state $\mathbf{x}$ and of selecting behaviour $\mathbf{u}$, so to find the total cost over the period $t(0)$ to $t(\mathrm{end})$ the cost must be integrated (which is mathematically equivalent to summing the instantaneous costs); the object of the exercise is to minimise the integral

$$\int_{t(0)}^{t(\mathrm{end})} C(\mathbf{x}, \mathbf{u})\, dt$$

It is important to realise that to minimise an integral it is necessary to be able to predict into the future; to evaluate the cost integral it is necessary to know all the future values of $\mathbf{x}$ and $\mathbf{u}$. In spite of this there are several techniques available which can be used in an attempt to find a solution to the problem of choosing the optimum behaviour sequence.

Dynamic programming

A strategy, in the meaning of the term used here, results in a trajectory being described in the vector space representing the animal's internal state from the initial state to the optimum or goal state. Different strategies, that is sequences of activities, produce different paths in this space; with each path can be associated a total cost from travelling

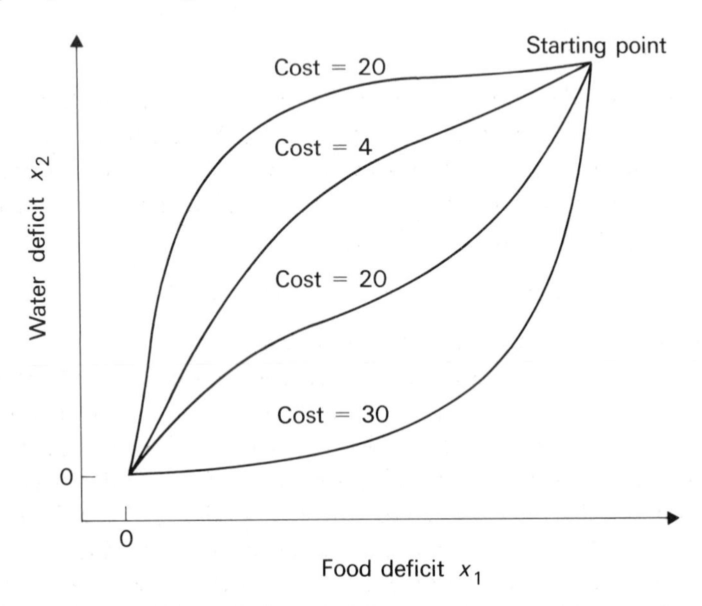

Fig. 13.6. Possible paths in causal factor state space. The overall cost (hypothetical) of each trajectory is indicated. (After Sibly & McFarland 1976.)

that way rather than any other, and the optimum strategy obviously produces the path with the lowest associated cost (Fig. 13.6). One way to decide on the best path is to evaluate the costs for all possible paths (the 'British Museum algorithm'). A more sophisticated approach is to use a technique called Dynamic Programming which is essentially an heuristic device for reducing the calculation required by the British Museum method to manageable proportions. Dynamic programming relies on the proposition that for any point on the optimum path between two points A and B the optimum pathway to B lies along the same trajectory AB. The problems appropriate to the present discussion are called 'final value' problems because the precise nature of the goal state is known in advance. The way of solving these problems is to choose the best *last* step to the goal since we know that the animal must take this if it is on the optimum trajectory, then to work out the best penultimate step and so on back to the starting state. Although the method is reliable in principle it may be hard to use, mainly because the computation requirements are so great in any non-trivial case. It has not yet been used systematically in the type of study with which this chapter is concerned.

Pontryagin method

The most suitable alternative to dynamic programming is one due to Pontryagin, a Russian rocket scientist, who showed that minimising the total cost of a trajectory for a system such as this where the state variables can be controlled by behaviour variables was equivalent to continuously maximising a quantity

$$H = \text{plant equations} - \text{cost function}$$

H is sometimes referred to as a Hamiltonian function.

In this equation the cost function part is the same thing as that described above in section 13.3.2, namely the relationship between the instantaneous cost and states and rates of behaviour. The plant equations represent the way in which internal states are influenced by the behaviour selected. Thus the process of minimising H can be thought of as trading off the cost of being in a certain state against the cost of performing the behaviour required to change the state, using the plant equations to determine what behaviour this will be. Keeping down the cost of behaviour means paying the cost of being in a certain state for longer. So a hungry and thirsty Barbary dove should try to reduce its deficit as fast as possible without incurring too high a cost from pecking very fast (e.g. by failing to see a predator creeping up behind it); the effect that a particular rate of pecking will have on the food deficit is determined by the plant equations. Two considerations are important for applying the method to behaviour.

(1) Behaviour cannot be performed at a negative rate.

(2) In the case of, say, feeding, the environment affects the change in state produced by each unit of feeding behaviour; in this case the yield of food per peck is determined by the availability of food.

For the mathematical argument required to find the optimal solution in the Barbary dove example the reader is referred to Sibly and

McFarland (1976). Intuitively it can be seen that if the cost of being in a state far removed from the optimum is high, then it is worth incurring a high cost from behaving at a high rate in order to move the internal state away from the danger area. However, as the state of the animal approaches optimum it is less worthwhile to incur the risks of a high rate of behaviour. What should be seen when an animal is recovering from, say, food deprivation is an initial rapid intake of food followed by a steady decrease in rate of intake as satiation is approached, assuming that there is no digestive bottleneck which prevents the animal from bringing its state to the goal point in a single meal.

It can be shown that if cost functions are quadratic as assumed by Sibly and McFarland then this negatively accelerated curve will be exponential in form. Such exponential curves have indeed been found in several animals (Bousfield 1933, McFarland & McFarland 1968, Meyer & Pudel 1972, McCleery 1977a) in laboratory feeding and drinking experiments including both feeding and drinking in the Barbary dove (McCleery in prep.). I have also shown that alternative decelerating curves predicted by other cost functions do not fit the data so well in rat feeding (McCleery 1977a) and drinking (McCleery 1977b).

The control law which an optimally designed animal should use therefore is to reduce all its deficits exponentially. This assumes that the animal can behave infinitely fast, or at least that its behaviour is never limited by external constraints, and can do several things at a time. In real life this is not possible of course; there is a maximum rate at which any particular behaviour can be performed, and only one activity can be performed at once. The effect of these constraints is that there is a finite combination of amounts of each activity that can be performed in a given time, shown diagrammatically in Fig. 13.7. The points in the shaded triangle represent combinations of two activities which can be carried out in a given time; for obvious reasons this is called a triangular constraint. Because of the relationship between rate of behaviour and consequences for internal state another similar triangle could be drawn representing the states into which the animal can get in a given time from a given initial state. If the availabilities of both food and water in this example were the same then this triangle would be the same shape as Fig. 13.7; but if one behaviour produced bigger returns in terms of change in internal state than the other then the corresponding axis would be stretched. Under these conditions Sibly and McFarland decided that evaluating the Pontryagin equation was too difficult to do analytically and they resorted to a graphical solution,

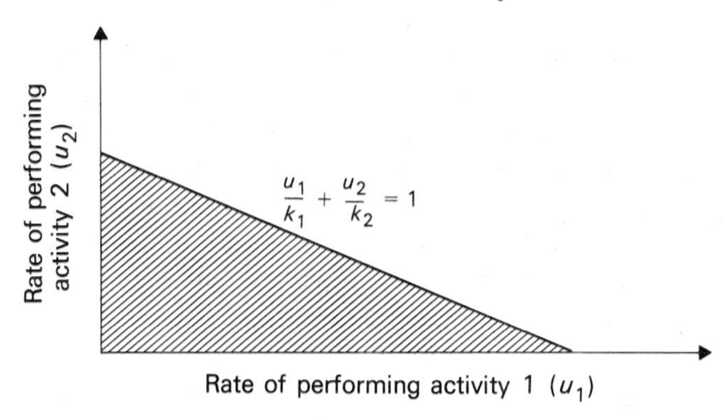

Fig. 13.7. The possible rates (u_1 and u_2) of performing activities 1 and 2 lie in a control region, represented by the shaded triangle. (From Sibly & McFarland 1976.)

shown in Fig. 13.8a. This figure is a series of contour lines representing lines of equal advantage or values of the Pontryagin function H. The two dimensional space has no direct relationship with the vector spaces described in section 13.4.1; it is simply a means of solving the equation. The axes of the space have no intuitive meaning; they are the parts of the Pontryagin equation which can be altered by the animal's behaviour, i.e. the terms in the equation containing u_1 (x-axis) or u_2 (y-axis). In this space the triangular control region defined above can be positioned so that the contours pass through it. The optimal solution is to choose the combination of behaviours which instantaneously cuts most contour lines, which means confers the best advantage. This may not be the same thing as following the line of maximum slope, for as can be seen in the figure this would not give the instantaneous maximum value of H as required by the Pontryagin method. In the state of affairs depicted in Fig. 13.8a, it can be seen that it is best to choose one behaviour and do it at the maximum possible rate. Which behaviour should be chosen depends partly on the position of the triangle which itself depends on internal state, and partly on the length of each side of the triangle, which means on the maximum rate at which the behaviour can be performed and the availability of the relevant commodity. Sibly and McFarland were able to show that this dependence is multiplicative, and amounts to saying 'choose the behaviour for which the product deficit × availability (called incentive by Sibly 1975) × maximum rate of behaviour is greatest'. The prediction, then, is that the animals in this experiment should, if they are optimising their behaviour *and* the cost functions are quadratic,

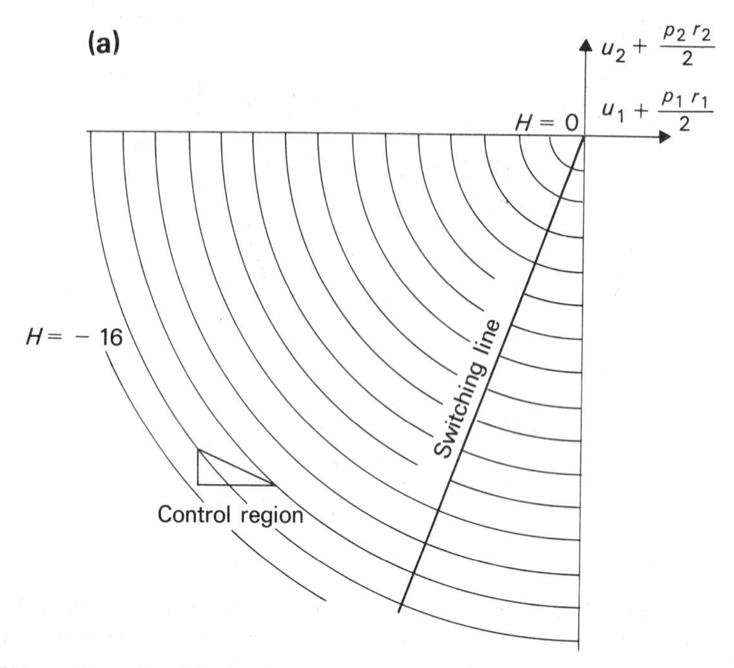

Fig. 13.8a. Graphical solution of the Pontryagin equation. Contours of equal advantage in a space whose axes represent the parts of the equation which can be altered by changing the rate of behaviour 1 (x-axis) or behaviour 2 (y-axis). The triangle represents the control region (Fig. 13.7), and the optimal control is that point in the control region (i.e. combination of rates u_1 and u_2) crossing the most contours.

choose the behaviour for which the product deficit × incentive is greatest.

The way in which following this control law will move the origin of the triangular control region with respect to the rather curious space shown in Fig. 13.8a is very hard to calculate but it can be said that the triangle will appear to move towards the origin as shown (Fig. 13.8b) by approaching the line which divides places where feeding is most advantageous from those where drinking is best; this line should then be followed to the goal point. For an optimally designed animal we should expect that the tendencies for different activities should correspond to the functional needs of the animal, so that the activity which is most advantageous is the one for which tendency is greatest. Thus the line dividing places where feeding and drinking are equally advantageous should correspond in causal terms with the line dividing states where feeding tendency is greatest from those where drinking tendency is

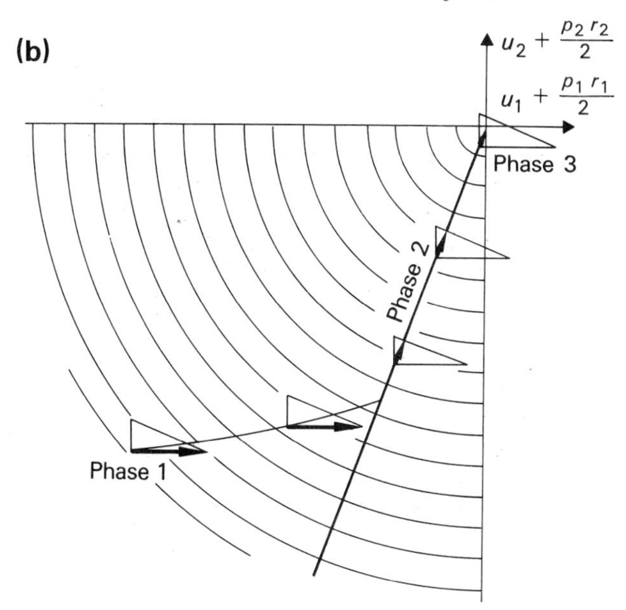

Fig. 13.8b. The optimal trajectory for the case shown in Fig. 13.8a. The optimal control at different points on the trajectory is shown by the thick arrows. (From Sibly & McFarland 1976.)

greatest. It has been shown that this line can be mapped using a suitable technique (Sibly 1975, Sibly & McCleery 1976).

Sibly and McFarland therefore make two predictions about the choices an animal will make under the constraints mentioned on page 393. (i) The animal should choose that behaviour for which the product deficit × incentive is greatest; (ii) animals should behave so that their internal state approaches the line dividing states where drinking is the dominant tendency from those where feeding is the dominant tendency, and this line should then be followed to the satiation point.

13.4.5 *Experimental results*

When a Barbary dove is working for food and water in a Skinner box as described above then a graph of its food and water intake plotted against each other can be taken as an approximation of the changes in internal state produced by its behaviour (Fig. 13.9a). The animal is taken to be choosing the behaviour which is 'dominant' in the sense of McFarland and Lloyd (1973), that is the behaviour which is resumed after an interruption of 1 minute during which no feeding or drinking is possible. It has been shown that in this deficit space the points where

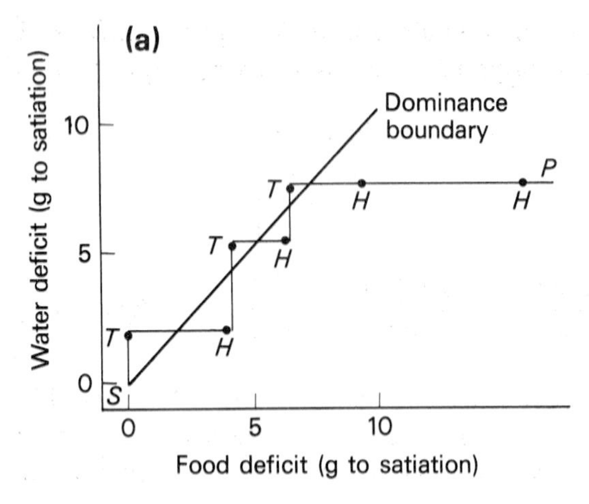

Fig. 13.9a. The motivational trajectory during recovery from deprivation shown in a space with hunger and thirst axes, both measured in grams needed to satiate. H and T indicate points where feeding or drinking were dominant. A dominance boundary was constructed to pass through the satiation point S leaving all the H points on one side and all the T points on the other.

feeding behaviour is dominant can be divided from the points where drinking is dominant by a straight line called the 'dominance boundary'; it is assumed that dominant behaviour is that for which tendency is greatest, so points on the boundary are deficit states (points in deficit space) where feeding tendency is equal to drinking tendency. If behaviour is really chosen according to the rule described above, then on the dominance boundary

$$\text{feeding tendency} = \text{drinking tendency}$$
$$k_1 x_1 r_1 = k_2 x_2 r_2$$

where k_1 = maximum feeding rate
x_1 = food deficit
r_1 = availability of food rewards ('incentive')
$k_2 x_2 r_2$ are the same thing for drinking

Sibly showed that when the availability of food rewards was changed the apparent position of the boundary in deficit space changed as predicted; if r_1 is increased then feeding tendency ought to be greater than drinking tendency for lower values of food deficit than was the case before the change, so the boundary should appear to rotate towards the water deficit axis in Fig. 13.9a.

The slope of the boundary in deficit space is x_2/x_1 so the change of slope to the new value x_2'/x_1' is given by $x_2 x_1^{-1}/x_2' x_1'^{-1}$. At the boundary

$$k_1 x_1 r_1 = k_2 x_2 r_2$$

So
$$\frac{x_2}{x_1} = \frac{k_1 r_1}{k_2 r_2}$$

Hence
$$\frac{x_2' x_1'^{-1}}{x_2 x_1^{-1}} = \frac{k_1' r_1' (k_2' r_2')^{-1}}{k_1 r_1 (k_2 r_2)^{-1}} = \frac{r_1' k_1'}{r_1 k_1} \quad \text{since } k_2 r_2 = k_2' r_2'$$

In other words the ratio of the boundary slopes at different reward rates should be proportional to the ratio of the reward rates. The expected relationship is the dashed line of Figure 13.9b; the actual

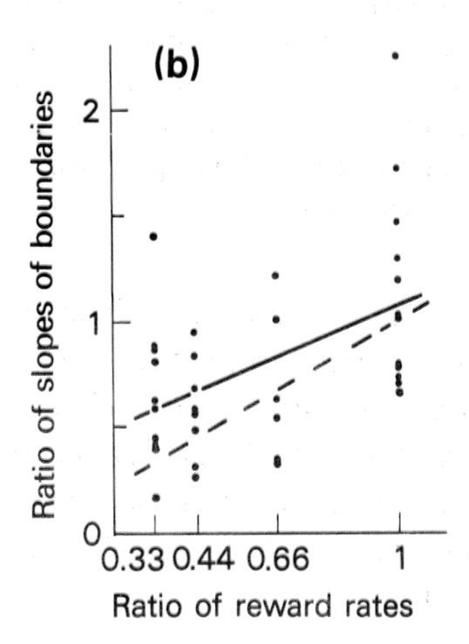

Fig. 13.9b. Ratio of slopes of dominance boundaries found at different relative reward rates. Dashed line: expected result if the feeding tendency is proportional to deficit times reward rate. Solid line: least squares fit to the data points, not significantly different from predicted. (From Sibly 1975.)

results (solid line) are not significantly different; this confirms the first prediction of Sibly and McFarland (see above).

Sibly and McCleery (1976) showed that when all other causal factors apart from deficits were held constant the dominance boundary appeared straight in deficit space, implying that feeding tendency was proportional to food deficit and drinking tendency to water deficit, as one would expect. When deficits were equivalent (subject to a suitable scaling being found) animals did follow the boundary line by alternating between behaviours until satiation is reached (Fig. 13.9a), confirming the second prediction of Sibly and McFarland (see above). Similar experiments by Larkin and McFarland (1978) in a more natural experimental environment have confirmed this result.

In summary, Sibly and McFarland predict that if an animal's behaviour is optimally organised, and if the cost functions are quadratic in state and rate of behaviour, then when food and water are readily available they will be taken at an exponentially decreasing rate; when they are not so readily available then the animal will choose the behaviour for which the product deficit × incentive is greatest. Such experimental evidence as there is confirms these predictions.

The immediate result of this exercise is to show that if the intuitively reasonable quadratic cost function is postulated then two predictions of the Pontryagin method are borne out by experimental results; the validity of the method is not in itself tested, since had it failed the test the conclusion would simply have been that the cost functions were wrong. In the longer term it does show that the technique can be used for real animals, and it also shows what kind of information is needed to test these ideas.

13.5 Another example of PMP – reproductive strategies in eusocial insects

Another interesting application of the Pontryagin maximum principle is the study by Macevicz and Oster (1976) of the way in which annual eusocial insects produce reproductives. Eusocial insects are an attractive subject for optimality studies: the ergonomics of colony development is well enough known to be quantified, and the social insects may be evolutionarily static since basic colony structure has remained unchanged for a long time (Wilson 1975). Macevicz and Oster take as their criterion of fitness of a colony the number of reproductives (queens) at the end of a breeding season. Their cost function is thus extremely simple for it corresponds to this number (but notice that it will be maximised rather than minimised). The genetic composition of reproductives and their relationship to the queen and workers are not taken into account; the maximisation of the number of reproductives is taken to be a goal of the colony as a whole, ignoring conflicts of interest between queen and workers (Trivers & Hare 1976) (section 1.5). To this extent the model could be criticised as being group selectionist; it is of considerable technical interest however, and complicating it by adding kinship considerations would not greatly affect the conclusions drawn. The controls available to the workers are formulated in a slightly different way from that of section 13.4; **u** which was the behaviour vector, is here taken to be the proportion of resources devoted to enlarging the number of workers, so that $1-\mathbf{u}$ of the total output is used for making queens, so there are only two 'behaviours' in the model, namely help make more workers and help make more queens. The dynamics, or plant equations, which determine the income of resources to the colony are rather complicated. The income, or the return function giving the amount of resources produced by the colony when it is in a given state, will partly depend on the numbers of workers and queens present and on the net metabolism and associated losses of energy in the colony, most of which could be measured directly. But it will also depend on such things as (1) the spatial distribution of feeding sites; (2) the ratio at which food can be obtained at each feeding site; (3) energetics of flight, and carrying capacity of each worker; (4) the time required to service a food source. These are factors which could be measured by detailed observations of the time and energy budgets of individual workers. Macevicz and Oster use a model based on such observations made on bumblebees by Oster (1976). Using a technique similar to that of Sibly and McFarland (1976) they were able to show that the optimum strategy is a 'bang-bang' control in which the colony concentrates entirely on making workers until one generation from the end of the season when all resources are thrown into making reproductives; no 'mixed' strategy would do better than this. The conclusion turns out to be relatively insensitive to the precise form of the dynamics, but using a particular form the optimum time of the switch can be calculated; it is generally one worker-lifetime from the end of the breeding season (which is assumed to be a predictable event). Comparison with the behaviour of real social insects shows that several species do conform to this general strategy; for example polistine wasps (Eberhard 1969), hornets (Ishay *et al.* 1967) and bumblebees (Plowright & Jay 1968). However in some cases reproductives are made throughout the season. The conditions under which Macevicz and Oster's model predict this are rather restricted and they decline to speculate about it too far.

Although not aiming at the same degree of detail in the prediction of behaviour sequences as the work of Sibly and McFarland this study reveals the same major features. As with the dove study the cost function is an informed guess, though it could be argued that it is not a very contentious one; the account of the dynamics is arguably more complete since it is based on detailed field work. An interesting similarity in the results is the finding that the optimum control is an all or nothing choice of the options rather than a graded response. As with the dove study, what has been shown is that if the animals are using the cost function guessed by the experimenters, and if they are optimising then the results are consistent with the available observations on real insects.

13.6 Retrospect

Studies like these may seem a little unconvincing, partly because of the sheer difficulty of understanding what has been done and the assumptions on which it is based, and partly because in many cases, for example that of the social insects, the result might seem obvious without the elaborate mathematics. It is not clear that quantitative theories could be produced by a more intuitive approach, though it might be questioned just how quantitative are predictions of the optimality models detailed here. The most important feature of these studies is probably the way in which they point up the type of work required before a rigorously functional explanation of behaviour can be given, a point which some recent popular writers on social behaviour would do well to take to heart. In the first place very detailed understanding of the basic machinery of the animal is required; secondly an accurate measure of the risk actually incurred by the animal from being in particular states and behaving in particular ways is needed. In order to test particular optimality theories properly a complete account of the sequence of behaviour adopted by individual animals over relatively long periods is required.

In spite of this gloom, the final message is intended to be one of confidence that if the technical difficulties involved in gathering such information can be solved then there are methods available which should enable a complete functional explanation of the phenomena of adaptation to be given.

13.7 The inverse optimality approach – courtship in the smooth newt

The smooth newt, *Triturus vulgaris*, performs its courtship on the floor of the ponds and streams which it inhabits in the spring and summer. During courtship the male is obliged to remain near the female throughout the proceedings since otherwise there is a danger that she will leave, or start courting with another male. Courtship is thus an activity which excludes all others; in particular if the male breaks off to breathe air from the surface of the pond he is unlikely to complete courtship successfully. The male may need oxygen from the surface since newts ascend to breathe with a frequency depending on their activity level, water temperature and the partial pressure of oxygen in the water and the atmosphere.

13.7.1 *Newt courtship*

A detailed study of newt courtship has been made by Halliday (1974), while a more general description is given by Houston *et al.* (1977). Courtship is initiated by the male, who approaches the female, sniffs her and attempts to take up a position in front of her, the orientation phase in Fig. 13.10. Provided she does not swim away, she will begin to advance towards the male who responds by retreating in front of her whipping his tail (retreat display). If she continues her advance the male will then turn and move slowly away from her (creep). After moving 5–10 cms he stops and quivers his tail and the female, who has followed him, may touch his tail with her nose. He responds by depositing a spermatophore on the substrate and then moving on. The effect of his subsequent movements is to position the female's cloaca over the spermatophore and if it touches her it is drawn up. For present purposes, the important features of courtship are the transitions from one part of the sequence to the next, some of which depend on the female doing the correct thing, while others depend on factors internal to the male (Halliday 1976). The timing of these transitions may have a decisive effect on whether the outcome is successful or not.

What makes this behaviour a particularly good candidate for an optimality analysis is the fact that there already exists a fairly detailed model of the relationship between internal state and behaviour, backed up by experimental evidence (Houston *et al.* 1977). Furthermore since the outcome of courtship is very obviously related to fitness, at least in the short term, the costs and benefits involved are easier to understand. However, since the model and the experimental evidence are based on laboratory behaviour it is doubtful whether enough evidence exists to enable the construction of cost functions directly; for this reason the inverse optimality approach seemed more appropriate. It is claimed on the basis of decision theory that some function (the objective function) is being maximised by the newt; the difficulty is to find it. When it

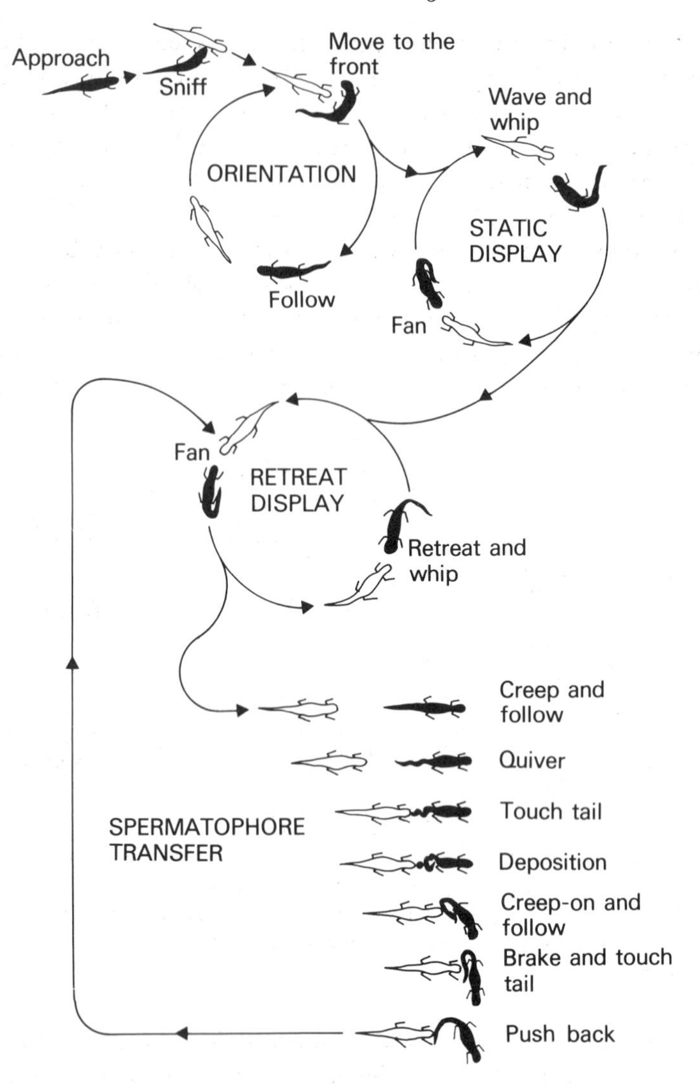

Fig. 13.10. The sexual behaviour of the smooth newt, *Triturus vulgaris*. (From Halliday 1974.)

has been found it will not necessarily be the case that by maximising the objective function fitness is maximised under laboratory conditions, though it is hoped that it would be in the wild.

13.7.2 *Causal models of newt courtship*

The basis of the causal explanation for newt courtship is similar to that propounded for the Barbary dove example. It is assumed that the animal's internal state can be represented as a vector space $\mathbf{x}$, and that the behaviour can be represented in another vector space $\mathbf{u}$, so that the trajectory in the state space $\mathbf{x}$ is made up of small elements x_i each corresponding to a behaviour state u_i. The behaviour states u_i are themselves determined by the internal state $\mathbf{x}$ (Fig. 13.4).

The model proposed by Houston *et al.* is shown in Fig. 13.11. It uses the symbols of mathematical control theory (see Toates 1975, McFarland 1971). The outputs of the model are the different behaviours represented by the arrow heads at the right hand side of the figure. The boxes labelled T_1, T_2 etc. represent thresholds and have the property that when the variable represented by the arrow leading into the

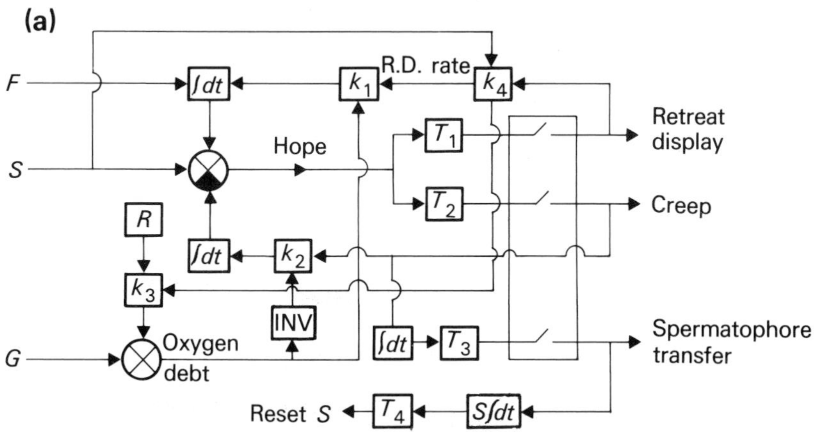

Fig. 13.11a. A control model for the courtship of the smooth newt. S is initial amount of sperm; F, initial sexual state of the female; G, initial value of oxygen debt; R, increase of oxygen debt with time; T_1-T_3, thresholds; k_1-k_4, modifiable parameters. (After Houston *et al.* 1977.)

box exceeds a certain value then the arrow leading out of the box is activated. The variable called 'hope' represents the male's assessment of the readiness of the female to accept his spermatophore. Initially it can be seen that hope depends on the state of readiness of the female, F, and the sperm supply of the male, S. At the beginning of the sequence the male shows retreat display, during which the value of hope increases (via boxes k_4 and k_1) until it reaches threshold T_2, at which point there is a switch to creep. During creep hope decreases (negative feedback via box k_2) possibly because the male cannot see the female and hence does not know whether she is responding or not. When a certain amount of creep has been performed (T_3, note the integrator between this box and the arrow representing creep) the male proceeds to spermatophore

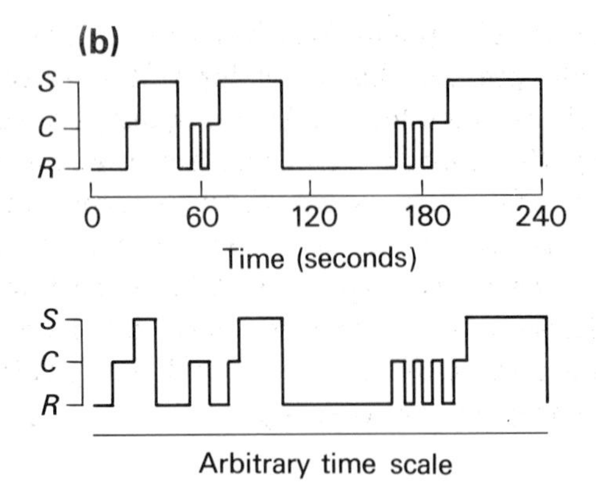

Fig. 13.11b. Comparison of mean durations of the courtship activities of 22 newts (top) with that of the control model. *R*, retreat display; *C*, creep; *S*, spermatophore deposition. (From Houston *et al.* (1977.)

deposition; in real life he must receive a tail touch from the female. If hope decays too quickly during creep and drops below T_1 before the tail touch then the male turns back to retreat display. The changes in hope during a sequence are shown in Fig. 13.11c.

The other important factor is oxygen debt, which is supposed by the model to have an effect on the rate at which hope changes as a result of retreat display (via box k_1) and creep (via box k_2). Since oxygen debt increases throughout the session its effect is to make it more likely that the male will get through to spermatophore deposition as time goes on.

The model gives quite faithful replicas of courtship (Fig. 13.11b),

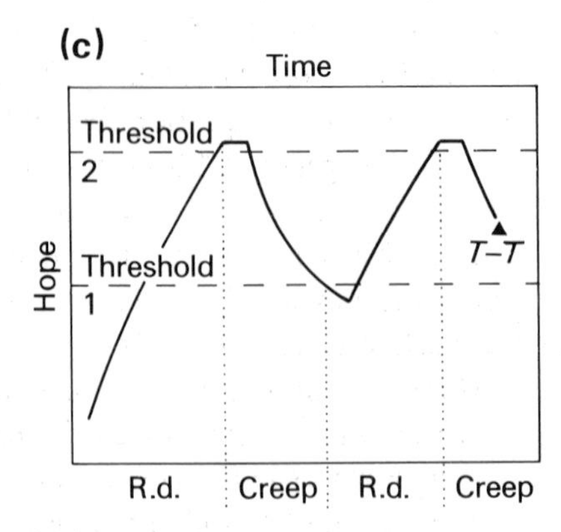

Fig. 13.11c. The trajectory of hope against time in the control model shown in Fig. 13.11a, showing the way in which the thresholds operate. *r.d.*, retreat display; *T.T.*, tail touch. (From Houston *et al.* 1977.)

and some of its specific predictions have been tested directly. For example, the model predicts that courtship will finish earlier when the initial oxygen debt is made bigger. This was tested experimentally by Halliday (1977a) who showed that courtship was faster than normal when the gas over the water was high in nitrogen, and slower than normal when it was high in oxygen, presumably because of the effect that this had on the initial oxygen debt. The most important state variables are therefore hope, oxygen debt and sperm supply. The behaviour variables are rate of retreat display, u_1, rate of creep, u_2, and rate of spermatophore transfer, u_3.

13.7.3 *Inverse optimality*

The work described in this section is due to Dempster (in press), and has been preliminarily described by McFarland (1977).

The solution

Using the model of Houston *et al.* (1977) described in the previous section, which gives an account of the trajectory in state space produced by different activities, Dempster set out to find a function which is maximised by this trajectory corresponding to the Pontryagin equation for H described in section 13.4.4. The causal model has two state variables, 'hope' and oxygen debt. However hope is a hypothetical variable which does not necessarily correspond to any physiologically measurable entity whereas the oxygen debt is a real variable whose value can be determined independently. This led Dempster to use only oxygen debt as a state variable and to identify hope as the costate function, which is part of the plant equation (see Sibly & McFarland 1976, McFarland & Houston in press). This costate function is defined in terms of the objective function, and means, strictly, the change in total future cost produced by a small change in state. For present purposes the important thing is that if the costate function is known then it is possible to work out the objective function because the relationship between them is known by definition. By using the equation for hope, derived from the control model of Houston *et al.*, as a costate function, Dempster was able to derive an objective function that is being maximised by the model (and therefore, one hopes, by the behaviour of real newts).

Dempster's reconstructed Pontryagin came out as follows:

$$H \quad \underset{\text{plant equations}}{-\lambda(\theta - x^{-p}u_2)} + \underset{\text{objective function}}{e^{-\phi S t}Su_1x + Su_3x}$$

where x = oxygen debt
u_1 = rate of retreat display t = time
u_2 = rate of creep S = spermatophore count
u_3 = rate of spermatophore transfer λ = costate function
ϕ, θ and p are constants

Dempster's claim is that the optimally designed newt should be maximising the objective function above, subject to the plant equation.

Interpretation

The plant equation, relating changes in oxygen debt ($\dot{x}$) to behaviour u, can be seen to have two parts, viz.:

$$\dot{x} = \theta - x^{-p}u_2$$

The first bit, θ, simply corresponds to the rate at which oxygen is used up during retreat display and spermatophore deposition. The second part means that oxygen debt increases less rapidly during creep; in simulation of this model (Houston 1977) it actually *decreases* during creep for all reasonable values of the parameters. This might be due to skin respiration, exceeding the rate at which oxygen is used in creep which is less strenuous—a proposition which can be tested empirically, though this has not yet been done.

The objective function is the thing that the newt is alleged to be maximising by its behaviour, as discussed in section 13.3.2. Dempster interprets it, not unreasonably, as meaning that the newt is maximising his chance of fertilising the female. The first bit of it he takes to represent an exponentially decaying likelihood that the female will leave as the retreat display proceeds. The presence of x (oxygen debt) in both terms could mean that females are actually selecting males who can stand up to a high oxygen debt, an idea further discussed by Halliday (1977a see 7.9.1). Besides oxygen debt the second term contains spermatophore count and rate of deposition, indicating that both these are to be maximised.

13.7.4 *The validity of the method*

In summary the study described here consists of taking a detailed model of the behaviour and internal state of an animal, making the assumption that its behaviour maximises something and then using quite sophisticated mathematics and 'green fingers' to produce the function which is maximised; this can be given a reasonably plausible biological interpretation and certainly suggests some further experiments. The approach differs from the one used in the work of Sibly and McFarland discussed in section 13.4 which the cost function was guessed. Apart from being a remarkable *tour de force*, what is the final verdict on this work?

The main worry is the mathematical problem of uniqueness. Although Dempster's solution seems reliable, there is no known method by which it can be shown to be the only possible, or even the only plausible solution. In other words there could easily be other mathematically valid solutions whose biological implications would be quite different. It is not at present clear how much depends on Dempster's more or less arbitrary rejection of hope as a state variable and its consequent identification as the costate function. Indeed Houston (1977) writes: 'Although the objective function has been described as a property of the individual organism it could just as well be thought of as a property of the individual investigator'. He concludes that it may have no relation to the action of selection. This makes the programme suggested by McFarland (1977) in which an animal could be assessed for how well adapted it was by seeing how well the cost function and the objective function match up seem problematic; failure to match could not be regarded as proof of bad adaptation since another, as yet undiscovered, objective function might match. It can be argued that if a good match were to be found then this was evidence for good adaptation, but even if we were able to say we had hit on the 'correct' objective function there are grounds for thinking that a perfect match would never be found (cf. Cody 1974a).

13.8 The way ahead

Although there have been successful applications of optimality theory to decision making in animals both at the level described here and at more general levels such as thermoregulatory strategy (Huey & Slatkin

1976), and although similar ideas are widely current at the time of writing, I think that we have only begun to take up the challenge to provide rigorous functional explanations in biology so clearly made by Williams (1966). While the basic principles described here will govern more successful and complete studies, the actual techniques used may be different. It seems quite clear, for example, that the direct approach of evaluating cost functions in the field rather than making a guess or using the inverse techniques described here will be the next major step. The study on herring gull behaviour during incubation mentioned earlier (section 13.3.4) is a first attempt at such an exercise, but instructive as this study is proving it has been found that even with a species whose behaviour and ecology have been so substantially worked on there are still formidable technical problems in obtaining the required information on risks and on the internal state of the animal.

With the ever increasing sophistication of computing machinery, and of biologists in its use, it may be that dynamic programming will be found to be a more appropriate tool than the Pontryagin maximum principle in some cases. Although the latter has a certain elegance and heuristic value, it can lead to equations for which there are no analytic solutions and numerical solution of these is in fact a very similar undertaking to dynamic programming. An attractive feature of dynamic programming from a theoretical point of view is the idea of sensitivity analysis, in which the departure from the optimal solution caused by making approximations in the calculations is evaluated (see Norman 1975 p. 77). This technique might also give us some insight into the law of diminishing returns involved in calculation of optimal solutions and may show us how natural selection could choose the best compromise between the truly optimum solution and a practical solution to the problem of decision and control.

In conclusion, the development of optimality models for the complete behaviour of animals from minute to minute is in an embryonic state even though it goes to the current limits of mathematical control theory, but it does seem to promise an eventual road to that biologist's Utopia, the Rigorous Functional Explanation.

Chapter 14
Optimal Tactics of Reproduction and Life-History

HENRY S. HORN

14.1 Introduction

In the game of life an animal stakes its offspring against a more or less capricious environment. The game is won if its offspring live to play another round. What is an appropriate tactical strategy for winning this game? How many offspring are needed? At what age should they be born? Should they be born in one large batch or spread out over a long lifespan? Should the offspring in a particular batch be few and tough or many and flimsy? Should parents lavish care on their offspring? Should parents lavish care on themselves to survive and breed again? Should the young grow up as a family, or should they be broadcast over the landscape at an early age to seek their fortunes independently?

Since even humans seldom think seriously about these questions, it would be silly to pretend that animals literally are scheming tacticians. However, we can still examine the tactical answers to these questions and discover how these answers vary from one environment to another. And we can still expect that potentially competitive species will sort themselves out among the available environments so that there is a very broad pattern of adaptive tactics in appropriate environments.

In this chapter I shall examine the theoretical answers to tactical questions about reproduction and life-history. For excellent reviews of the supporting data and for abundant references to the exquisite details of behaviour of particular species, see the papers by Stearns (1976, 1977), Southwood (1977), and Stubbs (1977). Other recent reviews of a more technical nature are by Southwood *et al.* (1974), Southwood (1976) and Giesel (1976). Unfortunately, most of the theoretical discussions of tactics of life-history are overloaded with turgid mathematical formalism. Hence, many of the important papers are unintelligible even to the authors of other important papers. Precious exceptions that are both simple and elegant are the papers by Charnov and Schaffer (1973) and Pianka and Parker (1975), and the short note by May (1977a).

Few of the results in this chapter are new. What is new is the simplicity of their derivation and the compact perspective on the following aspects of adaptive life histories. No formal mathematics are needed to get rigorous results that are free of the restrictive assumptions that traditional approaches entail. In fact the traditional approach has led to an apparent paradox in which different adaptive patterns are predicted depending on whether theoretical models favour growth and survival of populations or bet-hedging in the face of a capricious environment (Stearns 1976). On the one hand, variable environments favour early and copious breeding by lowering the chance that parents will survive to breed in another season; but on the other, variable environments favour spreading a more conservative reproductive output over a longer lifespan because of the high probability of the death of all offspring born in any one season. This paradox disappears when the models explicitly distinguish the opposite effects of juvenile versus parental mortality. The further distinction between adaptation for growth in open environments and persistence in crowded environments is reinforced by positive feedback in selection for both extremes of behaviour. In seasonal environments, adaptations to each season may produce a strategic pattern that is maladaptive to seasonality itself. Dispersal is important not only for denizens of changing habitats, but also for those that live in stable habitats. The study of both the theories and the facts of dispersal in stable and crowded populations has just begun.

The simple derivations have more than heuristic value. Stearns (1977) has recently assembled a lugubrious catalogue of ambiguities and philosophical difficulties that plague the theory of adaptive life-histories. Most of these complications affect the formal development of the theory more strongly than the ideas themselves. Once the ideas are separated from largely unnecessary formal restrictions, they may be explored, in theory and in fact, with fewer philosophical qualms. Stearns has also presented an encyclopaedic review of the places where the available facts fall short of testing theories. These gaps make an excellent shopping list for field work in the future.

14.2 r-selection and K-selection

When a small population is placed in a salubrious environment, it initially grows at an ever increasing pace as more and more individuals have families who have families; but eventually the population reaches the point where the environment simply lacks the capacity to support more, and it levels at a more or less constant number. Figure 14.1 shows this for a hypothetical population, with a dotted line showing the response of a population with a higher rate of growth, and a dashed line for a population with a higher tolerance of crowding. The population with a higher rate of growth has the numerical advantage at low population sizes or early in population growth. Conversely, higher crowding tolerance is favoured in crowded populations or late in population growth. The extremes that favour growth rate and crowding tolerance are called respectively r-selection and K-selection. These

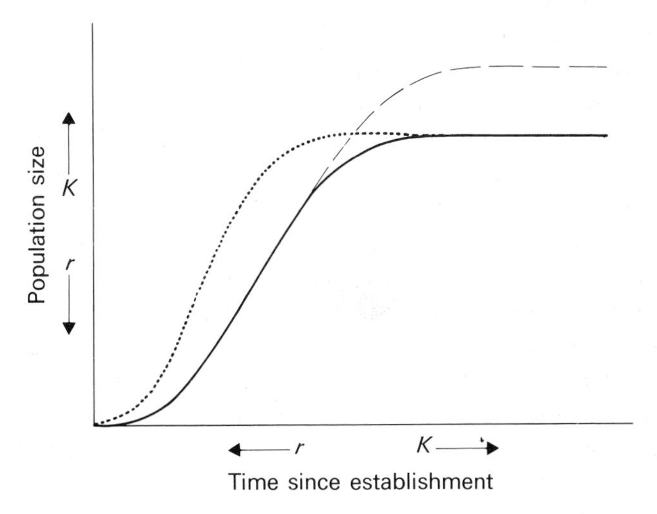

Fig. 14.1. Population growth in a benign but limited environment. The population grows multiplicatively until it reaches the limit of its crowding tolerance. The dotted line shows a population with a higher growth rate (higher r); the dashed line, one with higher tolerance of crowding (higher K). The axes show population sizes and times at which higher r or higher K is adaptive.

terms derive from the logistic equation of population growth in a limited environment, in which r is traditionally used to represent the rate of growth per capita when the population is sparse, and K is the Krowding tolerance, measured as the number or density of individuals at which further population growth is no longer possible. This equation is mathematically tractable and didactically useful even though it is unrealistic in detail. I personally think that the terms 'r-selection' and 'K-selection' are barbaric, but they have become commonplace since their introduction by MacArthur and Wilson (1967) to draw together ideas that had been discussed under many names by many people. I shall keep the jargon as a reminder of how abstract the concepts of r-selection and K-selection really are.

K-selection favours crowding tolerance via survival, competitive ability, and predator escape, all of which are made easier by large size and parental care, even at the expense of delayed and reduced breeding.

r-selection favours growth rate via fecundity, early breeding, dispersal, and rapid development, all of which are favoured by small size, even at the expense of greater sensitivity to environmental changes.

A benign and constant environment allows population growth to the point at which K-selection prevails. However, for r-selection to persist, the population must continually be reduced or put back to an early stage of population growth by predation, catastrophes of weather or climate, or the invasion of unoccupied areas. Patchy and insular habitats often provide a continuous supply of local environmental catastrophes and vacant areas for dispersal. Such habitats include true islands and archipelagos, ponds, mud puddles, water in pitcher plants or hollow stumps, open space in forests or in rocky intertidal shores, and edible bundles of food like carcasses, rotting logs, hosts for parasites, and small plants attacked by insects. Thus patchy habitats often combine r-selection with selection for dispersal.

14.3 The value and cost of early reproduction

In a classic paper Cole (1954) asserted that an annual animal could become reproductively equivalent to a potentially immortal perennial by adding only one further young to its litter. Cole proved this by a numerical approximation, and other people have contested his result for a variety of reasons, all of which turn out to be wrong.

I shall start by proving Cole's result for an unrealistic case, namely an immortal, parthenogenetic animal who grows to maturity in one year and then has either a single litter or an annual succession of litters, each of equal size. I shall then gradually add the effects of mortality among young, mortality among parents, variation in these mortalities, sexual reproduction, extended time to reach maturity, competition with other members of a growing population, and growth in body size and hence in potential litter size.

To prove Cole's result we need only to observe that if both young and adults are immortal and reproduce asexually one year after the last reproductive period, then there is no reproductive difference between an adult and any one of its offspring. Hence, adding one more youngster to the litter is a sufficient reproductive gain to offset the loss of all future reproduction by the parent. A visual representation of this proof is shown in Fig. 14.2.

If the young are subject to mortality so that only a fraction Y of them survive their first year, then in order for a parent to assure an additional surviving young at breeding time, the number of young added to the litter must be $1/Y$, since $(1/Y)(Y)=1$.

If parents are also mortal, so that only a fraction P of all parents survive from one annual breeding period to the next, then the probability of annual survival for any given parent is only P. This means that only P young, who survive to breed, need to be added to the litter

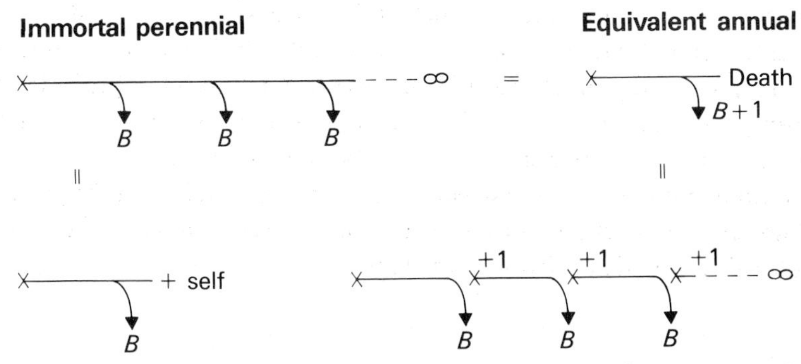

Fig. 14.2. Reproduction of an asexual immortal perennial compared with a more fecund annual. An individual is born at X and then bears B or more offspring at the times indicated by the arrows. That the patterns are reproductively identical is shown by rewriting the perennial's pattern in the format of the annual's and vice versa.

to equal the reproductive potential of the parent. So an annual that bears $B+P/Y$ young and then dies is reproductively equivalent to a perennial that bears B young each year for as long as it lives (Fig. 14.3). This is the conclusion that Charnov and Schaffer (1973) reached, but here it is proved with complete rigour, with fewer restrictive assumptions, and with virtually no mathematics. In particular it is not necessary to assume a stationary age distribution, or even constancy of B, P, and Y beyond the year in question.

We can use the result of Fig. 14.3 to generate adaptive relations between various aspects of life-histories. Massive, early breeding is favoured, even at the expense of death of the parent if a mortal reproductive effort can add more than P/Y young to the litter. This addition is made relatively less expensive if the litter is already large and if the

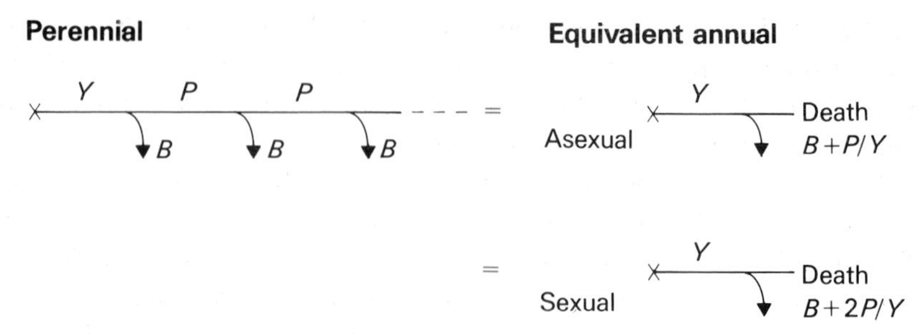

Fig. 14.3. Equivalent reproductive patterns for perennial and annual. This is like Fig. 14.2, but only a fraction Y of the juveniles survive their first season, and only a fraction P of parents survive subsequent seasons. An addition of P/Y young to the litter of an annual ($2P/Y$ for a sexual creature) is reproductively equivalent to potential immortality subject to a survival rate of P per year. The correction for sexual reproduction was derived by Donald Waller and Douglas Green (unpublished manuscript).

offspring are small and require little parental care. Low parental survival (P) and high youthful survival (Y) also lower the expense of the addition since they lower the value of P/Y. Massive reproductive effort will further lower parental survival, so that the selection for early and copious reproduction is self-reinforcing. So the following factors favour early, copious, and self-sacrificial breeding: large litter, small offspring, little parental care, low parental survival, and high survival of juveniles. By converse arguments the factors that favour restrained early breeding and thus survival for iterated breeding are: small litter, large young, parental care, high parental survival, and low survival of young. These lists are the same as those associated respectively with r-selection and K-selection.

For heuristic simplicity I have assumed an asexual population, but now the complications of sex are easy to deal with. For a sexual population even a surviving offspring at breeding time is not genetically equivalent to its parents. In the most extreme case, namely a large and highly outbred population, the young only carry half of the genes of each parent. However, this means that the additional number of young need only be doubled to make a single litter the genetically reproductive equivalent of potential immortality; the number of additional young is $2P/Y$ rather than just P/Y (Fig. 14.3). This is a special case of what Williams (1975) has called the cost of meiosis. My use of the cost of meiosis is independent of the controversy that surrounds the use of the concept to interpret the evolutionary origin of sexuality (Barash 1976b, Maynard Smith & Williams 1976, see also

Chapter 6). For a locally fragmented and inbred population, the number of additional young needed would be something between P/Y and $2P/Y$. All of the qualitative conclusions derived so far for asexual populations remain the same for sexual populations.

To discover the effect of variation in mortality, we must again separate the mortality of adults from that of juveniles. If the survival of parents is variable, the future is further discounted in a way that is closely analogous to lowered average survival of parents. For any survival rate other than 100%, the survival of a given individual is uncertain; if the survival rate is itself subject to unpredictable variation, this uncertainty is compounded. The increased uncertainty of survival to breed again favours early and copious breeding, just as lowered average survival does. Variations in survival of juveniles, especially if they die in batches, favours spreading the risk of reproductive failure among many batches; that is, conservative but iterated breeding. Thus variation in juvenile mortality also has an effect analogous to lowered juvenile mortality, though the line of reasoning is slightly different from the case of parental mortality.

If development is so rapid that maturity is reached in the interval between two reproductive seasons, then offspring of a young parent are already breeding by the next time that their parent breeds. The reproductive value of the parent's first offspring is compounded multiplicatively. However, if there is a very long period of development and adolescence before actual breeding, the reproductive value of the first batch of young is more nearly a simple addition to that of further young, rather than being compounded. Hence earlier and more exhaustive breeding, when it is favoured, is more highly favoured in species with rapid development to maturity than in those with slow development.

Stearns (1976) has provided an elegant test of these ideas by plotting a graph of the average number of seasons in which an individual breeds, against the ratio of average juvenile mortality to average adult mortality, for a variety of birds, mammals, fish, and insects. He found a strong positive correlation over about a sixfold range of both parameters; the species with relatively higher juvenile mortality indeed bred more often. However there was nearly a fourfold scatter of points, and Stearns pleads for more data to decide whether the cause of this scatter is biological or statistical. Southwood (1976) has gathered a more exhaustive and less critical set of relations, showing a correlation between size and generation time, both of which are inversely correlated with rate of population increase.

Although theory suggests that early breeding will often be exhaustive and that delayed breeding will often be repeated, other combinations are fully consistent with the causal machinery of the theory. For example, many birds and mammals have a limited clutch or litter, favouring repeated breeding. If they inhabit ephemeral habitats which favour early breeding, then birds and mammals might be expected to begin breeding as early as possible and to continue breeding for many seasons. The contrary combination of delayed and exhaustive breeding seems confined to beasts with rapid and potentially indeterminate growth in body size and thus in clutch. The most famous example is the sockeye salmon of the Pacific (Foerster 1968) which makes a self-sacrificial breeding effort at an age of three to seven years. Many examples are found among plants, including the spectacular case of a Philippine palm tree in Miami, Florida, which grew quietly for 44 years and then in about 4 months manufactured 300,000 fruits weighing a total of 600 kilograms and displayed on nearly 6 kilometers of stem (Tomlinson & Soderholm 1975). Other examples of unusually late breeding are combined with synchronous breeding of all members of a local population, for example every 13 or 17 years in cicadas (Lloyd & Dybas 1966) or about every 120 years in bamboo (Janzen 1976). In these cases the synchrony, which has been interpreted as a device for overwhelming predators, may be more important than the delay *per se*.

Growth of the population as a whole accentuates the reproductive value of the first batch of young over later batches. For a population that is growing at a per capita rate of r, young born T years in the future should be discounted by e^{-rT} because they will face e^{rT} times as many competitors as those born in the present. This discount for population growth is completely analogous to what economists call 'discounted present value' (Clarke 1976). It was an economic analogy that prompted Fisher (1958) to define the reproductive value of the future offspring of individuals of age x as:

$$\frac{e^{rx}}{l_x}\int_x^\infty b_t l_t e^{-rt} dt,$$

where r is the per capita growth rate of the population, b_t is the birth rate to individuals of age t, and l_t is the probability of survival from birth to age t. This recipe for reproductive value is used in many technical analyses of life-histories; so I shall explain it even though I shall not use it. Fisher's original definition was telegraphic, and most recent explications are opaque; so it is useful to rewrite Fisher's formula in a readily interpretable form:

$$\int_{x}^{\infty} b_t(l_t/l_x)\, e^{-r[t-x]}dt.$$

In this form reproductive value is clearly the sum into the future from age x of (reproductive output at a future age) (probability of reaching that future age from age x) (discount on future reproduction for growth in number of competitors in the population between age x and that future age). Pianka and Parker (1975) partition this reproductive value into current and future components and graphically analyse the tactical partitioning of reproductive effort into current reproduction versus survival for later reproduction. Their results are more detailed than those developed here, but not qualitatively different.

The effect of a declining population is slightly more complex. If the decline is due primarily to juvenile mortality, delayed breeding is favoured since future offspring will face fewer competitors than present offspring. If the decline of the population involves heavy mortality among adults, then the competitive value of delayed reproduction may be offset by the low probability of the potential parent's surviving a delay. If the population is steady, then age of reproduction *per se* is irrelevant, though any parent whose offspring carry some other competitive advantage like higher tolerance of crowding can augment this competitive advantage by breeding early.

The effect of parental growth in size and experience is easy to state in a qualitative fashion, but the recipes for calculating it are horrendous. The effect depends on whether growth in potential size or success of the litter exceeds the compound interest growth of fewer or smaller young invested in the population earlier. Simply put, delayed breeding and growth are favoured if the potential growth of a population of offspring within the body of the parent exceeds the potential growth of the population of offspring in the environment outside of the parent.

Schaffer and Elson (1975) have analysed the interaction of growth and mortality in determining adaptive timing of spawning by Atlantic salmon in different North American river systems from Maine to Ungava. These fish are born and spend their youth in rivers and streams, live through adolescence and adulthood in the sea, but return to their home rivers to spawn. In contrast with the Pacific salmon, their spawning is not exhaustive, and they may return to spawn again after a further period at sea. The mean age of first spawning is greater in the harsher rivers that forbode greater mortality for smaller fish. Commercial fishing increases the death rate of larger and hence older fish and thus favours earlier spawning. Conversely, rapid growth at sea favours delayed spawning. All of these observations are consistent with the theoretical expectations.

Perhaps the most important result of this section is that juvenile mortality and parental mortality have opposite effects on the prediction of adaptive tactics. Juvenile mortality biases toward K-selection, but adult mortality biases toward r-selection. There is really nothing subtle about this difference. Since the only reproductively effective offspring are those who reach maturity and breed, pre-reproductive mortality can be viewed as subtracting from the ultimately effective natality of the parent (Charnov & Schaffer 1973). In a sense, juvenile survival enters the analysis as though it were simply birth rather than survival itself. The effect of variation in mortality also differs for juvenile and parental mortality; and the effect of increased variation is analogous to lowering the respective average survivals. We obtain fully concordant results from an analysis of either average mortality or unpredictable variations in mortality; the paradox discussed by Stearns (1976) disappears. The dichotomy is rather whether environmental mortality falls more heavily on juveniles or adults.

Any increase in reproductive effort is likely to weaken a parent and to increase its subsequent risk of mortality, which in turn favours a further increase in current reproduction. Such self-reinforcing selection will be discussed when r and K selection are revisited. Because reproduction and survival are interdependent it is not always possible to design an optimal life history by exploring the effect of changes in reproductive behaviour. In addition to being mathematically difficult (Mirmirani & Oster 1978) or messy (Schaffer & Rosenzweig 1977), a general analysis may yield several sets of tactics that are better than anything close to them, or it may produce optimal tactics that cannot be attained. Such technical difficulties and biological ambiguities are details beyond resolution in the broad picture of this chapter, but they will surely provide interesting theoretical insights in the future.

14.4 Fluctuating reproduction in a fluctuating environment

If the quality of the environment fluctuates more or less regularly on a cycle that is not greatly shorter or longer than an animal's life-span, the appropriate strategy is reproduction during periods of boom but survival during periods of bust. Moreover, MacArthur (1968) used an ingenious graphical model to show that the rate of reproduction averaged over the whole environmental cycle is higher if there are extreme shifts between profligate breeding and dogged survival, than it is for a moderate response. MacArthur's analysis is repeated in Fig. 14.4.

There is an inevitable lag in the response of an animal to changes in its environment, particularly if the response requires a shift from physiological mobilization of resources for reproduction to storage for survival, or vice versa. The lag and the shift itself virtually guarantee that a population whose members are adapted to environmental fluctuations will continue to undergo drastic fluctuations in population size even during periods of relatively steady environmental conditions. These fluctuations in population size can be viewed as fluctuations in degree of crowding, to which the appropriate adaptation is yet more extreme shifts in reproductive behaviour, reinforcing a vicious cycle.

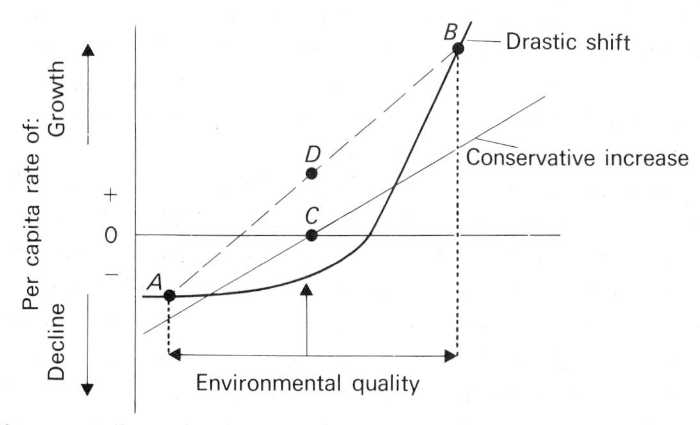

Fig. 14.4. Reproduction in a fluctuating environment. Two kinds of response to environmental quality are compared: a conservative increase in growth rate with increasing environmental quality, and a drastic shift from profligate breeding in good times to dogged survival in poor, at the expense of lowered growth in average conditions. In an average and constant environment (vertical arrow), the conservative response yields the higher rate of growth. However, if the environment shifts back and forth between sufficiently different extremes (horizontal arrows and dotted lines), while preserving the same average, then the average growth rate is a linear average of the extreme growth rates (e.g. of *A* and *B*), and it can be graphically calculated as the point at which a straight line between the extreme growth rates crosses the average environmental quality. The average growth rate for the drastic response (*D*) is higher than that of the conservative response (*C*) in a sufficiently variable environment (MacArthur 1968).

A peculiar pattern is common to empirical studies of cyclic populations of caterpillars (Wellington 1960), voles (Krebs *et al.* 1973), and locusts (summarized by Krebs 1972). As the population fluctuates in any given location, its composition alternates between primarily dispersive and primarily sedentary individuals. This shift in composition is itself enough to generate and to maintain the cycles of local population size. The data have yet to be considered from the opposite point of view; namely, to what extent are the cycles instrumental in selecting for the reproductive characteristics that generate them? More detailed theoretical discussions in the context of these examples can be found in Schaffer and Tamarin (1973) and May (1977b).

The famous examples of dramatic cycles in population size are small, ravenous herbivores. However, the arguments apply with equal force to nearly all of the species that inhabit a strongly seasonal environment and that grow to reproductive maturity in a single season. Furthermore, even the most mathematically simple models of population growth that include a lag in response show violent cyclic and even chaotic behaviour in a constant environment, as long as the potential rate of growth is sufficiently high (May 1976 pp. 11–17). Hence reproductive adaptations to a fluctuating environment may be important even in species that are not well known for regular cycles in population size.

14.5 Dispersal

In an environment that changes in quality from time to time, there are two ways to avoid the effects of hard times. One is to endure them locally; the other, to migrate to a better place. Migration or dispersal of young is most favourable when the pattern of environmental fluctuation varies from place to place, since there will often be another place where conditions are better (Fig. 14.5). However, if the environmental changes are highly correlated from place to place, then hard times in one place mean hard times in another, and endurance, even to the point of dormancy or diapause, is favoured over dispersal. Thus one expects dispersal as an adaptation to local vagaries of weather, competition, predation, and patchy changes in the plant community; but dormancy in response to seasonal and other broad climatological changes. A more detailed discussion is given by Gadgil (1971).

It is no surprise that dispersal should be advantageous when the local environment is deteriorating, but Hamilton and May (1977) have recently argued that it is adaptive for parents to enforce dispersal of

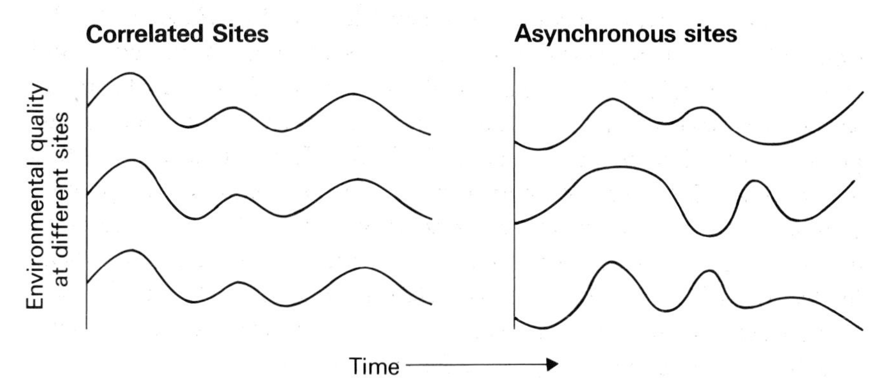

Fig. 14.5. Fluctuations in environmental quality at different sites. Where conditions are correlated from site to site, dispersers cannot improve their lot by moving; and local endurance, dormancy, or diapause is favoured. Where different sites fluctuate asynchronously, dispersal during bad times offers some chance of encountering better conditions elsewhere.

some of their young even in an environment that is crowded, but otherwise steadily healthy. Hamilton and May envision an environment that is crowded with sessile adults whose offspring compete within their own generation for the spaces vacated by the deaths of adults. Offspring who do not disperse must compete for their parent's site with both siblings and the dispersive offspring of other families. Therefore any parent whose entire family remains to compete for its space stands a finite chance of leaving no issue if any other members of the population have dispersive offspring; and it has no chance whatsoever to have more than one surviving offspring in the next generation. Dispersal of at least some offspring lowers the probability of total reproductive failure, and it gives a parent a more even chance in competition, with other parents of dispersive young, for placement of young in new vacancies. Hamilton and May discuss the complications that ensure when the strategy of dispersal that is optimal for parents is not optimal for their offspring, and they agonize over the restrictive assumptions of their model, but their main conclusion seems robust. It is advantageous for parents to enforce some dispersal by their offspring even when the local environment is not deteriorating and even when the process of dispersal itself incurs considerable mortality. Van Valen (1971) has presented a similar model showing the adaptive value of dispersal from local populations rather than families, over evolutionary spans of time rather than ecological.

14.6 r and K selection revisited

The major results of this chapter can be summarized in two diagrams that show the causal relations among environmental fluctuations, dynamics of populations, and aspects of natural history. Such diagrams are given for the extremes of *r*-selection and *K*-selection in Figs. 14.6

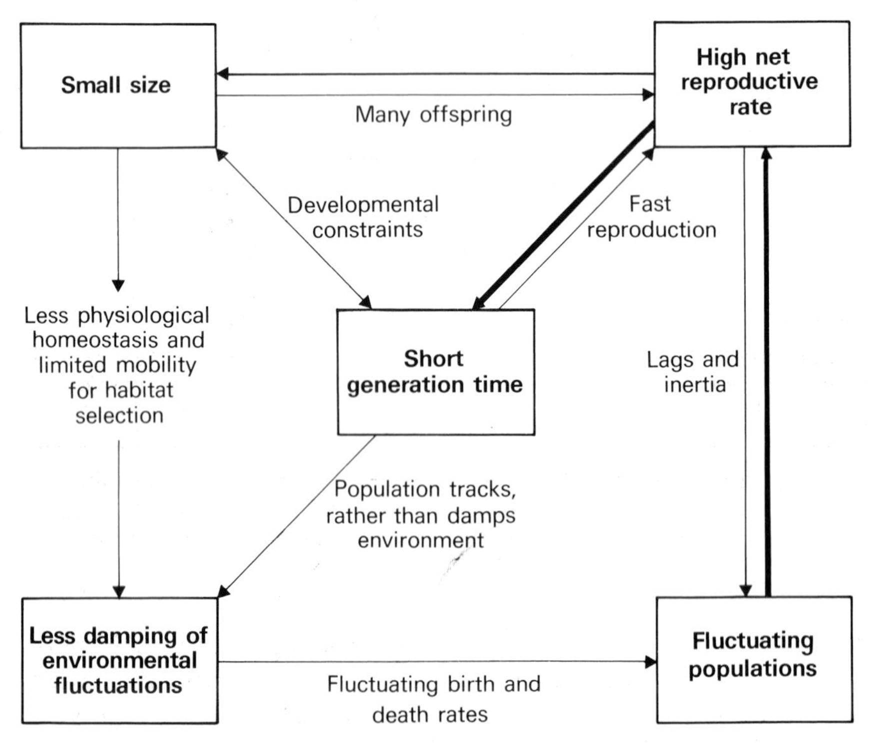

Fig. 14.6. Positive feedback loops reinforce *r*-selection. This is a summary of the relations among aspects of life-history, with arrows pointing from causes to effects that are discussed in the text. The heavy arrows represent actual selection. The idea of making this diagram is pilfered from Southwood (1977).

and 14.7. The diagrams also indicate the primary pathways by which selection favours rate of growth in one case and crowding tolerance in the other. Two additional relations in the diagram have not been discussed yet, but they are straightforward. Size is correlated with life-span or generation time, in part because smaller beasts live a more frenetic life and wear themselves out faster, and in part because it takes

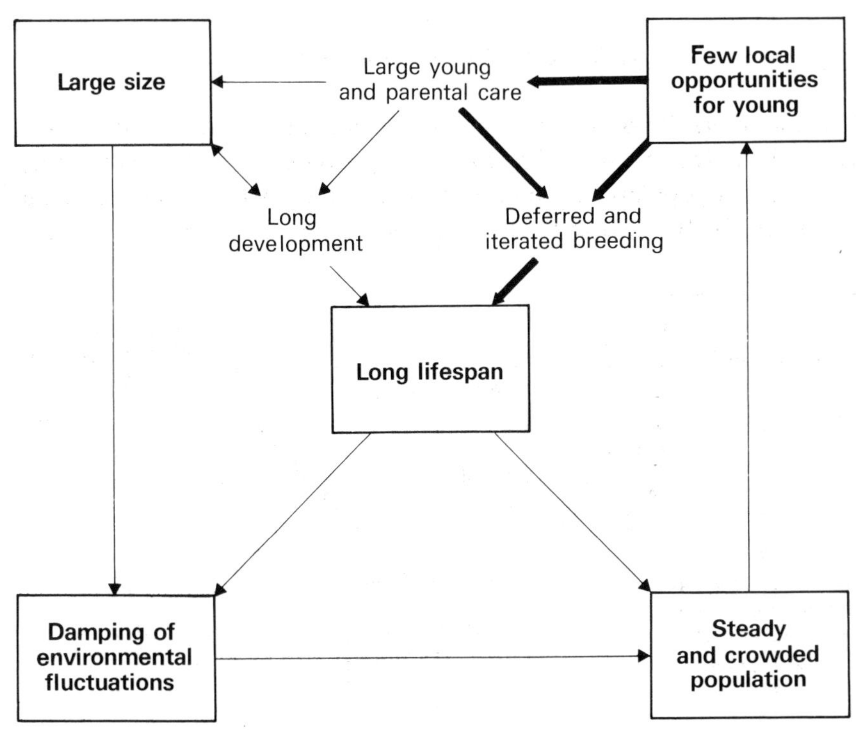

Fig. 14.7. Positive feedback loops reinforce *K*-selection. Some of the effects in this diagram are just the converse of those in Fig. 14.6, but the two diagrams are different in important respects that are discussed in the text.

longer to grow bigger. Small adult size implies greater sensitivity to environmental fluctuations, because small beasts cannot move easily to a new habitat when faced with local adversities, and though they may compensate by finding tiny hiding places where many physical fluctuations are damped, they are more sensitive to these physical fluctuations because of their small specific heat capacity and large ratio of surface to volume.

The first lesson of Figs. 14.6 and 14.7 is that *r*-selection and *K*-selection are both self-reinforcing. For example, selection in a fluctuating population favours a high reproductive rate, which is made easier by small size and a short generation time, which in turn reduces the ability of the population to damp environmental fluctuations, which increases the fluctuations in the population, further favouring a high reproductive rate, and so on. This positive feedback for both extremes of *r*-selection and *K*-selection suggests that animals in nature may show extreme characteristics more often than an intermediate set of characters. Perhaps this is a biological reason for the smouldering controversy over the mechanisms of population regulation in animals (Stubbs 1977, McLaren 1971).

Comparing Figs. 14.6 and 14.7, it is tempting to suggest that small animals are subject mainly to r-selection, and large animals to K-selection. There is enough truth in this suggestion for it to be a major tenet of modern ecology, but there is enough triviality in it to warrant scrutiny. Southwood and friends (1974) have noted the importance of relating the temporal scale of changes in the environment to the scale of the generation time of the animal. If environmental changes occur within a few generations, r-selection prevails. If the environment changes slowly over many generations, K-selection prevails. In complete ignorance of environmental effects, one can trivially state that most large animals will delay reproduction until they grow to be large. They may also be insensitive to environmental fluctuations that would destroy smaller animals, but they may be just as sensitive as the small beasts to longer term fluctuations that are commensurate with their lifespans. Stubbs (1977) has elegantly analysed data from a variety of animals, and found that mammals and birds are indeed dominated by K-selection, as befits large animals. However, she found that insects have some species whose life histories are dominated by r-selection, and some species dominated by K-selection.

The second lesson is that the diagrams of Figs. 14.6 and 14.7 are not simply the converse of one another. This hints that r-selection and K-selection are not simply opposite ends of a one-dimensional spectrum. In particular, the introduction of dispersal into population models has dramatically different consequences for r and K selection. Dispersal further reinforces r-selection because an abundance of small offspring is favourable for dispersal, and the mortality incurred by dispersal of young further increases the value of early and copious breeding. However, the introduction of dispersal into the diagram for K-selection may change its topology entirely. If dispersal provides youngsters with opportunities for colonization at some distance from their place of birth, then the fact that there are few local opportunities for them is less important, and the positive feedback loop that reinforces K-selection is broken. A whole new series of tactical questions is opened. Should many small young be broadcast at random? Should fewer young be sent out to seek their fortunes with a heavy grubstake from their parents? Should offspring remain at home and engage in territorial feuds with neighbouring clans? Should offspring look for vacancies or fight to create them? These questions need tactical answers before we can guess what mixture of tactical responses to K-selection and r-selection is the appropriate strategy for individuals in a crowded and steady population.

Figure 14.8 shows the relation between r and K selection at yet a further level of abstraction. In a given locality, population growth will turn r-selection into K-selection unless predation or regular catastrophes keep populations well below the capacity of the environment.

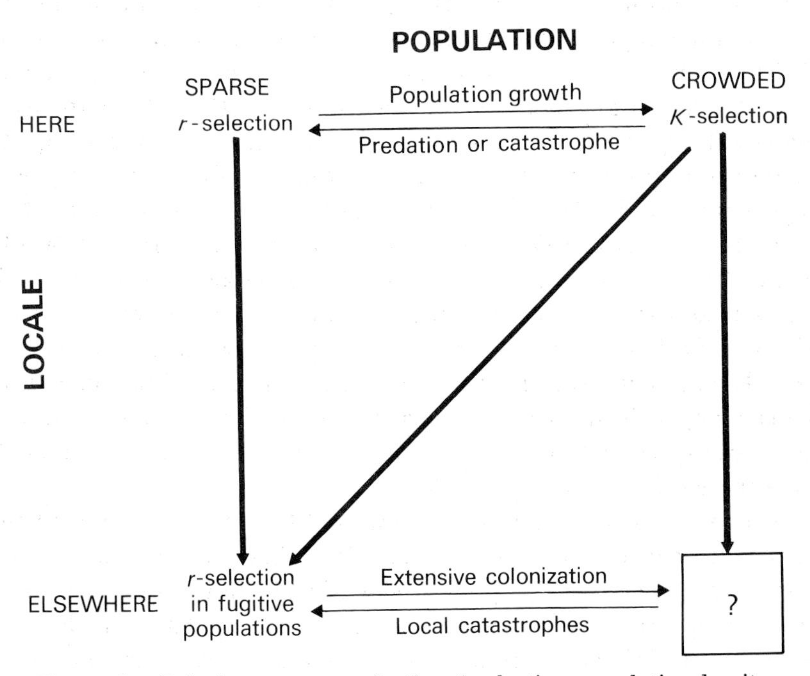

Fig. 14.8. Relations among r-selection, k-selection, population density, and dispersal. The heavy arrows represent dispersal. The big question mark indicates a critical gap in current theories.

Then the tactics of reproduction *per se* are less interesting than the adaptive response to predators or the ways of surviving or recovering from catastrophe (Wilbur *et al.* 1974). If a parent ensures the success of its descendants by sending them off to colonize newly vacant places, then r-selection continues indefinitely, but again the tactical machinery of dispersal may be of as much interest as reproduction itself. This may be why Southwood's (1977) list of synonyms of r-selection includes terms like fugitive species, opportunists, pioneers, and super tramps.

Dispersers from a crowded population may land in an uncrowded environment, in which case the adaptive life history may include elements favoured by r-selection and K-selection. However, if the dispersers land in a crowded population in another locality, adaptive tactics have yet to be explored by behavioural ecologists. It is not even clear whether dispersal may be favoured in globally crowded populations of motile species. Although the specific model of Hamilton and May (1977) does not apply, similar ideas could be developed in this context. In particular, in a crowded population of sedentary, long-lived adults, vacant territories due to recent deaths will be rare and scattered. Dispersal, especially in the form of itinerant adolescence, could expose youngsters to a large number of potential vacancies, even at the expense of encountering young competitors who are also looking for those same openings.

Woolfenden and Fitzpatrick (1978b) present a novel and exemplary analysis of dispersal in Florida scrub jays, mapping the dispersal of young of known parentage in a patchy and crowded environment. Male and female offspring both remain on their parents' territory, help to defend it, and assist in the feeding of the next batch of siblings. Females tend to spend a year or two helping while exploring widely in search of suitable mates and territories. Males stay at home for as much as three to five years, often extending their family's territory at the expense of adjacent families. A male helper then tends to 'bud off' his own territory from his father's, or to move quickly into any nearby vacancy. Thus in the face of the same crowded environment, males and females find breeding territories by very different patterns of dispersal (see also 3.5.2 and 9.2).

The tactically adaptive form of dispersal depends on the context. If adults can actively select habitats, then adults should disperse and leave well-provisioned offspring in appropriate places. If adults cannot move about or select habitats, they must broadcast young over both appropriate and inappropriate habitats. The first case is consistent with K-selection; the second entails very low survival of young, and therefore accords more with r-selection. The two cases have very different consequences for habitat selection, population dispersion, kinship structure, outbreeding, and even the adaptive significance of sex.

Wilbur (1977) has discussed these alternatives in the context of an engaging juxtaposition of data from salamanders and milkweeds. His four species of salamanders breed in ponds of varying size and persistence. Females of each species disperse their eggs in different ways, from one big bunch through several bunches, to singly or in pairs. The differences in dispersion are apparently adapted to patterns of larval survival in ponds that differ in persistence and predatory regime. The parents of milkweed seedlings cannot move about, but they can adjust the dispersal of their offspring through the number and timing of seed pods, and the size and number of seeds in each pod. Wilbur's six species of milkweeds show the same range of adjustments in dispersion of offspring as do his salamanders. However, he did not document any differences in the milkweeds' habitats to which these reproductive attributes might be adaptive.

There is a whole series of tactical questions about dispersal that have barely been asked, let alone been answered. Should adults disperse? Should young disperse? Should the tactics of parents and offspring agree? How should the tactics of dispersal vary with the level of crowding within a local population? I began with questions; I have ended with questions. It is a measure of progress that the two lists of questions are different.

References

The number(s) of the section(s) in which each reference appears is given on a separate line following the reference.

Alcock, J. 1975. *Animal Behaviour, an Evolutionary Approach.* Sunderland, Massachusetts, Sinauer.
5.7

Alcock, J. 1978. (In press.) The evolution of intraspecific diversity in male reproductive strategies in some bees and wasps. In Blum, M. S. & Blum, N. A. (eds.), *Reproductive Competition, Mate Choice and Sexual Selection.* New York, Academic Press.
8.4.1

Alexander, R. D. 1974. The evolution of social behaviour. *A. Rev. Ecol. Syst.*, 5, 325–383.
1.5.1, 3.1, 9.6, 9.8, 9.9.2

Alexander, R. D. 1975. Natural selection and specialized chorusing behaviour in acoustical insects. In Pimentel, D. (ed.), *Insects, Science and Society*, pp. 35–77. New York, Academic Press.
11.2.2

Alexander, R. D. & Sherman, P. W. 1977. Local mate competition and parental investment in social insects. *Science*, 196, 494–500.
1.5.1

Alexander, R. McN. 1967. *Functional Design in Fishes.* London, Hutchinson.
13.2.3

Allee, W. C. 1926. Studies in animal aggregations: causes and effects of bunching in land isopods. *J. exp. Zool.*, 45, 255–277.
3.4

Allee, W. C., Emerson, A. E., Park, O., Park, T., & Schmidt, K. P. 1949. *Principles of Animal Ecology.* Philadelphia, Saunders. 837 pp.
8.3.1

Allen, J. A. 1972. Evidence for stabilizing and apostatic selection by wild blackbirds. *Nature*, 237, 348–9.
5.3.2

Allen, J. A. & Clarke, B. 1968. Evidence for apostatic selection by wild passerines. *Nature*, 220, 501–2.
2.2.5, 5.3.2, 7.8

Altmann, S. A. 1956. Avian mobbing behaviour and predator recognition. *Condor*, 58, 241–253.
5.7

Altmann, S. A. & Altmann, J. 1970. *Baboon Ecology, African Field Research.* Chicago, University of Chicago Press.
3.4

Anderson, L. T. 1973. An analysis of habitat preference in mice as a function of prior experience. *Behaviour*, 47, 302–339.
12.5

Anderson, P. K. 1964. Lethal allele in *Mus musculus*, local distribution and evidence for isolation of demes. *Science*, 145, 177–178.
12.8

Andrew, R. J. 1961a. The motivational organisation controlling the mobbing calls of the blackbird (*Turdus merula*). I. Effects of flight on mobbing calls. *Behaviour*, 17, 224–246.
5.7

Andrew, R. J. 1961b. The motivational organisation controlling the mobbing calls of the blackbird (*Turdus merula*). II. The quantitative analysis of changes in the motivation of calling. *Behaviour*, 17, 288–321.
5.7

Andrew, R. J. 1961c. The motivational organisation controlling the mobbing calls of the blackbird (*Turdus merula*). III. Changes in the intensity of mobbing due to changes in the effect of the owl or to the progressive waning of mobbing. *Behaviour*, 18, 25–43.
5.7

Andrew, R. J. 1961d. The motivational organisation controlling the mobbing calls of the blackbird (*Turdus merula*). III. A general discussion of the calls of the blackbird and certain other passerines. *Behaviour*, 18, 161–176.
5.5.1, 5.7

Armstrong, E. A. 1947. *Bird Display and Behaviour: an Introduction to the Study of Bird Psychology*. London, Lindsay Drummond.
5.6

Armstrong, E. A. 1955. *The Wren*, London, Collins.
3.4

Armstrong, E. A. 1956. Distraction display and the human predator. *Ibis*, 98, 641–654.
5.6

Armstrong, E. A. 1973. *A Study of Bird Song*. New York, Dover.
10.5.4

Arnold, S. J. 1976. Sexual behaviour, sexual interference and sexual defense in the salamanders *Ambystoma maculatum*, *Ambystoma tigrinum* and *Plethodon jordani*. *Z. Tierpsychol.*, 42, 247–300.
7.10, 7.10.1

Baerends, G. P. & Baerends van Roon, J. M. 1950. An introduction to the study of the ethology of cichlid fishes. *Behav. Suppl.*, 1, 1–242.
10.3.2

Baker, H. G. 1955. Self-compatibility and establishment after 'long-distance' dispersal. *Evolution*, 9, 347–348.
6.2

Baker, H. G. 1959. Reproductive methods as factors in speciation in flowering plants. *Cold Spring Harbor Symp. Quant. Biol.*, 24, 177–191.
6.2

Baker, R. R. 1968. A possible method of evolution of the migratory habit in butterflies. *Phil. Trans. R. Soc. (Lond.) Ser. B*, 253, 309–341.
8.4.1

Baker, R. R. 1978. *The Evolution of Animal Migration*. London, English University Press.
8.3.1

Bantock, C. R. & Bayley, J. A. 1973. Visual selection for shell size in *Cepaea*. *J. Anim. Ecol.*, 42, 247–261.
5.3.3

Bantock, C. R. & Harvey, P. H. 1976. Colour polymorphism and selective predation experiments. *J. Biol. Educ.*, 8, 323–329.
5.3.3, 5.8

Bantock, C. R., Bayley J. A. & Harvey, P. H. 1976. Simultaneous selective predation of two features in a mixed sibling species population. *Evolution*, 29, 636–649.
5.3.3

Barash, D. P. 1974. The evolution of marmot societies: a general theory. *Science, N.Y.*, 185, 415–420.
3.5.1

Barash, D. P. 1975. Evolutionary aspects of parental behaviour of the alpine accentor. *Wilson Bulletin*, 87, 367–373.
5.6

Barash, D. P. 1976a. Social behaviour and individual differences in free-living Alpine marmots (*Marmota marmota*). *Anim. Behav.*, 24, 27–35.
5.5.3

Barash, D. P. 1976b. What does sex really cost? *Amer. Natur.* 110, 894–897.
14.3

Baron, R. A., Byrne, D. & Griffitt, W. 1974. *Social Psychology.* Borton, Allyn and Bacon.
10.5.2

Bastock, M. 1956. A gene mutation which changes a behaviour pattern. *Evolution*, 10, 421–439.
7.9.2

Bastock, M. 1967. *Courtship, A Zoological Study.* London, Heinemann.
7.9.2, 10.3.4

Bateman, A. J. 1948. Intra-sexual selection in *Drosophila. Heredity*, 2, 349–368.
7.1.2, 8.1

Bawa, K. S. & Opler, P. A. 1975. Dioecism in tropical forest trees. *Evolution*, 29, 167–179.
6.2

Bengtsson, B. O. 1978. Avoid inbreeding: at what cost? *J. theoret. Biol.* (in press).
6.2

Benson, W. W. 1971. Evidence for the evolution of unpalatibility through kin selection in the Heliconiinae (Lepidoptera). *Amer. Natur.*, 105, 213–226.
5.8

Berkowitz, L. 1975. *A Survey of Social Psychology.* Illinois, the Dryden Press.
10.5.2

Bernstein, R. 1971. The Ecology of Ants in the Mojave Desert: their interspecific relationships, resource utilization and diversity. Ph.D. dissertation. University of California, Los Angeles.
4.3.2

Bernstein, R. 1975. Foraging strategies of ants in response to variable food density. *Ecology*, 56, 213–219.
4.3.2

Berry, R. J. & Davis, P. E. 1970. Polymorphism and behaviour in the Arctic Skua. *Proc. Roy. Soc. B.*, 175, 255–267.
7.6.2

Bertram, B. C. R. 1975. Social factors influencing reproduction in wild lions. *J. Zool. Lond.*, 177, 463–482.
3.1, 3.5.1 3.5.2, 3.5.3, 3.5.4, 3.5.7, 9.8

Bertram, B. C. R. 1976. Kin selection in lions and in evolution. In Bateson, P. P. G. & Hinde, R. A. (eds.), *Growing Points in Ethology*, pp. 281–301. Cambridge. Cambridge University Press.
3.5.3, 3.5.4, 3.5.5, 7.10

Bertram, B. C. R. (In press). Serengeti predators and their social systems. In Norton-Griffiths, M. & Sinclair, A. R. E. (eds.), *Serengeti: The Dynamics of the Ecosystem.* Chicago, University of Chicago Press.
3.3.2

Beukema, J. J. 1968. Predation by the three-spined stickleback *Gasterosteus aculeatus* L, the influence of hunger and experience. *Behaviour*, 31, 1–126.
2.4.2

Birkhead, T. R. 1977. The effect of habitat and density on breeding success in common guillemot (*Uria aalge*). *J. Anim. Ecol.*, 46, 751–764.
11.2.3

Bischof, N. 1975. Comparative ethology of incest avoidance. In Fox, R. (ed.), *Biosocial Anthro-*

pology, pp. 37–67. New York, Malaby Press.
3.5.2

Black, R. 1971. Hatching success in the three-spined stickleback (*Gasterosteus aculeatus*) in relation to changes in behaviour during the parental phase. *Anim. Behav.*, 19, 532–541.
11.2.2

Blair, W. F. 1974. Character displacement in frogs. *Amer. Zool.*, 14, 1119–1125.
7.2.5

Blest, A. D. 1963. Longevity, palatability and natural selection in five species of New World saturniid moths. *Nature*, 197, 1183–1186.
5.2

Boch, R. 1956. Die Tänze der Bienen bei nahen und fernen Trachtquellen. *Z. vergl. Physiol.*, 38, 136–167.
4.2.3

Bourlière, F. 1963. Specific feeding habits of African carnivores. *Afr. wild Life*, 17, 20–27.
3.3.3

Bousfield, W. A. 1933. Certain quantitative aspects of the feeding behaviour of cats. *J. gen. Psychol.*, 8, 446–454.
13.4.4

Brackbill, H. 1958. Titmouse mother's helper. *Baltimore Evening Sun*. (June 18, 1958). Not seen. Cited by Skutch, 1961.
9.6, 9.9.2

Branch, G. M. 1975. Mechanisms reducing intraspecific competition in *Patella* spp., migration, differentiation and territorial behaviour. *J. Anim. Ecol.*, 44, 575–600.
11.2.5

Breder, C. M. 1976. Fish schools as operational structures. *Fishery Bull. Fish Wildl. Serv. U.S.*, 74, 471–502.
3.4

Breder, C. M. & Rosen D. E. 1966. *Modes of Reproduction in Fishes*. New York, Natural History Press.
6.4

Brian, A. D. 1952. Division of labour and foraging in *Bombus agrorum* Fabricus. *J. Anim. Ecol.*, 21, 223–240.
4.2.2

Britten, R. J. 1977. The sources of variation in evolution. In Duncan, R. & Weston-Smith, M. (eds.), *The Encyclopaedia of Ignorance, Life Sciences and Earth Sciences*. Pergamon Press, Oxford.
1.5

Brooke, M. de L. 1978. Some factors affecting laying date, incubation and breeding success of the Manx Shearwater (*Puffinus puffinus* Brünnich). *J. Anim. Ecol.* 47, 477–495.
10.3.4

Brooks, R. J. & Falls, J. B. 1975. Individual recognition by song in white-throated sparrows. I. Discrimination of songs of neighbors and strangers. *Can. J. Zool.*, 53, 879–888.
11.4.1

Brower, J. V. Z. 1958a. Experimental studies of mimicry in some North American butterflies. Part I. The Monarch *Danaus plexippus*, and the Viceroy, *Limenitis archippus archippus*. *Evolution*. 12, 32–47.
5.2

Brower, J. V. Z. 1958b. Experimental studies of mimicry in some North American butterflies Part II. *Battus philenor* and *Papio troilus*, *P. polyxenes* and *P. glaucus*. *Evolution*, 12, 123–136.
5.2

Brower, J. V. Z. 1958c. Experimental studies of mimicry in some North American butterflies. Part III. *Danaus gilippus berenice* and *Limenitis archippus floridensis*. *Evolution*, 12, 273–285.
5.2

Brower, L. P. & Brower J. V. Z. 1964. Birds, butterflies and plant poisons, a study in ecological

chemistry. *Zoologica*, New York, 49, 137–159.
5.2

Brown, J. L. 1964. The evolution of diversity in avian territorial systems. *Wilson Bull.*, 76, 160–169.
1.1, 11.2

Brown, J. L. 1969. The buffer effect and productivity in tit populations. *Amer. Natur.* 103, 347–354.
11.3.1

Brown, J. L. 1972. Communal feeding of nestlings in the Mexican jay (*Aphelocoma ultramarina*), interflock comparisons. *Anim. Behav.*, 20, 395–403.
3.5.3

Brown, J. L. 1974. Alternate routes to sociality in jays with a theory for the evolution of altruism and communal breeding. *Am. Zool.*, 14, 63–80.
9.3, 9.8

Brown, J. L. 1975a. Helpers among Arabian babblers, *Turdoides squamiceps*. *Ibis*, 117, 243–244.
9.3

Brown J. L. 1975b. *The Evolution of Behaviour*. New York, W. W. Norton and Co. Inc.
9.5, 12.4.1

Brown, J. L. In press. Communal breeding in birds. *A. Rev. Ecol. Syst.*
9.4

Brown, J. L. & Balda, R. P. 1977. The relationship of habitat quality to group size in Hall's babbler, *Pomatostomus halli*. *Condor*, 79, 312–320.
9.4

Brown, J. L. & Orians, G. H. 1970. Spacing patterns in mobile animals. *A. Rev. Ecol. Syst.*, 1, 239–262.
11.1

Bruce, H. M. 1966. Smell as an exteroceptive factor. *J. Anim. Sci.*, Suppl. 25, 83–89.
7.10

Bruning, D. F. 1974a. Social structure and reproductive behaviour in the Greater Rhea. *Living Bird*, 13, 251–294.
9.7

Bruning, D. F. 1974b. Social structure and reproductive behaviour in the Argentine gray rhea, *Rhea americana albescens*. Unpublished thesis, University of Colorado. 159 pp.
9.7

Buechner, H. K. & Schloeth, R. 1965. Ceremonial mating behaviour in Uganda kob (*Adenota kob thomasi* Neumann). *Z. Tierpsychol.*, 22, 209–225.
7.10, 11.2.2

Bulmer, M. C. 1972. Multiple niche polymorphism. *Amer. Natur.*, 106, 254–257.
12.8

Burgess, J. W. 1976. Social spiders. *Sci Amer.*, 234(3), 100–106.
10.3.2

Bush, G. L. 1969a. Mating behaviour, host specificity, and the ecological significance of sibling species in frugivorous flies of the genus *Rhagoletis* (Diptera: Tephritidae). *Amer. Natur.*, 103, 669–672.
12.8

Bush, G. L. 1969b. Sympatric host race formation and speciation in frugivorous flies of the genus *Rhagoletis* (Diptera: Tephritidae). *Evolution*, 23, 237–251.
12.8

Bush, G. L. 1974. The mechanism of sympatric host race formation in the true fruit flies (Tephritidae). In White, M. J. (ed.), *Genetic Mechanisms of Speciation in Insects*. Sydney, Australian and New Zealand Book Co.
12.8

Cain, A. J. 1953. Visual selection by tone of *Cepaea nemoralis*. *J. Conchol.* 23, 333–336.
5.3.3

Cain, A. J. & Sheppard, P. M. 1954.

Natural selection in *Cepaea. Genetics*, 39, 89–116.
5.3.1

Cain, A. J., Sheppard, P. M. & King, M. B. 1968. Studies on *Cepaea*. I. The genetics of some morphs and varieties of *Cepaea nemoralis* (C.) *Phil. Trans. R. Soc.* (Lond). Ser. B. 253, 383–396.
5.3.3

Calder, W. A. 1973. Microhabitat selection during nesting of hummingbirds in the Rocky Mountains. *Ecology*, 54, 127–134.
12.6.1

Cameron, A. W. 1964. Competitive exclusion between the rodent genera *Microtus* and *Clethrionomys. Evolution*, 18, 630–634.
12.4.2

Campanella, P. J. 1975. The evolution of mating systems in temperate zone dragonflies (Odonata: Anisoptera) II. *Libellula luctuosa* (Burmeister). *Behaviour*, 54, 278–310.
11.2

Campanella, P. J. & Wolf, L. L. 1974. Temporal leks as a mating system in a temperate zone dragonfly (Odonata: Anisoptera). I. *Plathemis lydia* (Drury). *Behaviour*, 51, 49–87.
11.2, 11.2.2

Caraco, T. & Wolf, L. L. 1975. Ecological determinants of group sizes of foraging lions. *Amer. Natur.* 109, 343–352.
3.6

Carl, E. 1971. Population control in arctic ground squirrels. *Ecology*, 52, 395–413.
11.3

Carpenter, F. L. & MacMillen, R. E. 1976a. Threshold model of feeding territoriality and test with a Hawaiian Honeycreeper. *Science*, 194, 639–642.
11.2.1

Carpenter, F. L. & MacMillen, R. E. 1976b. Energetic cost of feeding territories in an Hawaiian honeycreeper. *Oecologia*, 26, 213–224.
11.2.1

Carter, M. A. 1967. Selection in mixed colonies of *Cepaea nemoralis* and *Cepaea hortensis. Heredity*, 22, 117–139.
5.3.3

Catchpole, C. K. 1972. A comparative study of territory in the reed warbler (*Acrocephalus scirpaceus*) and the sedge warbler (*A. schoenobaenus*). *J. Zool. Lond.*, 166, 213–231.
11.2.5

Catchpole, C. K. 1977. Aggressive responses of male sedge warblers (*Acrocephalus schoenobaenus*) to playback of species song and sympatric song, before and after pairing. *Anim. Behav.*, 25, 489–496.
11.2.5

Chadab, R. & Rettenmeyer, C. W. 1975. Mass recruitment by army ants. *Science*, 188, 1124–1125.
4.3.1

Charnov, E. L. 1976a. Optimal foraging: attack strategy of a mantid. *Amer. Natur.* 110, 141–151.
2.2.2

Charnov, E. L. 1976b. Optimal foraging: the marginal value theorem. *Theor. Popul. Biol.*, 9, 129–136.
2.3.3, 8.4.1

Charnov, E. L. & Schaffer, W. M. 1973. Life history consequences of natural selection: Cole's result revisited. *Amer. Natur.* 107, 791–793.

Charnov, E. L. & Krebs, J. R. 1975. The evolution of alarm calls: altruism or manipulation? *Amer. Natur.* 109, 107–112.
5.5.1, 10.2.2

Charnov, E. L., Orians, G. H. & Hyatt, K. 1976a. The ecological implications of resource depression. *Amer. Natur.* 110, 247–259.
2.4.3, 11.2.1

Charnov, E. L., Maynard Smith, J.

& Bull, J. J. 1976b. Why be a hermaphrodite? *Nature*, 263, 125–126.
6.2

Chauvin, R. 1950. Le transport de proies chez les fourmis. Y-a-t-il entr'aide? *Behaviour*. 2, 249–256.
4.3.1

Cherrett, J. M. 1968. The foraging behavior of *Atta cephalotes* L. (Hymenoptera, Formicidae). *J. Anim. Ecol.*, 37, 387–403.
4.3.3

Cherrett, J. M. 1972. Chemical aspects of plant attack by leaf-cutting ants. In Harborne, J. B. (ed.), *Phytochemical Ecology*, pp. 13–24. Academic Press, London and New York.
4.3.3

Clarke, B. C. 1962a. Balanced polymorphism and the diversity of sympatric species. *Syst. Assoc. Publ.*, 4, 47–70.
5.3.2, 5.3.3

Clarke, B. C. 1962b. Natural selection in mixed populations of two polymorphic snails. *Heredity*, 17, 319–345.
5.3.3

Clarke, B. C. 1969. The evidence for apostatic selection. *Heredity*, 24, 347–352.
5.3.3

Clarke, C. A., Dickson, C. G. C. & Sheppard, P. M. 1963. Larval colour pattern in *Papilio demodocus*. *Evolution*, 17, 130–137.
5.3.1

Clarke, C. J. 1976. *Mathematical Bioeconomics*. New York, Wiley.
14.3

Clarke, T. A. 1970. Territorial behaviour and population dynamics of a pomacentrid fish, the Garibaldi, *Hypsypops rubicunda*. *Ecol. Monogr.*, 40, 189–212.
11.3.1

Clutton-Brock, T. H. 1974. Primate social organisation and ecology. *Nature*, 250, 539–542.
1.3.2

Clutton-Brock, T. H. 1977. Some aspects of intraspecific variation in feeding and ranging behaviour in primates. In Clutton-Brock, T. H. (ed.), *Primate Ecology, Studies of feeding and ranging behaviour in lemurs, monkeys and apes*. Academic Press, London.
1.3.2

Clutton-Brock, T. H. & Harvey, P. H. 1976. Evolutionary rules and primate societies. In Bateson, P. P. G. & Hinde, R. A. (eds.), *Growing Points in Ethology*, pp. 195–237. Cambridge, Cambridge University Press.
3.5.1, 3.5.3, 5.4

Clutton-Brock, T. H. & Harvey, P. H. 1977. Primate ecology and social organisation. *J. Zool., Lond.* 183, 1–39.
1.3.1, 1.3.2, 3.5.5, 5.4, 7.5.2

Cody, M. L. 1971. Finch flocks in the Mohave desert. *Theor. Pop. Biol.*, 2, 142–148.
2.4.1, 2.4.3

Cody, M. L. 1974a. Optimisation in ecology. *Science*, 183, 1156–1164.
2.4.3, 2.7, 3.3.1, 13.7.4

Cody, M. L. 1974b. *Competition and the Structure of Bird Communities*. Princeton University Press, Princeton.
11.4.1, 12.2.1, 12.4.2

Cody, M. L. & Cody, C. B. J. 1972. Territory size, clutch size and food in populations of wrens. *Condor*, 74, 473–477.
11.2.1

Cole, L. C. 1954. The population consequences of life history phenomena. *Quart. Rev. Biol.*, 29, 103–137.
14.3

Connell, J. H. 1961. The influence of interspecific competition and other factors on the distribution of the barnacle *Chthamalus stellatus*. *Ecology*, 42, 710–723.
12.4.2

Constantz, G. D. 1975. Behavioural ecology of mating in the male Gila Topminnow, *Poeciliopsis occidentalis* (Cyprinodentiformes: Poeciliidae). *Ecology*, 56, 966–973.
11.2.2

Cook, L. M. 1965. A note on apostasy. *Heredity*, 20, 631–636.
5.3.2

Cook, L. M. & Miller, P. 1977. Density-dependent selection on polymorphic prey—some data. *Amer. Natur.* 111, 594–598.
5.3.2

Cook, R. M. & Hubbard, S. F. 1977. Adaptive searching strategies in insect parasites. *J. Anim. Ecol.*, 46, 115–125.
2.3.5

Cook, R. M. & Cockrell, B. J. 1978. Predator ingestion rate and its bearing on feeding time and the theory of optimal diets. *J. Anim. Ecol.*, 47, 529–547.
2.3.4

Cooke, F., Finney, G. H. & Rockwell, R. F. 1976. Assortative mating in Lesser Snow Geese (*Anser caerulescens*). *Behav. Genet.*, 6, 127–140.
7.8

Cory, B. L. & Manion, J. J. 1955. Ecology and hybridisation in the genus *Bufo* in the Michigan-Indiana region. *Evolution*, 9, 42–51.
12.8

Coulson, J. C. 1966. The influence of the pair-bond and age on the breeding biology of the Kittiwake Gull *Rissa tridactyla*. *J. Anim. Ecol.*, 35, 269–279.
7.9.2

Coulson, J. C. 1968. Differences in the quality of birds nesting in the centre and on the edges of a colony. *Nature*, 217, 478–479.
3.2.6

Coulson, J. C. 1971. Competition for breeding sites causing segregation and reduced young production in colonial animals. In den Boer, P. J. & Gradwell, G. R. (eds.), *Proc. Adv. Study Inst. Dynamics Numbers in Pops. Oosterbeck* 1970, pp. 257–268. Wageningen, Pudoc.
10.3.4

Cowie, R. J. 1977. Optimal foraging in Great Tits (*Parus major*). *Nature*, 268, 137–139.
2.3.4, 8.4.1

Cowie, R. J. & Krebs, J. R. (In press). Optimal foraging in patchy environments. In *British Ecological Society Symposium on Population Dynamics*.
2.3.4, 2.3.5

Cox, C. R. & Le Boeuf, B. J. 1977. Female incitation of male competition: a mechanism of mate selection. *Amer. Natur.*, 111, 317–335.
7.4, 11.2.2

Craig, G. B. 1967. Mosquitoes, female monogamy induced by male accessory gland substance. *Science*, 156, 1499–1501.
7.10

Crisler, L. 1956. Observations of wolves hunting caribou. *J. Mammal.*, 37, 337–346.
3.2.2

Crisp, D. J. 1961. Territorial behaviour in barnacle settlement. *J. Exp. Biol.*, 38, 429–446.
11.1

Crook, J. H. 1964. The evolution of social organisation and visual communication in the weaver birds (Ploceinae). *Behaviour* Suppl. 10, 1–178.
1.3.1

Crook, J. H. 1965. The adaptive significance of avian social organisation. *Symp. Zool. Soc. Lond.*, 14, 181–218.
1.3.1, 11.2

Crook, J. H. 1970. The socio-ecology of primates. In Crook, J. H. (ed.), *Social Behaviour in Birds and Mammals*. London, Academic Press.
3.5.1, 11.2

Crook, J. H. & Gartlan, J. S. 1966. Evolution of primate societies. *Nature*, 210, 1200–1203.
1.3.1, 1.3.2, 3.2.3, 3.5.4

Crow, J. F. & Kimura, M. 1965. Evolution in sexual and asexual populations. *Amer. Natur.*, 99, 439–450.
6.1

Croxall, J. P. 1976. The composition and behaviour of some mixed-species bird flocks in Sarawak. *Ibis*, 118, 333–346.
3.4

Croze, H. J. 1970. Searching image in carrion crows. *Z. Tierpsychol. Beiheft.*, 5, pp. 86.
5.3.2, 11.2.3

Cuellar, O. 1976. Intraclonal histocompatibility in a parthenogenetic lizard, evidence of genetic homogeneity. *Science*, 193, 150–153.
6.1.1

Cullen, J. M. 1966. Reduction of ambiguity through ritualization. *Phil. Trans. Roy. Soc. B.*, 251, 363–374.
10.2.2

Cullen, J. M. 1972. Some principles of animal communication. In Hinde, R. A. (ed.), *Non Verbal Communication*, pp. 101–122. Cambridge, Cambridge University Press.
10.1.1

Curio, E. 1959. Verhaltensstudien am Trauerschnapper. *Z. Tierpsychol., Beiheft*, 3, 1–118.
5.7

Curio, E. 1975. The functional organization of anti-predator behaviour in the pied flycatcher, a study of avian visual perception. *Anim. Behav.*, 23, 1–115.
5.7, 5.8

Curio, E. 1976. *The Ethology of Predation*. Berlin, Springer-Verlag.
5.1, 5.3.2

Cushing, J. E. 1941. An experiment on olfactory conditioning in *Drosophila guttifera*. *Proc. Nat. Acad. Sci.*, 27, 496–499.
12.5

Darwin, C. 1871. *The Decent of Man, and Selection in Relation to Sex.* London, John Murray.
7.2.1

Darwin, C. 1872. *The Expression of the Emotions in Man and Animals.* London, John Murray.
10.1.5

Darwin, C. 1877. *The Different forms of Flowers on Plants of the Same Species.* London, John Murray.
6.2

Davidson, D. W. 1977. Foraging ecology and community organization in desert seed-eating ants. *Ecology*, 58, 725–737.
4.3.2

Davies, N. B. 1976a. Food, flocking and territorial behaviour of the pied wagtail (*Motacilla alba yarrellii*) in winter. *J. Anim. Ecol.*, 45, 235–254.
11.2, 11.2.1, 11.4

Davies, N. B. 1976b. Parental care and the transition to independent feeding in the young spotted flycatcher (*Musucicapa striata*). *Behaviour*, 59, 280–295.
1.5.1

Davies, N. B. 1977a. Prey selection and social behaviour in wagtails (Aves: Motacillidae). *J. Anim. Ecol.*, 46, 37–57.
2.2.1, 2.3.3

Davies, N. B. 1977b. Prey selection and the search strategy of the spotted flycatcher (*Muscicapa striata*), a field study on optimal foraging. *Anim. Behav.*, 25, 1016–1033.
2.3.6

Davies, N. B. 1978a. (In press). Parental meanness and offspring independence: an experiment with hand-reared great tits (*Parus major*). *Ibis*, 120.
1.5.1

Davies, N. B. 1978b. Territorial defence in the speckled wood butterfly (*Pararge aegeria*), the resident always wins. *Anim. Behav.*, 26, 138–147.
10.3.2, 11.3.1, 11.4, 11.4.2

Davies, N. B. & Green, R. E. 1976. The development and ecological significance of feeding techniques in the reed warbler (*Acrocephalus scirpaceus*). *Anim. Behav.*, 24, 213–229.
11.2.5

Davies, N. B. & Halliday, T. R. 1977. Optimal mate selection in the toad *Bufo bufo*. *Nature*, 269, 56–58.
7.9.4, 8.4.1

Davis, D. E. 1940. Social nesting habits of the smooth-billed Ani. *Auk*, 57, 179–218.
9.7

Davis, J. W. F. 1976. Breeding success and experience in the Arctic Skua, *Stercorarius parasiticus* (L.). *J. Anim. Ecol.*, 45, 531–535.
7.6.2

Davis, J. W. F. & O'Donald, P. 1976a. Sexual selection for a handicap, a critical analysis of Zahavi's model. *J. theor. Biol.*, 57, 345–354.
7.2.3

Davis, J. W. F. & O'Donald, P. 1976b. Territory size, breeding time and mating preference in the Arctic Skua. *Nature*, 260, 774–775.
7.6.2, 11.2.2

Davis, J. W. F. & O'Donald, P. 1976c. Estimation of assortative mating preferences in the Arctic Skua. *Heredity*, 36, 235–244.
7.6.2

Dawkins, M. 1971a. Perceptual changes in chicks, another look at the 'search image' concept. *Anim. Behav.*, 19, 566–574.
2.2.5

Dawkins, M. 1971b. Shifts of 'attention' in chicks during feeding. *Anim. Behav.*, 19, 575–582.
2.2.5

Dawkins, R. 1976. *The Selfish Gene*. Oxford University Press, Oxford.
1.4, 1.5, 3.5.3, 5.4, 5.5.1, 7.3, 10.1.1, 10.2.2

Dawkins, R. & Carlisle, T. R. 1976. Parental investment, mate desertion and a fallacy. *Nature*, 262, 131–133.
5.6, 6.4

Dawkins, R. & Dawkins, M. 1973. Decisions and the uncertainty of behaviour. *Behaviour*, 45, 83–103.
10.2.3

Day, M. G. 1968. Food habits of British stoats (*Mustela erminea*) and weasels (*Mustela nivalis*). *J. Zool. Lond*, 155, 485–497.
3.3.3

Deakin, M. A. B. 1966. Sufficient conditions for genetic polymorphism. *Amer. Natur.*, 100, 690–691.
12.8

De Benedictis, P. A., Gill, F. B., Hainsworth, F. R., Pyke, G. H. & Wolf, L. L. 1978. Optimal meal size in hummingbirds. *Amer. Natur.* 112: 301–316.
2.6

De Boer, J. N. & Heuts, B. A. 1973. Prior exposure to visual cues affecting dominance in the jewel fish *Hemichromis bimaculatus* Gill 1862 (Pisces, Cichlidae). *Behaviour*, 44, 299–321.
10.3.2

Dempster, M. A. H. (In press). Optimal temporal organisation of animal behaviour. In Dempster, M. A. H. & McFarland, D. J. (eds.), *Animal Economics*. London and New York, Academic Press.
13.7.3

Devine, M. C. 1977. Copulatory plugs, restricted mating opportunities and reproductive competition among male garter snakes. *Nature*, 267, 345–346.
8.3.3

Dexter, R. W. 1952. Extra-parental cooperation in the nesting of Chimney Swifts. *Wilson Bull.*, 64, 133–139.
9.8

Dill, L. M. (In press). An energy-based model of optimal feeding territory size. *Theor. Pop. Biol.*,
11.3.2

Douglass, R. J. 1976. Spatial interactions and microhabitat selections of two locally sympatric voles, *Microtus montanus* and *Microtus pennsylvanicus*. *Ecology*, 57, 346–352.
12.2.2

Douglas-Hamilton, I. 1973. On the ecology and behaviour of the Lake Manyara elephants. *E. Afr. Wildl. J.*, 11, 401–403.
3.5.3

Douthwaite, R. J. 1970. Some aspects of the biology of the Lesser Pied Kingfisher, *Ceryle rudis*. Unpublished Ph.D. Thesis, Makerere College. (Not seen, cited by Fry, 1972a.)
9.8

Downhower, J. F. & Armitage, K. B. 1971. The Yellow-bellied Marmot and the evolution of polygamy. *Amer. Natur.*, 105, 355–370.
7.9.4

Doyle, R. W. 1975. Settlement of planktonic larvae, a theory of habitat selection in varying environments. *Amer. Natur.*, 109, 113–126.
12.6.1, 12.6.2

Drent, R. H. 1970. Functional aspects of incubation in the Herring gull (*Larus argentatus*, Pont.). *Behaviour* Suppl. 17, 1–132.
13.3.2

Driver, P. M. & Humphries, D. A. 1969. The significance of the high-intensity alarm call in captured passerines. *Ibis*, 111, 243–244.
5.5.1

Dunbar, R. I. M. & Dunbar, E. P. 1974. Social organization and ecology of the Klipspringer (*Oreotragus oreotragus*) in Ethiopia. *Z. Tierpsychol.*, 35, 481–493.
3.2.2

Dunford, C. 1977. Kin selection for ground squirrel alarm calls. *Amer. Natur.* 111, 782–785.
5.5, 5.5.3

Dunn, E. 1977. Predation by weasels (*Mustela nivalis*) on breeding tits (*Parus* spp.) in relation to the density of tits and rodents. *J. Anim. Ecol.*, 46, 634–652.
11.2.3, 11.2.5

Eberhard, M. J. 1969. The social biology of polistine wasps. *Misc. Publ. Mus. Zool.* Univ. Michigan, 140, 1–101.
13.5

Ebersole, J. P. 1977. The adaptive significance of interspecific territoriality in the reef fish *Eupomacentrus leucostictus*. *Ecology*, 58, 914–920.
11.2.5

Edmunds, M. 1974. *Defence in Animals*. Harlow, Essex, Longman.
5.1, 5.3

Edwards, W. 1954. The theory of decision making. *Psychol. Bull.*, 51, 380–417. (Reprinted in Edwards, W. & Tversky, A. (eds.) 1967. *Decision Making*. Penguin Books.)
13.2.1

Ehrlich, P. R. & Raven, R. H. 1967. Butterflies and plants. *Sci. Am.*, 216(6), 105–113.
5.2

Eibl-Eibesfeldt. 1952. Nahrungserwerb und Beuteschema der Erdkröte (*Bufo bufo* L.). *Behaviour*, 4, 1–36.
5.2

Eidmann, H. 1935. Zur Kenntnis der Blattschneiderameise *Atta sexdens* ins besondere ihre Oekologie. *Z. angew. Ent.*, 22, 385–436.
4.3.3

Eisenberg, J. F., Muckenhirn, N. A. & Rudran, R. 1972. The relation between ecology and social structure in primates. *Science, N.Y.*, 176, 863–874.
3.2.3

Elgerd, O. I. 1967. *Control Systems Theory*. New York, McGraw Hill.
13.4.2

Elner, R. W. & Hughes, R. N. 1978. Energy maximisation in the diet of the Shore Crab, *Carcinus maenas* (L.). *J. Anim. Ecol.* 47, 103–116.
2.2.1, 2.2.3

Elliot, P. F. 1975. Longevity and the evolution of polygamy. *Amer. Natur.*, 109, 281–287.
7.9.4

Elton, R. A. & Greenwood, J. J. D. 1970. Exploring apostatic selection. *Heredity*, 25, 629–633.
5.3.2

Emerson, A. E. 1949. The organisation of insect societies. In Allee, W. C. *et al.* (eds.), *Principles of Animal Ecology*, W. B. Saunders Co., Philadelphia, pp. 419–435.
1.5.1

Emlen, J. M. & Emlen, M. G. R. 1975. Optimal choice in diet: test of a hypothesis. *Amer. Natur.* 109, 427–435.
2.8

Emlen, S. T. 1971. The role of song in individual recognition in the indigo bunting. *Z. Tierpsychol.*, 28, 241–246.
11.4.1

Emlen, S. T. 1976a. Altruism in Mountain Bluebirds? *Science*, 191, 808–809.
9.6

Emlen, S. T. 1976b. Lek organization and mating strategies in the bullfrog. *Behav. Ecol. Sociobiol.*, 1, 283–313.
8.1, 11.2.2

Emlen, S. T. & Oring, L. W. 1977. Ecology, sexual selection, and the evolution of mating systems. *Science*, 197, 215–223.
7.5, 7.5.2, 7.5.3, 7.5.4, 8.1, 11.2.2

Erickson, C. J. & Zenone, P. G. 1976. Courtship differences in male ring doves: avoidance of cuckoldry? *Science*, 192, 1353–1354.
7.9.3, 10.3.4

Estes, R. D. 1976. The significance of breeding synchrony in the wildebeest. *E. Afr. Wildl. J.*, 14, 135–152.
3.2.1

Estes, R. D. & Goddard, J. 1967. Prey selection and hunting behaviour of the African wild dog. *J. Wildlife Management*, 31, 52–70.
3.3.4, 5.4, 9.8

Evans, H. E. 1977. Extrinsic versus intrinsic factors in the evolution of insect sociality. *Bioscience.* 27, 613–617.
4.1, 4.6

Evans, P. R. 1976. Energy balance and optimal foraging strategies in shorebirds: some implications for their distributions and movements in the non-breeding season. *Ardea*, 64, 117–139.
2.6

Ewer, R. F. 1968. *Ethology of Mammals.* London, Elek Science.
11.4.1

Falconer, D. S. 1960. *Introduction to Quantitative Genetics.* Edinburgh, Oliver & Boyd.
7.2.4

Falls, J. B. & Brooks, R. J. 1975. Individual recognition by song in white-throated sparrows. II. Effects of location. *Can. J. Zool.*, 53, 1412–1420.
11.4.1

Farr, J. A. 1976. Social facilitation of male sexual behaviour, intrasexual competition, and sexual selection in the Guppy, *Poecilia reticulata* (Pisces, Poeciliidae). *Evolution*, 30, 707–717.
7.9.1

Felsenstein, J. 1974. The evolutionary advantage of recombination. *Genetics*, 78, 737–756.
6.1

Fischer, R. B. 1958. The breeding biology of The Chimney Swift, *Chaetura pelagica. New York State Museum and Science Service Bulletin*, 368, 1–141.
9.8

Fisher, R. A. 1930. *The Genetical Theory of Natural Selection.* Oxford, The Clarendon Press.
1.1, 5.2, 6.1, 6.3, 7.2.2, 10.3.1, 12.8

Fisher, R. A. 1958. *The Genetical Theory of Natural Selection* (2nd. edition). New York, Dover.
5.2, 11.2.2, 14.3

Floody, O. R. & Arnold, A. P. 1975. Uganda Kob (*Adenota Kob thomasi*): territoriality and the spatial distributions of sexual and agonistic behaviours at a territorial ground. *Z. Tierpsychol.*, 37, 192–212.
11.2.2

Foerster, R. E. 1968. The Sockeye Salmon, *Oncorhynchus nerka. Fish. Res. Board Can. Bull.*, 162, 1–422.
14.3

Ford, E. B. 1940. Polymorphism and taxonomy. In Huxley, J. (ed.), *The New Systematics*, pp. 493–513. Oxford, Clarendon Press.
5.3

Ford, E. B. 1945. *Butterflies.* London, Collins.
5.2

Ford, E. B. 1964. *Ecological Genetics.* London, Methuen.
5.3

Foxon, G. E. H. 1964. Blood and respiration. In Moore, J. A. (ed.), *Physiology of the Amphibia*, Vol. 1. pp. 151–209. New York, Academic Press.
7.2.5

Frame, L. H. & Frame, G. W. 1976. Female African wild dogs emigrate. *Nature*, 263, 227–229.
3.5.1, 3.5.2, 3.5.3, 3.5.4, 3.5.5

Frankie, G. W. 1976. Pollination of widely dispersed trees by animals in Central America, with an emphasis on bee pollination systems. In Burley, J. & Styles, B. T. (eds.), *Tropical Trees, Variation, Breeding and Conservation*, pp. 151–159. Academic Press, New York.
4.5

Franklin, W. L. 1974. The social behaviour of the vicuña. In Geist, V. & Walther, F. (eds.), *The Behaviour of Ungulates and its relation to Management*, Vol. 1, pp. 477–487. Switzerland, IUCN, Morges.
3.5.4

Fretwell, S. D. 1972. *Populations in a Seasonal Environment.* Princeton University Press, Princeton.
8.3.1, 11.3.1

Fretwell, S. D. & Lucas, H. L. 1970. On territorial behaviour and other factors influencing habitat distribution in birds. *Acta Biotheoretica*, 19, 16–36.
8.3.1, 12.6.3

Fricke, H. W. 1975. Sozialstruktur und ökologische. Spezialisierung bei verwandten Fischen (Pomacentridae). *Z. Tierpsychol.*, 39, 492–520.
1.3.1

Frisch, K. von. 1967. *The Dance Language and Orientation of Bees.* Belknap Press, Cambridge, MA.
4.2.3, 4.5

Frith, H. J. 1962. The Mallee-Fowl. Sydney, Angus and Robertson.
6.4

Fry, C. H. 1972a. The social organization of Bee-eaters (Meropodae) and cooperative breeding in hot-climate birds. *Ibis*, 114, 1–14.
9.1, 9.4, 9.8, 9.9.2

Fry, C. H. 1972b. The biology of African Bee-eaters. *Living Bird*, 11, 75–112.
9.4, 9.8

Fry, C. H. 1975. Cooperative breeding in Bee-eaters and longevity as an attribute of group-breeding birds. *Emu*, 74, 308–309.
9.9.2

Fry, C. H. 1977. The evolutionary significance of cooperative breeding in birds. pp. 127–136. In Stonehouse B. & Perrins, C. M. (eds.) *Evolutionary Ecology*, Macmillan, London.
9.8

Gadgil, M. 1971. Dispersal: population consequences and evolution. *Ecology*, 52, 253–261.
14.5

Gadgil, M. 1972. Male dimorphism as a consequence of sexual selection. *Amer. Natur.*, 106, 574–580.
7.7, 11.2.2

Gale, W. F. 1971. An experiment to determine substrate preference of the fingernail clam *Sphaerium transversum* (Say). *Ecology*, 52, 367–370.
12.2.3

Gartlan, J. S. 1968. Structure and function in primate society. *Folia primatol.*, 8, 89–120.
3.2.2

Gass, C. L., Angehr, G. & Centa, J. 1976. Regulation of food supply by feeding territoriality in the rufous hummingbird. *Can. J. Zool.*, 54, 2046–2054.
11.2.1

Gaston, A. J. 1976. Factors affecting the evolution of group territories of babblers (*Turdoides*) and long-tailed tits. Unpublished dissertation. Oxford University, 233 pp.
9.3, 9.4, 9.6, 9.9.2

Geist, V. 1971. *The Mountain Sheep.* Chicago, University Press.
10.3.1

Geist, V. 1974a. On fighting strategies in animal combat. *Nature*, 250, 354.
11.4.2

Geist, V. 1974b. On the relationship of social evolution and ecology in ungulates. *Am. Zool.*, 14, 205–220.
3.5.2

Ghiselin, M. T. 1969. The evolution of hermaphroditism among animals. *Q. Rev. Biol.*, 44, 189–208.
6.2

Gibb, J. A. 1954. Feeding ecology of tits, with notes on the treecreeper and goldcrest. *Ibis*, 96, 513–543.
12.2.3

Gibb, J. A. 1957. Food requirements and other observations on captive tits. *Bird Study*, 4, 207–215.
12.2.3

Gibb, J. A. 1960. Populations of tits and goldcrests and their food supply in pine plantations. *Ibis*, 102, 163–208.
2.1, 12.2.3

Gibson, D. O. 1974. Batesian mimicry without distastefulness? *Nature*, 250, 77–79.
5.2

Giesel, J. T. 1970. On the maintenance of a shell pattern and behaviour polymorphism in *Acmaea digitalis*, a limpet. *Evolution*, 24, 98–119.
5.3.1

Giesel, J. T. 1976. Reproductive strategies as adaptations to life in temporally heterogeneous environments. *A. Rev. Ecol. Syst.*, 7, 57–79.
14.1

Gilbert, C. H. 1918. Contribution to the life history of the sockeye salmon. No. 5. Appendix. *Report of the Commissioner of Fisheries, Province of British Columbia*, 26–52.
12.5

Gilbert, L. E. & Raven, P. E. 1975. (eds.). *Coevolution of Animals and Plants.* Austin, Univ. Texas Press.
5.1

Gill, F. B. & Wolf, L. L. 1975. Economics of feeding territoriality in the golden-winged sunbird. *Ecology*, 56, 333–345.
11.2.1

Gilliard, E. T. 1969. *Birds of Paradise and Bower Birds.* London, Weidenfeld and Nicolson.
12.4.2

Glas, P. 1960. Factors governing density in the chaffinch (*Fringilla coelebs*) in different types of wood. *Arch. Néerl. Zool.*, 13, 466–472.
12.4.1

Goodhart, C. B. 1962. Variation in a colony of the snail *Cepaea nemoralis* (L.). *J. Anim. Ecol.*, 31, 207–237.
5.3.3

Goodhart, C. B. 1963. Area effects and non-adaptive variations between populations of *Cepaea* (Mollusca). *Genetics*, 18, 459–465.
5.3.3

Goss-Custard, J. D. 1970. Feeding dispersion in some overwintering wading birds. In Crook, J. H. (ed.), *Social Behaviour in Birds and Mammals*, pp. 3–34. London, Academic Press.
11.2.4

Goss-Custard, J. D. 1976. Variation in the dispersion of redshank *Tringa totanus* on their winter feeding grounds. *Ibis*, 118, 257–263.
11.2.4

Goss-Custard, J. D. 1977a. Optimal foraging and the size selection of worms by redshank *Tringa totanus*. *Anim. Behav.*, 25, 10–29.
2.2.3

Goss-Custard, J. D. 1977b. Predator responses and prey mortality in the redshank *Tringa totanus* (L.) and a preferred prey *Corophium volutator* (Pallas). *J. Anim. Ecol.*, 46, 21–36.
2.2.4

Gould, J. L. 1976. The dance-language controversy. *Q. Rev. Biol.*, 51, 211–244.
4.2.3, 10.2.2

Gould, L. L. & Heppner, F. 1974. The vee formation of Canada geese. *Auk*, 91, 494–506.
3.4

Grant, P. R. 1969. Experimental studies of competitive interaction in a two-species system. I. *Microtus* and *Clethrionomys* species in enclosures. *Can. J. Zool.*, 47, 1059–1082.
12.4.2

Grant, P. R. 1970. Experimental studies of competitive interaction in a two-species system. II. The Behaviour of *Microtus*, *Clethrionomys* and *Peromyscus* species. *Anim. Behav.*, 18, 411–426.
12.4.2

Grant, P. R. 1971. Experimental studies of competitive interaction in a two-species system. III. *Microtus* and *Peromyscus* species in enclosures. *J. Anim. Ecol.*, 40, 323–350.
12.4.2

Grant, P. R. 1972. Convergent and divergent character displacement. *Biol. J. Linn. Soc.*, 4, 39–68.
7.2.5, 12.4.2

Grant, P. R. 1975. Population performance of *Microtus pennsylvanicus* confined to woodland habitat, and a model of habitat occupancy. *Can. J. Zool.*, 53, 1447–1465.
8.3.1

Grant, P. R., Grant, B. R., Smith, J. N. M., Abbott, I. J. & Abbott, L. K. 1976. Darwin's finches: population variation and natural selection. *Proc. Nat. Acad. Sci., U.S.A.*, 73, 257–261.
2.7, 12.7

Grant, V. 1958. The regulation of recombination in plants. *Cold Spring Harbor Symp. Quant. Biol.*, 23, 337–363.
6.2

Green, R. H. 1971. A multivariate statistical approach to the Hutchinsonian niche: bivalve molluscs of central Canada. *Ecology*, 52, 543–556.
12.2.1

Greenewalt, C. H. 1968. *Bird Song: Acoustics and Physiology*. Washington, Smithsonian Institution Press.
10.3.2

Greenwood, P. J., Harvey P. H. & Perrins, C. M. 1978. Inbreeding and dispersal in the Great Tit. *Nature*, 271, 52–54.
3.5.2

Grey Walter, W. 1953. *The Living Brain*. London, Duckworth.
10.5.3

Grimes, L. G. 1976. The occurrence of cooperative breeding behaviour in African birds. *Ostrich*, 47, 1–15.
9.1

Guthrie, R. D. 1971. A new theory of mammalian rump patch evolution. *Behaviour*, 38, 132–145.
5.4

Haldane, J. B. S. 1953. Animal populations and their regulation. *Modern Biology*, 15, 9–24.
5.1

Haldane, J. B. S. & Spurway, H. 1954. A statistical analysis of communication in *Apis melifera* and a comparison with communication in other animals. *Insectes Sociaux*, 1, 247–283.
10.2.2

Hall, K. R. L. 1965. Behaviour and ecology of the wild Patas monkey, *Erythrocebus patas*, in Uganda. *J. Zool. Lond*, 148, 15–87.
3.2.2

Hall-Craggs, J. 1969. The aesthetic content of bird song. In Hinde, R. A. (ed.), *Bird Vocalizations*, pp. 367–381. Cambridge, Cambridge University Press.
10.5.4

Halliday, T. R. 1974. The sexual behaviour of the smooth newt, *Triturus vulgaris* (Urodela, Salamandridae). *J. Herpetol.*, 8, 277–292.
7.9.1, 13.7.1

Halliday, T. R. 1976. The libidinous newt. An analysis of variations in the sexual behaviour of the smooth newt, *Triturus vulgaris*. *Anim. Behav.*, 24, 398–414.
7.9.1, 7.9.2, 13.7.1

Halliday, T. R. 1977a. The effects of experimental manipulation of breathing behaviour on the sexual behaviour of the smooth newt, *Triturus vulgaris*. *Anim. Behav.*, 25, 39–45.
7.9.1, 13.7.2

Halliday, T. R. 1977b. The courtship of European Newts: an evolutionary perspective. pp. 185–232. In Taylor, D. H. & Guttman, S. I. (eds.), *The Reproductive Biology of Amphibians*. New York, Plenum.
7.2.3, 7.2.5

Hamilton, W. D. 1963. The evolution of altruistic behaviour. *Amer. Natur.*, 97, 354–356.
5.5, 5.5.1

Hamilton, W. D. 1964. The genetical theory of social behaviour. I, II. *J. theor. Biol.*, 7, 1–52.
1.5, 1.5.1, 1.6, 3.5.3, 5.5.1, 5.8, 9.1, 9.9.2

Hamilton, W. D. 1967. Extraordinary sex ratios. *Science*, 156, 477–488.
1.5.1, 6.3, 10.3.1

Hamilton, W. D. 1971. Geometry for the selfish herd. *J. theor. Biol.*, 31, 295–311.
3.2.5, 3.2.6, 5.4, 5.5.1

Hamilton, W. D. 1972. Altruism and related phenomena, mainly in social insects. *A. Rev. Ecol. Syst.*, 3, 193–232.
4.1

Hamilton, W. D. & May, R. M. 1977. Dispersal in stable habitats. *Nature*, 269, 578–581.
14.5, 14.6

Hammer, C. 1941. *Biological and Ecological Investigations on Flies associated with Pasturing Cattle and their Excrement*. Copenhagen: Bianco Lunos Bogtrykkeri A/S, 257 pp.
8.2

Harcourt, A. H., Stewart, K. S. & Fossey, D. 1976. Male emigration and female transfer in wild mountain gorilla. *Nature*, 263, 226–227.
3.5.2, 3.5.3

Harris, M. 1976. *Cows, Pigs, and Wars and Witches. The Riddles of Culture*. Fontana.
10.3.2

Harrison, C. J. O. 1969. Helpers at the nest in Australian passerine birds. *Emu*, 69, 30–40.
9.1

Hartley, P. H. T. 1950. An experimental analysis of interspecific recognition. *Symp. Soc. Exp. Biol.*, 4, 313–336.
5.7

Hartshorne, C. 1973. *Born to Sing.* Bloomington and London, Indiana University Press.
10.5.4

Harvey, P. H. 1976. Factors influencing the shell pattern of *Cepaea nemoralis* (L.) in east Yorkshire: a test case. *Heredity*, 36, 1–10.
5.3.3

Harvey, P. H., Birley, N. & Blackstock, T. H. 1975. The effect of experience on the selective behaviour of song thrushes feeding on artificial populations of *Cepaea* (Held.) *Genetica*, 45, 211–216.
5.3.3

Harvey, P. H., Jordan, C. A. & Allen, J. A. 1974. Selection behaviour of wild blackbirds at high prey densities. *Heredity*, 32, 401–404.
5.3.2

Hassell, M. P. & May, R. M. 1974. Aggregation of predators and insect parasites and its effect on stability. *J. Anim. Ecol.*, 43, 567–594.
2.3.1, 2.3.5

Hazlett, B. 1968. Size relations and aggressive behaviour in the hermit crab *Clibanarius vitatus*. *Z. Tierpsychol.*, 25, 608–614.
10.3.2

Hay, D. E. & McPhail, J. D. 1975. Mate selection in three-spined sticklebacks. *Can. J. Zool.*, 53, 441–450.
7.8

Healey, M. C. 1967. Aggression and self regulation of population size in deermice. *Ecology*, 48, 377–392.
11.3.1

Heatwole, H. 1965. Some aspects of the association of Cattle Egrets with cattle. *Anim. Behav.*, 13, 79–83.
3.4

Hébrant, F. 1967. Etude d'influence du poids des individus, et de l'humidité du milieu sur la consommation d'oxygene d'ouvriers de *Cubitermes exiguus* Methot (Isoptera, Termitidae). Compt. Rend. 5th Congr. U.I.E.I.S., Toulouse, 1965, pp. 107–115.
4.4

Heinrich, B. 1974. Thermoregulation in bumblebees. I. Brood incubation by *Bombus vosnesenskii* queens. *J. comp. physiol.*, 88, 129–140.
4.4

Heinrich, B. 1976. The foraging specializations of individual bumblebees. *Ecol. Monogr.*, 46, 105–128.
4.2.2

Heinrich, B. & Raven, P. H. 1972. Energetics of pollination. *Science*, 176, 497–602.
4.2.1

Heinrich, B., Mudge, P. & Deringis, P. 1977. (In press). Laboratory analysis of flower constancy in foraging bumblebees: *Bombus ternarius* and *B. terricola*. *Behav. Ecol. and Sociobiol.*
4.2.2

Herrera, C. M. 1975. Trophic diversity of the Barn Owl *Tyto alba* in continental western Europe. *Ornis. Scand.*, 5, 181–191.
2.2.3

Herrnstein, R. J. & Loveland, D. H. 1975. Maximising and matching on concurrent ratio schedules. *J. Exp. Anal. Behav.*, 24, 107–116.
2.3.2

Hewson, R. & Healing, T. D. 1971. The stoat *Mustela erminea* and its prey. *J. Zool. Lond*, 164, 239–244.
3.3.3

Hilden, O. 1965. Habitat selection in birds. *Ann. Zool. Fenn.*, 2, 53–75.
12.5, 12.6.3

Hill, J. L. 1974. *Peromyscus:* effect of early pairing on reproduction. *Science*, 186, 1042–1044.
6.2

Hinde, R. A. 1954a. Factors governing the changes in strength of a partially inborn response, as shown by the mobbing behaviour of the chaffinch (*Fringilla coelebs*) I. The nature of the response, and an examination of its course. *Proc. Roy. Soc. B.*, 142, 306–331.
5.7

Hinde, R. A. 1954b. Factors governing the changes in strength of a partially inborn response, as shown by the mobbing behaviour of the chaffinch (*Fringilla coelebs*). II. The waning of the response. *Proc. Roy. Soc. B.*, 142, 331–358.
5.7

Hinde, R. A. 1956. The biological significance of the territories of birds. *Ibis*, 98, 340–369.
11.2

Hinde, R. A. 1959. Behaviour and speciation in birds and lower vertebrates. *Biol. Rev.*, 34, 85–128.
12.8

Hinde, R. A. 1970. *Animal Behaviour. A Synthesis of Ethology and Comparative Psychology.* New York, McGraw Hill.
10.3.3

Hinde, R. A. 1974. *Biological Bases of Human Social Behaviour.* McGraw-Hill, New York.
1.5.1

Hirons, G. J. M. 1976. A population study of the tawny owl (*Strix aluco*) and its main prey species in woodland. Unpublished D.Phil. Thesis. Oxford University.
11.2.1

Hirth, D. H. & McCullough, D. R. 1977. Evolution of alarm signals in ungulates with special reference to white-tailed deer. *Amer. Natur.*, 111, 31–42.
5.4, 5.5.3

Hodgson, J. R. 1972. Local distribution of *Microtus montanus* and *Microtus pennsylvanicus* in south western Montana. *J. Mammal.*, 42, 323–337.
12.2.2

Hodos, W. & Trumbule, G. H. 1967. Strategies of schedule preference in chimpanzees. *J. exp. Anal. Behav.*, 10, 503–514.
13.2.2

Hogan-Warburg, A. J. 1966. Social behaviour of the ruff, *Philomachus pugnax* (L.). *Ardea*, 54, 109–229.
11.2.2

Hölldobler, B. 1971. Communication between ants and their guests. *Sci. Amer.*, 224(3), 86–93.
10.4

Hölldobler, B. 1976a. Tournaments and slavery in a desert ant. *Science*, 192, 912–914.
11.4.1, 11.4.2

Hölldobler, B. 1976b. Recruitment behaviour, home range orientation and territoriality in harvester ants, *Pogonomyrmex*. *Behav. Ecol. and Sociobiol.*, 1, 3–44.
4.3.2

Hölldobler, B. & Wilson, E. O. 1970. Recruitment trails in the harvester ant *Pogonomyrmex badius*. *Psyche.*, 77, 385–399.
4.3.2

Hölldobler, B. & Wilson, E. O. 1977a. Weaver ants: social establishment and maintenance of territory. *Science*, 195, 900–902.
11.4.1

Hölldobler, B. & Wilson, E. O. 1977b. Colony-specific territorial pheromone in the African weaver ant *Oecophylla longinoda* (Latreille). *Proc. Natl. Acad. Sci. U.S.A.*, 74, 2072–2075.
11.4.1

Holm, C. H. 1973. Breeding sex ratios, territoriality and reproductive success in the red-winged blackbird (*Agelaius phoeniceus*). *Ecology*, 54, 356–365.
11.2.2

Holm, S. N. 1966. The utilization and management of bumblebees for red clover and alfalfa seed production. *Ann. Rev. Entomol.*, 11, 151–182.
4.2.2

Honigberg, B. M. 1970. Protozoa associated with termites and their role in digestion, in *Biology of Termites*, Vol. II, Krishna, K. & Weesner, F. M. (eds.), pp. 1–36. Academic Press, New York and London.
4.4

Hoogland, J. L. & Sherman, P. W. 1976. Advantages and disadvantages of Bank Swallow (*Riparia riparia*) coloniality. *Ecol. Monogr.*, 46, 33–58.
3.2.2, 3.2.3, 5.7, 9.8

Horn, H. S. 1968. The adaptive significance of colonial nesting in the Brewer's Blackbird *Euphagus cyanocephalus*. *Ecology*, 49, 682–694.
11.2.1, 11.2.3, 11.2.4

Houston, A. I. 1977. Models of Animal Motivation. Unpublished D.Phil. Thesis. University of Oxford.
13.4.3, 13.7.3, 13.7.4

Houston, A. I., Halliday, T. R. & McFarland, D. J. 1977. Towards a model of the courtship of the smooth newt, *Triturus vulgaris*, with special emphasis on problems of observability in the simulation of behaviour. *Med. & Biol. Eng. & Comput.*, 15, 49–61.
13.4.1, 13.7.1, 13.7.2, 13.7.3

Houston, D. C. 1974. Food searching in griffon vultures. *E. Afr. Wildl. J.*, 12, 63–77.
3.3.1

Howard, E. 1920. *Territory in Bird Life*. London, John Murray.
11.2

Howard, J. C. 1977. H-2 and mating preferences. *Nature*, 266, 406–408.
7.9.2

Howard, R. D. (In press)a. Factors influencing early embryo mortality in bullfrogs. *Ecology*.
11.2.2

Howard, R. D. (In press)b. The evolution of mating strategies in Bullfrogs, *Rana catesbeiana*. *Evolution*.
7.5.1, 7.7, 7.9.2, 10.3.2, 11.2.2

Howe, H. F. 1977. (In press). Nestling sex ratio adjustment among common grackles. *Science*. 198: 744–46.
6.3

Howland, C. I., Janis, I. L. & Kelley, H. H. 1953. *Communication and Persuasion*. New Haven, Yale University Press.
10.5.2

Hrdy, S. B. 1974. Male-male competition and infanticide among the Langurs (*Presbytis entellus*) of Abu, Rajasthan. *Folia primatol.*, 22, 19–58.
3.5.4, 3.5.6, 7.10

Hrdy, S. B. 1976. Care and exploitation of non-human primate infants by conspecifics other than the mother. *Adv. Study Behav.*, 6, 101–158.
3.5.4, 9.8

Hubbard, S. F. & Cook, R. M. 1978. (In press). Optimal foraging by parasitoid wasps. *J. Anim. Ecol.* 47, 593–604.
2.3.5

Huettel, M. D. & Bush, G. L. 1972. The genetics of host selection and its bearing on sympatric speciation in *Procecidochares* (Diptera: Tephritidae). *Entomol. Exp. Appl.*, 15, 465–480.
12.8

Huey, R. B. & Slatkin, M. 1976. Costs and benefits of lizard thermoregulation. *Q. Rev. Biol.*, 51, 363–384.
13.2.3, 13.3.4, 13.8

Hughes, R. N. 1978. (In press). Optimal diets and the energy maximisation premise: the effects

of recognition time and learning. *Amer. Natur.*
2.2.2, 2.2.5

Huxley, J. S. 1938. The present standing of the theory of sexual selection. In de Beer, G. R. (ed.), *Evolution: Essays on Aspects of Evolutionary Biology Presented to Professor E. S. Goodrich on his Seventieth Birthday.* Oxford, The Clarendon Press, pp. 11–42.
7.2.1

Huxley, J. S. 1966. Ritualization of behaviour in animals and men. *Phil. Trans. Roy. Soc. Lond. Ser. B.*, 251, 249–271.
10.2.1

Inouye, D. W. 1976. Resource Partitioning and Community Structure: a Study of Bumblebees in the Colorado Rocky Mountains. Unpublished Ph.D. Dissertation. University of North Carolina, Chapel Hill.
4.2.2

Ishay, J., Bytinsky-Salz, H. & Shulov, A. 1967. Contributions to the bionomics of the oriental hornet (*Vespa orientalis*). *Israel. J. Ent.*, 2, 45–106.
13.5

James, F. C., 1971. Ordinations of habitat relationships among breeding birds. *Wilson Bull.*, 83, 215–236.
12.2.1

Janzen, D. H. 1976. Why bamboos wait so long to flower. *A. Rev. Ecol. Syst.*, 7, 347–391.
14.3

Jarman, P. J. 1974. The social organisation of antelope in relation to their ecology. *Behaviour*, 48, 215–267.
1.3.1, 3.2.4, 3.5.1, 3.5.2, 3.5.3, 7.5.2, 11.2

Järvinen, O. 1976. Apostatic selection in small natural populations. *Hereditas*, 82, 127–129.
5.3.2

Jenni, D. A. 1974. Evolution of polyandry in birds. *Amer. Zool.*, 14, 129–144.
6.4

Jenssen, T. A. 1973. Shift in structural habitat of *Anolis opalinus* due to congeneric competition. *Ecology*, 54, 863–869.
12.4.2

Jellis, R. 1977. *Bird Sounds and their Meaning.* London, B.B.C. Publications.
10.4

Johnson, L. K. 1974. The Role of Agonstic Behaviour in the Foraging Strategies of *Trigona* Bees. Unpublished Ph.D. Dissertation. University of California, Berkeley.
4.2.1

Johnson, L. K. & Hubbell, S. P. 1974. Aggression and competition among stingless bees: field studies. *Ecology*, 55, 120–127.
4.2.1

Johnson, L. K. & Hubbell, S. P. 1975. Contrasting foraging strategies and coexistence of two bee species on a single resource. *Ecology*, 56, 1398–1406.
4.2.1

Jolly, A. 1966. *Lemur Behaviour: a Madagascar Field Study*, Chicago, Chicago University Press.
10.3.2

Jolly, A. 1972. *The Evolution of Primate Behavior.* New York: Macmillan.
9.8

Jones, J. S. 1973a. The genetic structure of a southern peripheral population of the snail *Cepaea nemoralis*. *Proc. Roy. Soc. Lond., Ser. B.*, 183, 371–384.
5.3.3

Jones, J. S. 1973b. Ecological genetics and natural selection in molluscs. *Science*, 182, 546–552.
5.3.3

Jones, J. S., Leith, B. H. & Rawlings, P. 1977. Polymorphism in *Cepaea:* a problem with too

many solutions? *A. Rev. Ecol. Syst.*, 8, 109–143.
5.3.3

Jones, R. B. & Nowell, N. W. 1974. The urinary aversive pheromone of mice: species, strain and grouping effects. *Anim. Behav.*, 22, 187–191.
11.4.1

Jones, R. E. 1977a. Search behaviour: a study of three caterpillar species. *Behaviour*, 60, 237–259.
2.4.1

Jones, R. E. 1977b. Movement patterns and egg distribution in cabbage butterflies. *J. Anim. Ecol.*, 46, 195–212.
2.4.1

Kamil, A. C. (In press). Systematic foraging for nectar by Amakihi *Loxops virens*. *J. Comp. Physiol. Psychol.*
2.4.3

Kaufmann, J. H. 1974. Social ethology of the whiptail wallaby, *Macropus parryi*, in northeastern New South Wales. *Anim. Behav.*, 22, 281–369.
3.5.1

Kawai, M. 1965. Newly acquired pre-cultural behavior of the natural troop of Japanese monkeys of Koshima islet. *Primates*, 6, 1–30.
3.5.7

Kear, J. 1962. Food selection in finches with special reference to interspecific differences. *Proc. Zool. Soc. Lond.*, 138, 163–204.
2.7

Kenward, R. E. 1978. Hawks and doves: attack success and selection in goshawk flights at woodpigeons. *J. Anim. Ecol.*, 47, 449–460.
3.2.2

Kettlewell, H. B. D. 1955a. Selection experiments on industrial melanism in the Lepidoptera. *Heredity*, 9, 323–335.
12.6.1

Kettlewell, H. B. D. 1955b. Recognition of appropriate backgrounds by the pale and black phases of Lepidoptera. *Nature*, 175, 943–944.
12.6.1

Kettlewell, H. B. D. 1956. Further selection experiments on industrial melanism in the Lepidoptera. *Nature*, 175, 934.
5.3.1, 12.6.1

Kettlewell, H. B. D. 1958. Survey of the frequencies of *Biston betularia* and its melanic forms in Great Britain. *Heredity*, 12, 51–75.
12.6.1

Kettlewell, H. B. D. 1968. Industrial melanism in moths and its contribution to our knowledge of evolution. *Proc. Roy. Inst. G.B.*, 36, 1–14.
5.3.1

Kettlewell, H. B. D. & Conn, D. L. T. 1977. Further background-choice experiments on cryptic Lepidoptera. *J. Zool. Lond.*, 181, 371–376.
12.6.1

Kiester, A. R. & Slatkin, M. 1974. A strategy of movement and resource utilization. *Theor. Pop. Biol.*, 6, 1–20.
8.3.1

Kiester, A. R., Gorman, G. C. & Arroyo, D. C. 1975. Habitat selection behaviour of three species of *Anolis* lizards. *Ecology*, 56, 220–225.
12.2.1

Kikkawa, J. 1968. Ecological association of bird species and habitats in Eastern Australia; similarity analysis. *J. Anim. Ecol.*, 37, 143–165.
12.2.1

Kislalioglu, M. & Gibson, R. N. 1976. Prey 'handling time' and its importance in food selection by the 15-spined stickleback *Spinachia spinachia* (L.). *J. exp. mar. Biol. Ecol.*, 25, 151–158.
2.2.1

Kitchen, D. W. 1974. Social behavior and ecology of the pronghorn. *Wildl. Monogr.*, 38, 1–96.
11.2.2

Kleerekoper, H., Timms, A. M., Westlake, G. F., Davy, F. B., Malar, T. & Anderson, V. M. 1970. An analysis of locomotor behaviour of the Goldfish (*Carasius auratus*). *Anim. Behav.*, 18, 317–330.
2.4.1

Kleiman, D. G. 1967. Some aspects of social behaviour in the Canidae. *Am. Zool.*, 7, 365–372.
3.5.7

Kleiman, D. G. & Eisenberg, J. F. 1973. Comparisons of candid and felid social systems from an evolutionary perspective. *Anim. Behav.*, 21, 637–659.
3.1, 3.5.4, 3.5.7

Klingel, H. 1965. Notes on the biology of the plains zebra, *Equus quagga boehmi* Matschie. *E. Afr. Wildl. J.*, 3, 86–88.
3.5.3, 3.5.4

Klopfer, P. H. 1963. Behavioural aspects of habitat selection: the role of early experience. *Wilson Bull.*, 75, 15–22.
12.5

Klopfer, P. H. 1967. Behavioural stereotypy in birds. *Wilson Bull.*, 79, 290–300.
12.7

Kluyver, H. N. & Tinbergen, L. 1953. Territory and regulation of density in titmice. *Arch. Néerl. Zool.*, 10, 265–286.
11.3.1, 12.4.1

Knapton, R. W. & Krebs, J. R. 1974. Settlement patterns, territory size, and breeding density in the song sparrow (*Melospiza melodia*). *Can. J. Zool.*, 52, 1413–1420.
11.3.2

Knight-Jones, E. W. 1953. Decreased discrimination during settling after prolonged planktonic life in larvae of *Sprirorbis borealis* (Serpulidea). *J. Marine Biol. Ass. U.K.*, 32, 337–345.
12.6.2

Knox, P. B. 1967. Apomixis: seasonal and population differences in a grass. *Science*, 157, 325–326.
6.1.2

Konishi, M. 1973. Locatable and nonlocatable acoustic signals for barn owls. *Amer. Natur.*, 107, 775–785.
5.5

Konishi, M. & Nottebohm, F. 1969. Experimental studies in the ontogeny of avian vocalizations. In Hinde, R. A. (ed.), *Bird Vocalizations*, pp. 29–48. Cambridge, Cambridge University Press.
10.5.4

Körner, I. 1939. Zeitgedächtnis and Alarmierung bei den Bienen. *Z. vergl. Physiol.*, 27, 445–459.
4.2.3

Kortlandt, A. & Kooij, M. 1963. Protohominid behaviour in primates. *Symp. zool. Soc. Lond.*, 10, 61–88.
3.2.3

Köster, F. 1971. Zum Nistverhalten des Ani *Crotophaga ani Bonn. zool. Beitr.*, 22, 4–27.
9.6, 9.7

Krebs, C. J. 1972. *Ecology*. New York, Harper & Row.
14.4

Krebs, C. J., Gaines, M. S., Keller, B. L., Myers, J. H. & Tamarin, R. H. 1973. Population cycles in small rodents. *Science*, 179, 35–44.
14.4

Krebs, J. R. 1970. Regulation of numbers in the great tit (Aves: Passeriformes). *J. Zool. Lond*, 162, 317–333.
11.3

Krebs, J. R. 1971. Territory and breeding density in the great tit, *Parus major* L. *Ecology*, 52, 2–22.
11.1, 11.2.3, 11.2.5, 11.3.1, 11.3.2, 11.4.1, 12.6.3

Krebs, J. R. 1973a. Behavioural aspects of predation. In Bateson,

P. P. G. & Klopfer, P. H. (eds.), *Perspectives in Ethology*. pp. 73–111. New York, Plenum Press.
5.3.2

Krebs, J. R. 1973b. Social learning and the significance of mixed-species flocks of chickadees (*Parus* spp). *Can. J. Zool.*, 51, 1275–1288.
3.4

Krebs, J. R. 1974. Colonial nesting and social feeding as strategies for exploiting food resources in the Great Blue Heron (*Ardea herodias*). *Behaviour*, 51, 99–134.
3.3.1, 3.3.2

Krebs, J. R. 1976. Habituation and song repertoires in the great tit. *Behav. Ecol. Sociobiol.*, 1, 215–227.
11.4.1

Krebs, J. R. 1977a. Song and territory in the Great tit. In Stonehouse, B. & Perrins, C. M. (eds.), *Evolutionary Ecology*. pp. 47–62. London. Macmillan.
10.3.2, 11.4.1

Krebs, J. R. 1977b. The significance of song repertoires: the Beau Geste hypothesis. *Anim. Behav.*, 25, 475–478.
10.4, 11.4.1

Krebs, J. R. & Cowie, R. J. 1976. Foraging strategies in birds. *Ardea*, 64, 98–116.
2.5

Krebs, J., Ashcroft, R. & Webber, M. 1978. Song repertoires and territory defence in the great tit (*Parus major*). *Nature*, 271, 539–542.
11.4.1

Krebs, J. R., Erichsen, J. T., Webber, M. I. & Charnov, E. L. 1977. Optimal prey selection in the Great Tit (*Parus major*). *Anim. Behav.*, 25, 30–38.
2.2.3

Krebs, J. R., Kacelnik, A. & Taylor, P. J. 1978. (In press). Optimal sampling by foraging birds: an experiment with great tits (*Parus major*). *Nature*.
2.3.2, 2.5

Krebs, J. R., MacRoberts, M. H. & Cullen, J. M. 1972. Flocking and feeding in the Great Tit *Parus major*—an experimental study. *Ibis*, 114, 507–530.
3.3.1

Krebs, J. R., Ryan, J. C. & Charnov, E. L. 1974. Hunting by expectation or optimal foraging? A study of patch use by chickadees. *Anim. Behav.*, 22, 953–964.
2.3.4, 2.3.5

Kroodsma, D. E. 1977. Correlates of song organization among North American wrens. *Amer. Natur.*, 111, 995–1008.
Part 2, Introduction.

Kruijt, J. P. & Hogan, J. A. 1967. Social behaviour on the lek in black grouse, *Lyrurus tetrix tetrix* (L.). *Ardea*, 55, 203–240.
11.2.2

Kruuk, H. 1964. Predators and anti-predator behaviour of the black headed gull *Larus ridibundus*. *Behaviour Suppl.*, 11, 1–129.
3.2.3, 3.2.6, 5.7, 11.2.3

Kruuk, H. 1972. *The Spotted Hyena*. Chicago and London, University of Chicago Press.
3.2.1, 3.2.3, 3.2.4, 3.2.5, 3.3.1, 3.3.2, 3.3.3, 3.3.4, 3.5.1, 3.5.2, 3.5.3, 3.5.6, 11.2

Kruuk, H. 1975. Functional aspects of social hunting by carnivores. In Baerends, G., Beer, C. & Manning, A. (eds.), *Function and Evolution in Behaviour: Essays in honour of Professor Niko Tinbergen, F.R.S.*, pp. 119–141. Oxford: Oxford University Press.
1.3.1, 3.3.3, 3.5.7

Kruuk, H. & Turner, M. 1967. Comparative notes on predation by lion, leopard, cheetah, and wild dog in the Serengeti area, East Africa. *Mammalia*, 31, 1–27.
3.3.3

Kulesza, G. 1975. Comment on 'Niche, habitat, and ecotope'. *Amer. Natur.*, 109, 476–479.
12.1

Lack, D. 1968. *Ecological Adaptations for Breeding in Birds*. London, Methuen.
1.3.1, 5.8, 6.4, 7.3, 7.6.1

Lack, D. 1971. *Ecological Isolation in Birds*. Oxford, Blackwell Scientific.
12.3, 12.6.3, 12.8

Lack, D. & Southern, H. N. 1949. Birds on Tenerife. *Ibis*, 91, 607–626.
12.8

Lamotte, M. 1951. Recherches sur la structure genetique des populations naturelles de *Cepaea nemoralis* (L.). *Bull. Biol. Fr. Belg. Suppl.*, 35, 1–239.
5.3.3

Lancaster, J. B. 1971. Play-mothering: the relations between juvenile females and young infants among free-ranging Vervet Monkeys (*Cercopithecus aethiops*). *Folia Primatol.*, 15, 161–182.
9.8

Larkin, S. & McFarland, D. J. 1978. (In press). The cost of changing from one activity to another. *Anim. Behav.*
13.4.5

Lawton, J. H., Beddington, J. R. & Bonser, P. 1974. Switching in invertebrate predators. In Usher, M. B. & Williamson, M. H. (eds.), *Ecological Stability*. pp. 141–158. London, Chapman and Hall.
2.2.5

Lea, S. E. G. 1976. Titration of schedule parameters by pigeons. *J. exp. Anal. Behav.*, 25, 43–54.
43–54.
13.2.2

Le Boeuf, B. J. 1972. Sexual behaviour in the Northern Elephant Seal *Mirounga angustirostris*. *Behaviour*. 41, 1–26.
7.4

Le Boeuf, B. J. 1974. Male-male competition and reproductive success in Elephant Seals. *Amer. Zool.*, 14, 163–176.
7.4

Le Boeuf, B. J. & Peterson, R. S. 1969. Social status and mating activities in Elephant Seals. *Science*, 163, 91–93.
7.4

Lerner, I. M. 1954. *Genetic Homeostasis*. Edinburgh, Oliver & Boyd.
6.2

Levene, H. 1953. Genetic equilibrium when more than one ecological niche is available. *Amer. Natur.*, 87, 331–333.
12.8

Levins, R. 1968. *Evolution in Changing Environments*. Princeton, Princeton University Press.
12.6.2, 12.7

Levins, R. & MacArthur, R. 1966. The maintenance of genetic polymorphism in a spatially heterogeneous environment: variations on a theme by Howard Levene. *Amer. Natur.*, 100, 585–589.
12.8

Leyhausen, P. 1965. The communal organisation of solitary mammals. *Symp. Zool. Soc. Lond.*, 14, 249–263.
11.1

Light, S. F. 1934. The distribution and biology of the common dry-wood termite, *Kalotermes minor*, pp. 201–224. In Kafoid, C. A. *et al.* (eds.), *Termites and Termite Control*.
4.4

Lill, A. 1968a. An analysis of sexual isolation in the domestic fowl: I. The basis of homogamy in males. *Behaviour*, 30, 107–126.
7.8

Lill, A. 1968b. An analysis of sexual isolation in the domestic fowl: II. The basis of homogamy in females. *Behaviour*, 30, 127–145.
7.8

Lill, A. 1974. Sexual behavior of the lek-forming white-bearded manakin (*Manacus manacus trinitatis*). *Z. Tierpsychol.*, 36, 1–36.
11.2.2

Lill, A. & Wood-Gush, D. G. M. 1965. Potential ethological isolating mechanisms and assortative mating in the domestic fowl. *Behaviour*, 25, 16–44.
7.8

Lindauer, M. 1952. Ein Beitrag zur Frage der Arbeitsteilung im Bienenstradt. *Z. vergl. Physiol.*, 34, 299–345.
4.2.3

Linsley, E. G. 1958. The ecology of solitary bees. *Hilgardia*, 27, 543–599.
4.2.2

Linsley, E. G. & Cazier, M. A. 1972. Diurnal and seasonal behaviour patterns among adults of *Protoxaea gloriosa* (Hymenoptera, Oxaeidae). *Am. Mus. Novitates*, 2509, 1–25.
4.5

Littlejohn, M. J. 1969. The systematic significance of isolating mechanisms. In *Systematic Biology*. National Academy of Sciences, Washington, D.C. 459–482.
7.2.5

Lloyd, M. & Dybas, H. S. 1966. The periodical cicada problem. *Evolution*, 20, 133–149 and 466–505.
14.3

Logan, F. A. 1965. Decision making by rats. *J. comp. physiol. Psychol.*, 59, 1–12.
13.2.2, 13.2.3

Loop, M. S. & Scoville, S. A. 1972. Response of newborn *Eumeces inexpectatus* to prey-object extracts. *Herpetologica*, 28, 254–256.
5.2

Lorenz, K. 1966. *On Aggression*. London, Methuen.
10.3.1, 11.4.2

Low, R. M. 1971. Interspecific territoriality in a pomacentrid reef fish, *Pomacentrus flavicauda* Whitley. *Ecology*, 52, 648–654.
11.2.5

Lüscher, M. 1951. Significance of 'fungus gardens' in termite nests. *Nature*, 167, 34–35.
4.4

Lüscher, M. 1955. Der Sauerstoffverbrauch bei Termiten und die Ventilation des Nestes bei *Macrotermes natalensis* (Haviland). *Acta Trop.*, 12, 289–307.
4.4

Lüscher, M. 1961. Air conditioned termite nests. *Sci. Am.*, 205, (1) 138–145.
4.4

MacArthur, R. H. 1968. Selection for life tables in periodic environments. *Amer. Natur.*, 102, 381–383.
14.4

MacArthur, R. H. 1972. *Geographical Ecology*. Harper & Row.
2.7

MacArthur, R. H. & Pianka, E. R. 1966. On the optimal use of a patchy environment. *Amer. Natur.*, 100, 603–609.
1.1, 2.2.2

MacArthur, R. H. & Wilson, E. O. 1967. *The Theory of Island Biogeography*. Princeton, Princeton University Press.
14.2

MacArthur, R. H., MacArthur, J. W. & Preer, J. 1962. On bird species diversity. II. Prediction of bird census from habitat measurements. *Amer. Natur.*, 96, 167–174.
12.2.1

Macdonald, D. W. 1977. The behavioural ecology of the red fox *Vulpes vulpes*. A study of social organisation and resource exploitation. Unpublished D.Phil. thesis. Oxford.
10.3.2, 11.4.1

Macdonald, D. W. & Henderson, D. G. 1977. Aspects of the behaviour and ecology of mixed-species bird flocks in Kashmir. *Ibis*, 119, 481–493.
3.4

Macevicz, S. & Oster, G. 1976. Modelling social insect populations II: Optimal reproductive strategies in annual eusocial insect colonies. *Behav. Ecol., Sociobiol.*, 1, 265–282.
13.5

MacKay, T. F. C. & Doyle, R. W. 1978. An ecological genetic analysis of the settling behaviour of a marine polychaete. I. Probability of settlement and gregarious behaviour. *Heredity*, 40, 1–12.
12.2.3, 12.6.1

MacKay, T. F. C. & Doyle, R. W. (In prep.). An ecological genetic analysis of the settling behaviour of a marine polychaete. II. Changes in dominance.
12.6.1

MacLean, G. L. 1973a. The Sociable Weaver, Part II: Nest architecture and social organization. *Ostrich*, 44, 191–218.
9.4

MacLean, G. L. 1973b. The Sociable Weaver, Part III: Breeding biology and moult. *Ostrich*, 44, 219–240.
9.4

MacRoberts, M. H. & MacRoberts, B. R. 1976. Social organization and behavior of the Acorn Woodpecker in central coastal California. *Orn. Monog.*, 21, 1–115.
9.3, 9.4

Mader, W. J. 1975. Biology of the Harris' Hawk in Southern Arizona. *Living Bird*, 14, 59–85.
9.4

Malcolm, J. R. & van Lawick, H. 1975. Notes on wild dogs (*Lycaon pictus*) hunting zebras. *Mammalia*, 39, 231–240.
3.2.5, 3.3.3

Manning, J. T. 1975. Male discrimination and investment in *Asellus aquaticus* (L.) and *A. meridianus* Racovitsza (Crustacea: Isopoda). *Behaviour*, 55, 1–14.
7.9.3, 8.4.1

Marler, P. 1955. Characteristics of some animal calls. *Nature*, Lond., 176, 6–8.
5.5, 5.7

Marler, P. 1957. Specific distinctiveness in the communication signals of birds. *Behaviour*, 11, 13–39.
5.5

Marler, P. 1959. Developments in the study of animal communication. In Bell, P. R. (ed.), *Darwin's Biological Work. Some Aspects Reconsidered.* New York, Wiley.
10.1.5

Marler, P. R. 1968. Visual systems. In Sebeok, T. A. (ed.), *Animal Communications*, pp. 103–126. Bloomington and London, Indiana University Press.
10.1.5

Marsh, N. & Rothschild, M. 1974. Aposomatic and cryptic Lepidoptera tested on the mouse. *J. Zool. Lond.*, 174, 89–122.
5.2

Martinez, D. R. & Klinghammer, E. 1970. The behavior of the whale *Orcinus orca*: a review of the literature. *Z. Tierpsychol.*, 27, 828–839.
3.3.2

Mather, K. 1955. Polymorphisms as an outcome of disruptive selection. *Evolution*. 9, 52–61.
12.8

Mathews, R. W. 1968. *Microstigmus comes:* sociality in a sphecid wasp. *Science*, 160, 787–788.
787–788.
4.6

Matthews, E. G. 1977. Signal-based frequency-dependent defense strategies and the evolution of mimicry. *Amer. Natur.*, 111, 213–222.
5.2

May, R. M. 1976. *Theoretical Ecology: Principles and Applications.* Oxford, Blackwell.
14.4

May, R. M. 1977a. Optimal life-

history strategies. *Nature*, 267, 394–395.
14.1

May, R. M. 1977b. Thresholds and breakpoints in ecosystems with a multiplicity of stable states. *Nature*, 269, 471–477.
14.4

May, R. M. & MacArthur, R. H. 1972. Niche overlap as a function of environmental variability. *Proc. Nat. Acad. Sci. U.S.A.*, 69, 1109–1113.
12.4.2

Maynard Smith, J. 1956. Fertility, mating behaviour and sexual selection in *Drosophila subobscura*. *J. Genet.*, 54, 261–279.
7.9.2

Maynard Smith, J. 1964. Group selection and kin selection. *Nature*, 201, 1145–1147.
6.1

Maynard Smith, J. 1965. The evolution of alarm calls. *Amer. Natur.*, 99, 59–63.
5.5.1

Maynard Smith, J. 1966. Sympatric speciation. *Amer. Natur.*, 100, 637–650.
12.8

Maynard Smith, J. 1968. Evolution in sexual and asexual populations. *Amer. Natur.*, 102, 469–473.
6.1

Maynard Smith, J. 1974a. *Models in Ecology*. Cambridge, Cambridge University Press.
11.3.2

Maynard Smith, J. 1974b. The theory of games and the evolution of animal conflicts. *J. theor. Biol.*, 47, 209–221.
8.1, 8.5

Maynard Smith, J. 1976a. Group Selection. *Q. Rev. Biol.*, 51, 277–283.
1.4, 5.5.1, 6.1, 11.3.1

Maynard Smith, J. 1976b. A short-term advantage for sex and recombination through sib-competition. *J. theor. Biol.*, 63, 245–258.
6.1.2

Maynard Smith, J. 1976c. Evolution and the theory of games. *Amer. Sci.*, 64, 41–45.
8.1, 10.3.1

Maynard Smith, J. 1976d. Sexual selection and the handicap principle. *J. theor. Biol.*, 57, 239–242.
7.2.3

Maynard Smith, J. 1977. Parental investment—a prospective analysis. *Anim. Behav.*, 25, 1–9.
5.6, 6.4

Maynard Smith, J. 1978. *The Evolution of Sex*. Cambridge, Cambridge University Press.
6.1, 6.1.2

Maynard Smith, J. & Parker G. A. 1976. The logic of asymmetric contests. *Anim. Behav.*, 24, 159–175.
8.4.2, 8.5, 10.3.1, 10.3.2, 11.4.2

Maynard Smith, J. & Price, G. R. 1973. The logic of animal conflict. *Nature*, 246, 15–18.
10.3.1, 11.4.2

Maynard Smith, J. & Ridpath, M. G. 1972. Wife sharing in the Tasmanian native hen, *Tribonyx mortierii*: a case of kin selection? *Amer. Natur.*, 106, 447–452.
3.5.3, 9.5

Maynard Smith, J. & Williams, G. C. 1976. Reply to Barash. *Amer. Natur.*, 110, 897.
14.3

Mayr, E. 1972. Sexual selection and natural selection. In Campbell, B. (ed.), *Sexual Selection and the Descent of Man*. Chap. 5. London, Heinemann.
7.3

Mayr, E. & Gilliard, G. T. 1952. Altitudinal hybridization in New Guinea honey-eaters. *Condor*, 54, 325–337.
12.8

McCleery, R. H. 1977a. On satiation curves. *Anim. Behav.*, 25, 1005–1015.
13.4.4

McCleery, R. H. 1977b. Temporal organisation of feeding and drink-

ing behaviour. Unpublished D.Phil. thesis. University of Oxford.
13.4.4

McDonald, J. F. & Ayala, F. J. 1974. Genetic response to environmental heterogeneity. *Nature*, 250, 572–574.
12.7

McFarland, D. J. 1970. Recent developments in the study of feeding and drinking in animals. *J. psychosom. Res.*, 14, 229–237.
13.4, 13.4.1

McFarland, D. J. 1971. *Feedback Mechanisms in Animal Behaviour*. London, Academic Press.
13.4.1, 13.7.2

McFarland, D. J. 1976. Form and function in animal behaviour. In Bateson, P. P. G. & Hinde, R. A. (eds.), *Growing Points in Ethology*. pp. 59–93. Cambridge & London, Cambridge University Press.
13.2.2

McFarland, D. J. 1977. Decision making in animals. *Nature*, 269, 15–21.
13.3.4, 13.7.3, 13.7.4

McFarland, D. J. (In press) 1978. The assessment of priorities by animals. In Dempster, M. & McFarland, D. J. (eds.), *Animal Economics*. London, Academic Press.
2.2.4

McFarland, D. J. & Houston, A. I. (In press). *The State Space Analysis of Behaviour*. London, Pitmans.
13.7.3

McFarland, D. J. & Lloyd, I. H. 1973. Time-shared feeding and drinking. *Q. J. exp. Psychol.*, 25, 48–61.
13.4.5

McFarland, D. J. & McFarland F. J. 1968. Dynamic analysis of an avian drinking response. *Med. Biol. Engng.*, 6, 659–668.
13.4.4

McFarland, D. J. & Sibly, R. M. 1975. The behavioural final common path. *Phil. Trans. R. Soc. Ser. B.*, 270, 265–293.
13.4, 13.4.1

McGuire, W. J. 1969. The nature of attitudes and attitude change. In Lindzey, G. & Aronson, E. (eds.), *Handbook of Social Psychology*, Vol. 3, pp. 136–314. Reading, Mass., Addison Wesley.
10.5.2

McLaren, I. A. (ed.), 1971. *Natural Regulation of Animal Populations*. New York, Atherton.
14.6

McNab, B. K. 1963. Bioenergetics and the determination of home range size. *Amer. Natur.*, 97, 133–140.
11.2.1

McPhail, J. D. 1969. Predation and the evolution of a stickleback (*Gasterosteus*). *J. Fish. Res. Bd. Canada*, 26, 3183–3208.
7.7

Mech, L. D. 1970. *The Wolf: The Ecology and Behavior of an Endangered Species*. New York, Natural History Press.
3.2.3, 3.2.5, 3.3.3, 3.5.1, 3.5.3, 3.5.4

Menge, B. A. 1972. Foraging strategy of a starfish in relation to actual prey availability and environmental predictability. *Ecol. Monogr.*, 42, 25–50.
2.2.3

Menzel, E. W. 1971. Communication about the environment in a group of young chimpanzees. *Folia primatol.*, 15, 220–232.
3.3.1

Menzel, R. 1968. Das Gedächtnis der Honigbiene für Spektralfarban. *Z. vergl. Physiol.*, 60, 82–102.
4.2.3

Meyer, J. E. & Pudel, V. 1972. Experimental studies on food intake in obese and normal human subjects. *J. psychosom. Res.*, 16, 305–308.
13.4.4

Michener, C. D. 1974. *The Social*

Behavior of the Bees. Cambridge, Mass. Belknap Press.
4.5, 4.6

Miller, M. R. 1975. Gut morphology of mallards in relation to diet quality. *J. Wildl. Manag.*, 39, 168–173.
2.2.5

Miller, R. S. 1964. Ecology and distribution of pocket gophers (Geomyidae) in Colorado. *Ecology*, 45, 256–272.
12.8

Milton, K. & May, M. L. 1976. Body weight, diet and home range area in primates. *Nature*, 259, 459–462.
11.2.1

Mirmirani, M. & Oster, G. 1978. (In press). Competition, kin selection, and evolutionarily stable strategies. *Theor. Pop. Biol.*
14.3

Missakian, E. A. 1973. Genealogical mating activity in free-ranging groups of rhesus monkeys (*Macaca mulatta*) on Cayo Santiago. *Behaviour*, 45, 225–241.
3.5.3

Mitton, J. B. & Koehn, R. K. 1975. Genetic organisation and adaptive response of allozymes to ecological variables in *Fundulus heteroclitus*. *Genetics*, 79, 97–111.
12.8

Mockford, E. L. 1971. Parthenogenesis in Psocids (Insecta: Psocoptera). *Am. Zool.*, 11, 327–339.
6.1.1

Mohnot, S. M. 1971. Some aspects of social changes and infant killing in the Hanuman Langur, *Presbytis entellus* (Primates: Cercopithecidae) in Western India. *Mammalia*, 35, 175–198.
3.5.4

Moment, G. B. 1962. Reflexive selection: a possible answer to an old puzzle. *Science*, 136, 262–263.
5.3.2

Moore, N. W. 1964. Intra- and interspecific competition among dragonflies. *J. Anim. Ecol.*, 33, 49–71.
11.3.1

Moore, W. S. 1976. Components of fitness in the unisexual fish *Poeciliopsis monacha-occidentalis*. *Evolution*, 30, 564–578.
6.1.1

Morris, D. 1952. Homosexuality in the ten-spined stickleback (*Pygosteus pungitius*). *Behaviour*, 4, 233–261.
10.4

Morris, D. 1957. 'Typical intensity' and its relation to the problem of ritualisation. *Behaviour*, 11, 1–12.
10.3.3

Morris, D. J. 1956. The feather postures of birds and the problem of the origin of social signs. *Behaviour*, 9, 75–114.
10.1.5, 10.2.3

Morris, R. D. 1969. Competitive exclusion between *Microtus* and *Clethrionomys* in the aspen parkland of Saskatchewan. *J. Mammal.*, 50, 291–301.
12.4.2

Morse, D. H. 1970. Ecological aspects of some mixed species foraging flocks of birds. *Ecol. Monogr.*, 40, 119–168.
12.4.2

Morton, E. S. 1977. On the occurrence and significance of motivation-structural rules in some bird and mammal sounds. *Amer. Natur.*, 111, 855–869.
10.3.2, 10.3.3

Moss, R. 1969. A comparison of red grouse (*Lagopus l. scoticus*) stocks with the production and nutrition value of heather (*Caluna vulgaris*). *J. Anim. Ecol.* 38, 103–122.
11.2.1

Moss, R. 1972. Effects of captivity on gut lengths in red grouse. *J. Wildl. Manag.*, 36, 99–104.
2.2.5

Moss, R., Miller, G. R. & Allen, S. E. 1972. Selection of heather by captive red grouse in relation to age of the plant. *J. Appl. Ecol.*, 9, 771–781.
2.2.4

Muller, H. J. 1932. Some genetic aspects of sex. *Amer. Natur.*, 66, 118–138.
6.1

Muller, H. J. 1964. The relation of recombination to mutational advance. *Mutation Res.*, 1, 2–9.
6.1

Murdoch, W. W. 1969. Switching in general predators: Experiments on predator specificity and stability of prey populations. *Ecol. Mongr.*, 39, 335–354.
5.3.2

Murdoch, W. W. & Oaten, A. 1975. Predation and population stability. *Adv. Ecol. Res.*, 9, 1–131.
2.2.5, 2.3.5, 5.3.2

Murdoch, W. W., Avery, S. & Snyth, M. E. B. 1975. Switching in predatory fish. *Ecology*, 56, 1054–1105.
2.2.5

Murie, J. O. 1971. Behavioural relationships between two sympatric voles (*Microtus*): relevance to habitat segregation. *J. Mammal.*, 52, 181–186.
12.4.2

Murray, B. G. 1971. The ecological consequences of interspecific territorial behavior in birds. *Ecology*, 52, 414–423.
11.2.5

Murton, R. K. 1971. The significance of a specific search image in the feeding behaviour of the woodpigeon. *Behaviour*, 40, 10–42.
3.3.1

Mykytowycz, R. 1968. Territorial marking by rabbits. *Sci. Am.*, 218(5), 116–126.
11.4.1

Nagel, E. 1961. *The Structure of Science*. London, Routledge & Kegan Paul.
13.2.3

Neill, S. R. St. J. & Cullen, J. M. 1974. Experiments on whether schooling by their prey affects the hunting behaviour of cephalapods and fish predators. *J. Zool. Lond.*, 172, 549–569.
3.2.4, 3.2.6, 3.2.7

Nelson, K. 1964. The temporal patterning of courtship behavior in the glandulocaudine fishes. *Behaviour*, 24, 90–146.
10.2.3

Newton, I. & Marquiss, M. 1978. (In press). Sex ratio among nestlings of the European Sparrowhawk. *Amer. Natur.*
6.3

Nice, M. M. 1941. The role of territory in bird life. *Amer. Midl. Nat.*, 26, 441–487.
11.2

Nisbet, I. C. T. 1973. Courtship-feeding, egg-size and breeding success in Common Terns. *Nature*, 241, 141–142.
7.9.2, 10.3.4

Nisbet, I. C. T. 1977. Courtship feeding and clutch size in Common Terns *Sterna hirundo*. In Stonehouse, B. & Perrins, C. M. (eds.), *Evolutionary Ecology*, pp. 101–109. London, Macmillan.
10.3.4

Noble, G. K. 1939. The role of dominance in the social life of birds. *Auk*, 56, 263–273.
11.1

Noirot, Ch. 1970. The nests of termites. In Krishna, K. & Weesner, F. M. (eds.), *Biology of Termites*, Vol. II. pp. 73–125. Academic Press, New York and London.
4.4

Norberg, R. A. 1977a. Occurrence and independent evolution of bilateral ear asymmetry in owls and implications in owl taxonomy. *Phil. Trans. Roy. Soc. Lond.*, 280, 375–408.
5.5

Norberg, R. A. 1977b. An ecological theory on foraging time and energetics and choice of optimal food-searching method. *J. Anim. Ecol.*, 46, 511–530.
2.6

Norman, J. M. 1975. *Elementary Dynamic Programming*. London, Edward Arnold.
13.8

O'Brien, W. J., Slade, N. A. & Vinyard, G. L. 1976. Apparent size as the determinant of prey selection by Bluegill sunfish (*Lepomis macrochirus*). *Ecology*, 57, 1304–1311.
2.2.3

O'Donald, P. 1963. Sexual selection and territorial behaviour. *Heredity*, 18, 361–364.
11.2.2

O'Donald, P. 1968. Natural selection by glow worms in a population of *Cepaea nemoralis*. *Nature*, 217, 194.
5.3.3

O'Donald, P. 1972. Sexual selection by variations in fitness at breeding time. *Nature*, 237, 349–351.
7.6.1

O'Donald, P. 1973a. Frequency-dependent sexual selection as a result of variations in fitness at breeding time. *Heredity*, 30, 351–368.
7.6.1

O'Donald, P. 1973b. Models of sexual and natural selection in polygamous species. *Heredity*, 31, 145–156.
7.8

O'Donald, P. 1974. Polymorphisms maintained by sexual selection in monogamous species of birds. *Heredity*, 32, 1–10.
7.7

O'Donald, P. 1977. Mating preferences and sexual selection in the Arctic Skua. II. Behavioural mechanisms of the mating preferences. *Heredity*, 39, 111–119.
7.6.2

O'Donald, P. & Davis, J. W. F. 1975. Demography and selection in a population of Arctic Skuas. *Heredity*, 35, 75–83.
7.6.2

O'Donald, P. & Davis, J. W. F. 1977. Mating preferences and sexual selection in the Arctic Skua. III. Estimation of parameters and tests of heterogeneity. *Heredity*, 39, 121–132.
7.6.2

Oettingen-Spielberg, T. 1949. Über das Wesen der Suchbiene. *Z. vergl. Physiol.*, 31, 454–489.
4.2.3

Oldham, C. 1929. Mollusca eaten by rabbits. *J. Conc.* 18, 335.
5.3.3

Olivier, R. C. D. & Laurie, W. A. 1974. Habitat utilization by hippopotamus in the Mara River. *E. Afr. Wildl. J.*, 12, 249–271.
3.5.5

Orians, G. H. 1961. The ecology of blackbird (*Agelaius*) social systems. *Ecol. Monogr.*, 31, 285–312.
11.3.1

Orians, G. H. 1969. On the evolution of mating systems in birds and mammals. *Amer. Natur.*, 103, 589–603.
1.1, 11.2.2

Orians, G. H. & Janzen, D. H. 1974. Why are embryos so tasty? *Amer. Natur.*, 108, 581–592.
5.8

Orians, G. H. & Willson, M. F. 1964. Interspecific territories of birds. *Ecology*, 45, 736–745.
11.2.5, 12.4.2

Orians, G. H., Orians, C. E. & Orians, K. J. 1977. Helpers at the nest in some Argentine Blackbirds. In Stonehouse, B. & Perrins, C. M. (eds.), *Evolutionary Ecology*, pp. 137–151. London, Macmillan Press.
9.4

Oster, G. 1976. Modelling social insect populations. I. Ergonomics of foraging and population growth

in bumblebees. *Amer. Natur.*, 110, 215–245.
13.5

Oster, G. & Heinrich, B. 1976. Why do bumblebees major? A mathematical model. *Ecol. Monogr.*, 46, 129–133.
2.5, 4.2.2

Otte, D. 1974. Effects and functions in the evolution of signalling systems. *A. Rev. Ecol. Syst.*, 5, 385–417.
10.4

Owens, N. W. & Goss-Custard, J. D. 1976. The adaptive significance of alarm calls given by shorebirds on their winter feeding grounds. *Evolution*, 30, 397–398.
5.5.3

Owings, D. H. & Coss, R. G. 1977. Snake mobbing by California ground squirrels: adaptive variation and ontogeny. *Behaviour*, 42, 50–69.
5.7

Packard, V. 1957. *The Hidden Persuaders.* London, Penguin.
10.5.1

Packer, C. 1975. Male transfer in olive baboons. *Nature*, 255, 219–220.
3.5.2, 6.2

Packer, C. 1977a. Reciprocal altruism in *Papio anubis*. *Nature*, 265, 441–443.
1.6

Packer, C. R. 1977b. Inter-troop Transfer and Inbreeding Avoidance in *Papio anubis* in Tanzania. Unpublished Ph.D. thesis, Cambridge University.
3.5.2

Page, G. & Whitacre, D. F. 1975. Raptor predation on wintering shorebirds. *Condor*, 77, 73–83.
11.2.4

Papageorgis, C. 1975. Mimicry in neotropical butterflies. *Am. Sci.*, 63, 522–532.
5.2

Parker, E. D. & Selander, R. K. 1976. The organization of genetic diversity in the parthenogenetic lizard *Cnemidophorus tesselatus*. *Genetics*, 84, 791–805.
6.1.1

Parker, G. A. 1970a. The reproductive behavior and the nature of sexual selection in *Scatophaga stercoraria* L. II. The fertilization rate and the spatial and temporal relationships of each sex around the site of mating and oviposition. *J. Anim. Ecol.*, 39, 205–228.
8.2, 8.3.1, 8.3.3, 8.4.2

Parker, G. A. 1970b. The reproductive behavior and the nature of sexual selection in *Scatophaga stercoraria* L. IV. Epigamic recognition and competition between males for the possession of females. *Behaviour*, 37, 113–139.
8.2

Parker, G. A. 1970c. Sperm competition and its evolutionary effect on copula duration in the fly *Scatophaga stercoraria*. *J. Insect Physiol.*, 16, 1301–1328.
7.10, 8.3.3, 8.4.1

Parker, G. A. 1970d. The reproductive behaviour and the nature of sexual selection in *Scatophaga stercoraria* L. (Diptera: Scatophagidae). VII. The origin and evolution of the passive phase. *Evolution*, 24, 774–788.
8.3.3

Parker, G. A. 1970e. Sperm competition and its evolutionary consequences in the insects. *Biol. Rev.*, 45, 525–568.
525–568.
7.10, 8.3.3

Parker, G. A. 1971. The reproductive behaviour and the nature of sexual selection in *Scatophaga stercoraria* L. (Diptera: Scatophagidae). VI. The adaptive significance of emigration from the dropping during the phase of genital contact. *J. Anim. Ecol.*, 40, 215–233.
8.3.3

Parker, G. A. 1974a. The reproductive behavior and the nature of sexual selection in *Scatophaga stercoraria* L. IX. Spatial distribution of fertilization rates and evolution of male search strategy within the reproductive area. *Evolution*, 28, 93–108.
8.3.1, 8.3.2, 8.3.3

Parker, G. A. 1974b. Courtship persistence and female-guarding as male time-investment strategies. *Behaviour*, 48, 157–184.
8.4.1, 8.6

Parker, G. A. 1974c. Assessment strategy and the evolution of fighting behaviour. *J. theor. Biol.* 47, 223–243.
10.3.2, 11.4.2

Parker, G. A. 1978. (In press). Sexual selection and sexual conflict. In Blum, M. S. & Blum, N. A. (eds.), *Reproductive Competition, Mate Choice and Sexual Selection.* New York, Academic Press.
8.4.1, 8.4.2, 8.6

Parker, G. A. & Stuart, R. A. 1976. Animal behavior as a strategy optimizer: evolution of resource assessment strategies and optimal emigration thresholds. *Amer. Natur.*, 110, 1055–1076.
8.4.1, 8.4.2, 8.5

Parry, V. 1973. The auxiliary social system and its effect on territory and breeding in Kookaburras. *Emu*, 73, 81–100.
9.4, 9.9.2

Parry, V. 1975. The auxiliary social system and its effect on territory and breeding in Kookaburras. *Emu*, 74, 311.
9.9.2

Parsons, J. 1971. Cannibalism in herring gulls, *Br. Birds*, 64, 528–537.
13.3.2

Partridge, L. 1974. Habitat selection in titmice. *Nature*, 247, 573–574.
12.2.3, 12.5

Partridge, L. 1976a. Field and laboratory observations on the foraging and feeding techniques of blue tits (*Parus caeruleus*) and coal tits (*Parus ater*) in relation to their habitats. *Anim. Behav.*, 24, 534–544.
2.7, 12.6.1

Partridge, L. 1976b. Individual differences in feeding efficiencies and feeding preferences of captive great tits. *Anim. Behav.*, 24, 230–240.
2.7

Partridge, L. (In press). Differences in behaviour between blue and coal tits reared under identical conditions. *Anim. Behav.*
12.6.1

Patterson, I. J. 1965. Timing and spacing of broods in the black-headed gull *Larus ridibundus*. *Ibis*, 107, 433–459.
1.1, 3.2.6, 11.2.3

Peitzmeier, J. 1949. Zur Ausbreitung der Parklandschaftspopulation der Misteldrossel in Niedersachsen. *Beitr. Naturk. Nieders.*, 2, 4–8.
12.5

Peek, F. W. 1972. An experimental study of the territorial function of vocal and visual display in the male red-winged blackbird (*Agelaius phoeniceus*). *Anim. Behav.*, 20, 112–118.
11.4.1

Perrins, C. M. 1965. Population fluctuation and clutch-size in the great tit (*Parus major* L.). *J. Anim. Ecol.*, 34, 601–647.
11.3.1, 11.4.1

Perrins, C. M. 1968. The purpose of high-intensity alarm calls in small passerines. *Ibis*, 110, 200–201.
5.5.1

Perrins, C. M. 1970. The timing of birds' breeding seasons. *Ibis*, 112, 242–253.
7.6.1

Peters, R. P. & Mech, L. D. 1975. Scent-marking in wolves. *Am. Sci.*, 63, 628–637.
11.4.1

Petit, C. 1958. Le déterminisme

génétique et psychophysiologique de la compétition sexuelle chez *Drosophila melanogaster*. *Bull, Biol. France et Belgique*, 92, 248–329.
7.8

Phillips, R. R. 1971. The relationship between social behaviour and the use of space in the benthic fish *Chasmodes bosquianus* Lacepede (Teleosti: Bleniidae). II. The effect of prior residency on social and enclosure behaviour. *Z. Tierpsychol.*, 29, 389–408.
10.3.2

Pianka, E. R. & Parker, W. S. 1975. Age-specific reproductive tactics. *Amer. Natur.*, 109, 453–464.
14.1, 14.3

Pimentel, D., Smith, G. J. C. & Soans, J. 1967. A population model of sympatric speciation. *Amer. Natur.*, 101, 493–504.
12.8

Pitelka, F. A. 1973. Cyclic pattern in lemming populations near Barrow, Alaska. In Britton, M. E. (ed.), *Alaskan Arctic Tundra*. Arctic Institute of North America technical paper no. 25, 199–215.
12.4.1

Pitelka, F. A., Holmes, R. T. & MacLean, S. F. Jr. 1974. Ecology and evolution of social organisation in arctic sandpipers. *Amer. Zool.*, 14, 185–204.
11.2.2

Pittendrigh, C. S. 1958. Adaptation, natural selection and behaviour. In Roe, A. & Simpson, G. G. (eds.), *Behaviour and Evolution*, pp. 390–419. Newhaven and London, Yale University Press.
13.3.1

Plath, O. E. 1934. *Bumblebees and Their Ways*. MacMillan, New York.
4.5

Plowright, R. & Jay, S. 1968. Caste differentiation in bumblebees (*Bombus* Latr. Hym.). I. The determination of female size. *Insectes. Soc.*, 15, 171–192.
13.5

Powell, G. V. N. 1974. Experimental analysis of the social value of flocking by starlings (*Sturnus vulgaris*) in relation to predation and foraging. *Anim. Behav.*, 22, 501–505.
3.2.2

Power, H. W. 1975. Mountain Bluebirds: experimental evidence against altruism. *Science*, 189, 142–143.
9.6

Prince, G. J. & Parsons, P. A. 1977. Adaptive behaviour of *Drosophila* adults in relation to temperature and humidity. *Aust. J. Zool.*, 25, 285–290.
12.3, 12.6.1

Prout, T. 1968. Sufficient conditions for multiple niche polymorphism. *Amer. Natur.*, 102, 493–496.
12.8

Pulliam, H. R. 1973. On the advantages of flocking. *J. theor. Biol.*, 38, 419–422.
3.2.2, 3.2.7

Pulliam, H. R. 1974. On the theory of optimal diets. *Amer. Natur.*, 108, 59–75.
2.2.2

Pulliam, H. R. 1975a. Diet optimization with nutrient constraints. *Amer. Natur.*, 109, 765–768.
2.2.4

Pulliam, H. R. 1975b. Coexistence of sparrows: a test of community theory. *Science*, 189, 474–476.
2.7

Pulliam, H. R. 1976. The principle of optimal behaviour and the theory of communities. In Bateson, P. P. G. & Klopfer, P. H. (eds.), *Perspectives in Ethology*, Vol. 2, pp. 311–332. New York, Plenum Press.
13.2.3

Pusey, A. E. (In press). Intercommunity transfer of chimpanzees in Gombe National Park. In Hamburg, D. A. (ed.), *Perspectives on Human Evolution* (*Great*

Apes), Vol. 6. New York, Staples Press/W. A. Benjamin.
3.5.2

Pyke, G. 1974. Studies in the Foraging Efficiency of Animals. Unpublished Ph.D. thesis, University of Chicago.
2.4.1, 2.4.2

Pyke, G. H., Pulliam, H. R. & Charnov, E. L. 1977. Optimal foraging: a selective review of theory and tests. *Q. Rev. Biol.*, 52, 137–154.
2.4.1

Rand, A. S. 1964. Ecological distribution in anoline lizards of Puerto Rico. *Ecology*, 45, 745–752.
12.2.1

Rand, A. S. & Rand, W. M. 1976. Agonistic behaviour in nesting Iguanas: a stochastic analysis of dispute settlement dominated by the minimisation of energy cost. *Z. Tierpsychol.*, 40, 279–299.
10.3.2

Rasa, O. A. E. 1976. Invalid care in the Dwarf Mongoose (*Helogale undulata rufula*). *Z. Tierpsychol.*, 42, 337–342.
3.5.6

Rasa, O. A. E. 1977. The ethology and sociology of the dwarf mongoose (*Helogale undulata rufula*). *Z. Tierpsychol.*, 43, 337–406.
3.2.2, 3.2.3, 3.5.1, 3.5.4, 3.5.6, 3.5.7

Rechten, C. 1978. Interspecific mimicry in birdsong: does the Beau Geste hypothesis apply? *Anim. Behav.*, 26, 305.
10.4

Rettenmeyer, C. W. 1963. Behavioral studies of army ants. *Kansas Univ. Science Bull.*, 44, 281–465.
4.3.1

Rettenmeyer, C. W. 1970. Insect mimicry. *Ann. Rev. Ent.*, 15, 43–74.
5.2

Richards, A. J. 1973. The origin of *Taraxacum* agamo-species. *Bot. J. Linn. Soc.*, 66, 189–211.
6.1.1

Richards, K. W. 1973. Biology of *Bombus polaris* Curtis and *B. hyperboreus* Schonherr at Lake Hazen, Northwest territories (Hymenoptera: Bombini). Quaest. *Entomol.* 9, 115–157.
4.5

Richardson, A. M. M. 1974. Differential climatic selection in natural populations of the land snail *Cepaea nemoralis*. *Nature*, 247, 572–573.
5.3.3

Ricklefs, R. E. 1975. The evolution of cooperative breeding in birds. *Ibis*, 117, 531–534.
9.8

Ridpath, M. G. 1972a. The Tasmanian native hen, *Tribonyx mortierii*. I. Patterns of behavior. *CSIRO Wildl. Res.*, 17, 1–51.
9.3, 9.5

Ridpath, M. G. 1972b. The Tasmanian native hen, *Tribonyx morterii*. II. The individual, the group, and the population. *CSIRO Wildl. Res.*, 17, 53–90.
3.5.3, 3.5.4, 9.3, 9.4, 9.5

Ridpath, M. G. 1972c. The Tasmanian native hen, *Tribonyx mortierii*. III. Ecology. *CSIRO Wildl. Res.* 17, 91–118.
9.3, 9.5

Riemann, J. G., Moen, J. M. & Thorson, B. J. 1967. Female monogamy and its control in houseflies. *J. Insect. Physiol.*, 13, 407–418.
7.10

Rissing, S. W. & Wheeler, J. 1976. Foraging responses of *Veromessor pergandei* to changes in seed production. *Pan-Pacific Entomol.*, 52, 63–72.
4.3.2

Robertson, D. R., Sweatman, H. P. A., Fletcher, E. A. & Cleland, M. G. 1976. Schooling as a mechanism for circumventing the

territoriality of competitors. *Ecology*, 57, 1208–1220.
3·3·4

Rockwell, R. F. & Cooke, F. 1977. Gene flow and local adaptation in a colonially nesting dimorphic bird: the lesser snow goose (*Anser caerulescens caerulescens*). *Amer. Natur.*, 111, 91–97.
12.8

Rockwood, L. L. 1975. Distribution, density, and dispersion of two species of leaf-cutting ants (*Atta*) in Guanacaste Province, Costa Rica. *Biotropica*, 73, 176–193.
4·3·3

Rockwood, L. L. 1976. Plant selection and foraging patterns in two species of leaf-cutting ants, *Atta*. *Ecology*, 57, 48–61.
4·3·3

Rogers, D. 1972. Random search and insect population models. *J. Anim. Ecol.*, 41, 369–383.
2.3.5, 2.4.1

Rohwer, S. 1977. Status signalling in Harris' sparrows: some experiments in deception. *Behaviour*, 61, 107–129.
10.3.2

Rohwer, S. 1978. Parent cannibalism of offspring and egg raiding as a courtship strategy. *Amer. Natur.* 112: 429–440.
10.4

Rohwer, S. & Rohwer, F. C. 1978. (In press). Status signalling in sparrows: experimental deceptions achieved. *Anim. Behav.*, 26.
10.3.2

Rohwer, S., Fretwell, S. D. & Tuckfield, R. C. 1976. Distress screams as a measure of kinship in birds. *Am. Midl. Natur.*, 96, 418–430.
5·5·1, 5·5·3

Rood, J. P. 1974. Banded mongoose males guard young. *Nature*, 248, 176.
3·5·6

Rood, J. P. 1975. Population dynamics and food habits of the banded mongoose. *E. Afr. Wildl. J.*, 13, 89–111.
3·2·3, 3·5·1, 3·5·3, 3·5·4, 3·5·5, 3·5·6

Rosen, R. 1967. *Optimality Principles in Biology*. London, Butterworth.
13·3·1

Ross, P. & Crews, D. 1977. Influence of the seminal plug on mating behaviour in the garter snakes. *Nature*, 267, 344–345.
8·3·3

Rothstein, S. I. 1973. The niche variation model—is it valid? *Amer. Natur.*, 107, 598–620.
12.7

Rothstein, S. I. (In press). Gene frequencies and selection for inhibitory traits, with special emphasis on the adaptiveness of territoriality. *Amer. Natur.*
11.2.1

Roughgarden, J. 1972. Evolution of niche width. *Amer. Natur.*, 106, 683–718.
12.7

Roughgarden, J. 1974. Species packing and the competition function with illustrations from coral reef fish. *Theoret. Pop. Biol.*, 5, 163–186.
12.4.2

Rovner, J. S. 1968. Territoriality in the Sheet-web spider *Linyphia triangularis* (Clerck) (Araneae, Linyphiidae). *Z. Tierpsychol.*, 25, 232–242.
11.4.2

Rowell, T. E., Hinde, R. A. & Spencer-Booth, Y. 1964. 'Aunt'-infant interaction in captive Rhesus Monkeys. *Anim. Behav.*, 12, 219–226.
9.8

Rowley, I. 1965a. The life history of the Superb Blue Wren, *Malurus cyaneus*. *Emu*, 64, 251–297.
9·3, 9·4, 9·5

Rowley, I. 1965b. White-winged Choughs. *Australian Natural History*, 15, 81–85.
9·9·2

Rowley, I. 1968. Communal species of Australian birds. *Bonn. Zool. Beitr.*, 19, 362–370.
9.1

Rowley, I. 1976. Cooperative breeding in Australian birds. *Proc. XVI Int. Orn. Congr.* (Canberra, Australia), pp. 657–666.
9.1

Royama, T. 1970. Factors governing the hunting behaviour and selection of food by the Great Tit (*Parus major* L.). *J. Anim. Ecol.*, 39, 619–668.
2.3.1

Rozin, P. 1976. The selection of foods by rats, humans and other animals. In Hinde, R. A., Beer, C. & Shaw, E. (eds.), *Adv. Study Behav.* Vol. 6. New York, Academic Press.
2.2.4

Rubinoff, I. & Kropach, C. 1970. Differential reactions of Atlantic and Pacific predators to sea snakes. *Nature*, 228, 1288–1290.
5.2

Rudran, R. 1973. Adult male replacement in one-male troops of purple-faced langurs (*Presbytis senex senex*) and its effect on population structure. *Folia primatol.*, 19, 166–192.
3.5.4

Sade, D. S. 1967. Determinants of dominance in a group of free-ranging rhesus monkeys. In Altmann, S. A. (ed.), *Social communication Among Primates*, pp. 99–114. Chicago, Chicago University Press.
3.5.2, 3.5.3

Sale, P. F. 1968. Influence of cover availability on depth preference of the juvenile maini, *Acanthurus triostegus sandvicensis*. *Copeia*, 68, 802–807.
12.3

Sargent, T. D. 1968. Cryptic moths: effects on background selections of painting the circumocular scales. *Science*, 159, 100–101.
12.6.1

Sauer, E. G. F. & Sauer, E. M. 1966. The behavior and ecology of the South African Ostrich. *Living Bird*, 5, 45–75.
9.7

Schaffer, W. M. & Elson, P. F. 1975. The adaptive significance of variations in life history among local populations of Atlantic Salmon in N. America. *Ecology*, 56, 577–590.
14.3

Schaffer, W. M. & Rosenzweig, M. L. 1977. Selection for optimal life histories. II. Multiple equilibria and the evolution of alternative reproductive strategies. *Ecology*, 58, 60–72.
14.3

Schaffer, W. M. & Tamarin, R. H. 1973. Changing reproductive rates and population cycles in lemmings and voles. *Evolution*, 27, 111–124.
14.4

Schaller, G. B. 1972. *The Serengeti Lion*. Chicago, Chicago University Press.
2.1, 3.2.2, 3.2.4, 3.2.5, 3.2.6, 3.3.2, 3.3.3, 3.3.4, 3.5.1, 3.5.2, 3.5.3, 3.5.4, 3.5.5, 3.5.6, 3.5.7, 3.6, 9.8

Schaller, G. B. & Lowther, G. R. 1969. The relevance of carnivore behaviour to the study of early hominids. *S. West. J. Anthrop.*, 25, 307–341.
3.3.3

Schneider, D. 1974. The sex-attractant receptor of moths. *Sci. Am.*, 231(1), 28–35.
7.10

Schneirla, T. C. & Topoff, H. R. 1971. *Army Ants*. W. H. Freeman and Co., San Francisco.
4.3.1

Schnetter, M. 1950. Veränderungen der genetischen Konstitution in näturlichen Populationen der polymorphen Bänderschnecken. *Verb. dt. Zool.* Marburg, 1950, 192–206.
5.3.3

Schoener, T. W. 1968. Sizes of feeding territories among birds. *Ecology*, 49, 123–141.
11.1, 11.2.1

Schoener, T. W. 1971. Theory of feeding strategies. *A. Rev. Ecol. Syst.*, 2, 369–404.
2.2.2, 13.3.1

Sealander, J. A. 1952. The relationship of nest protection and huddling to survival of *Peromyscus* at low temperature. *Ecology*, 33, 63–71.
3.4

Selander, R. K. 1964. Speciation in wrens of the genus *Campylorhynchus*. *Univ. Calif. Publ. Zool.*, 74, 1–224.
9.3, 9.4

Selander, R. K. 1972. Sexual selection and dimorphism in birds. In Campbell, B. (ed.), *Sexual Selection and the Descent of Man*. Chap. 8. London, Heinemann.
7.6

Semler, D. E. 1971. Some aspects of adaptation in a polymorphism for breeding colours in the Threespine Stickleback (*Gasterosteus aculeatus*). *J. Zool. Lond.*, 165, 291–302.
7.7

Seton, E. T. 1953. *The Lives of Game Animals*, 3. Mass, Branford, Newton Centre.
5.4

Shalter, M. D. 1978. Localization of passerine seeet and mobbing calls by goshawks and pygmy owls. *Z. Tierpsychol.*, 46, 260–267.
5.5

Shalter, M. D. & Schleidt, W. M. 1977. The ability of barn owls *Tyto alba* to discriminate and localize avian alarm calls. *Ibis*, 119, 22–27.
5.5

Shepher, J. 1971. Mate selection among second generation kibbutz adolescents and adults: incest avoidance and negative imprinting. *Arch. sex. Behav.*, 1, 293–307.
6.2

Sheppard, D. H., Klopfer, P. H. & Oelke, H. 1968. Habitat selection: differences in stereotypy between insular and continental birds. *Wilson Bull.*, 80, 452–457.
12.7

Sheppard, P. M. 1951. Fluctuations in the selective value of certain phenotypes in the polymorphic land snail *Cepaea nemoralis* (L.). *Heredity*, 5, 125–134.
5.3.1

Sherman, P. W. 1977. Nepotism and the evolution of alarm calls. *Science*, 197, 1246–1253.
3.5.3, 5.5.2, 5.5.3, 5.8

Shettleworth, S. J. 1972. The role of novelty in learned avoidance by unpalatable 'prey' by domestic chicks (*Gallus gallus*). *Anim. Behav.*, 20, 29–35.
5.2

Sibley, C. G. 1957. The evolutionary and taxonomic significance of sexual dimorphism and hybridization in birds. *Condor*, 59, 166–191.
12.8

Sibly, R. M. 1975. How incentive and deficit determine feeding tendency. *Anim. Behav.*, 23, 437–446.
13.4.1, 13.4.4, 13.4.5

Sibly, R. M. & McCleery, R. H. 1976. The dominance boundary method of determining motivational state. *Anim. Behav.*, 24, 108–124.
13.4.4, 13.4.5

Sibly, R. M. & McFarland, D. J. 1976. On the fitness of behaviour sequences. *Amer. Natur.*, 110, 601–617.
13.4, 13.4.1, 13.4.2, 13.4.4, 13.5, 13.7.4

Siegfried, W. R. & Underhill, L. G. 1975. Flocking as an anti-predator strategy in doves. *Anim. Behav.*, 23, 504–508.
3.2.2

Simon, C. A. 1975. The influence of food abundance on territory size

in the inguanid lizard, *Sceloporus jarrovi*. *Ecology*, 56, 993–998.
11.2.1

Simon, C. A. & Middendorf, G. A. 1976. Resource partitioning by an iguanid lizard: temporal and microhabitat aspects. *Ecology*, 57, 1317–1320.
12.2.1

Simon, H. A. 1956. Rational choice and the structure of the environment. *Psych. Rev.*, 63, 129–138.
13.2.1

Simpson, M. J. A. 1968. The display of the Siamese fighting fish, *Betta splendens*. *Anim. Behav. Monogr.*, 1, 1–73.
10.3.2

Sinclair, A. R. E. 1977. *The African Buffalo: A Study of Resource Limitation of Populations*. Chicago, Univ. of Chicago Press.
10.3.2, 11.1

Sinniff, D. B. & Jessen, C. R. 1969. A simulation model of animal movement patterns. *Adv. Ecol. Res.*, 6, 185–219.
2.4.1

Skaife, S. H. 1955. *Dwellers in Darkness*. Longmans, Green, New York.
4.4

Skutch, A. F. 1953. Life history of the Southern House Wren. *Condor*, 55, 121–149.
9.6

Skutch, A. F. 1959. Life history of the Groove-billed Ani. *Auk*, 76, 281–317.
9.7

Skutch, A. F. 1961. Helpers among birds. *Condor*, 63, 198–226.
9.1, 9.6, 9.9.2

Skutch, A. F. 1969. Life histories of Central American birds. Volume III. Golden naped Woodpecker. *Pacific Coast Avifauna*, 35, 479–517.
9.9.2

Slaney, P. A. & Northcote, T. G. 1974. Effects of prey abundance on density and territorial behavior of young rainbow trout (*Salmo gairdneri*) in laboratory stream channels. *J. Fish. Res. Board Can.*, 31, 1201–1209.
11.2.1

Smith, C. C. 1968. The adaptive nature of social organisation in the genus of tree squirrels *Tamiasciurus*. *Ecol. Monog.*, 38, 31–64.
11.2.1

Smith, D. G. 1976. An experimental analysis of the function of red-winged blackbird song. *Behaviour*, 56, 136–156.
11.4.1

Smith, J. N. M. 1974a. The food searching behaviour of two European thrushes. I. Description and analysis of search paths. *Behaviour*, 48, 276–302.
2.4.1

Smith, J. N. M. 1974b. The food searching behaviour of two European thrushes. II. The adaptiveness of the search patterns. *Behaviour*, 49, 1–61.
2.4.1, 2.4.2

Smith, J. N. M. & Sweatman, H. P. A. 1974. Food searching behaviour of titmice in patchy environments. *Ecology*, 55, 1216–1232.
2.5

Smith, J. N. M., Grant, P. R., Grant, B. R., Abbott, I. J. & Abbott, L. K. (In press). Seasonal variation in feeding habits of Darwin's ground finches. *Ecology*.
2.2.3

Smith, W. J. 1968. Message meaning analysis. In Sebeok, T. A. (ed.), *Animal Communication*, pp. 44–60. Bloomington and London, Indiana University Press.
10.1.5

Smith, W. J. 1977. *The Behaviour of Communicating: An Ethological Approach*. Harvard University Press.
10.1.5, 10.3.3

Smythe, N. 1970. On the existence of 'pursuit invitation' signals in

mammals. *Amer. Natur.*, 104, 491–494.
5.4

Smythe, N. 1977. The function of mammalian alarm advertising: social signals or pursuit invitation? *Amer. Natur.*, 111, 191–194.
5.4

Snow, D. W. 1954. The habitats of Eurasian tits (*Parus* spp). *Ibis*, 96, 565–585.
12.2.3

Soulé, M. & Stewart, B. R. 1970. The 'niche variation' hypothesis: a test and alternatives. *Amer. Natur.*, 104, 85–97.
12.7

Southern, H. N. 1970. The natural control of a population of tawny owls (*Strix aluco*). *J. Zool. Lond.*, 162, 197–285.
11.1, 11.3.1

Southwood, T. R. E. 1976. Bionomic strategies and population parameters. In May, R. M. (ed.), *Theoretical Ecology: Principles and Applications*. Oxford, Blackwell.
14.1, 14.3

Southwood, T. R. E. 1977. Habitat, the templet for ecological strategies? *J. Anim. Ecol.*, 46, 337–365.
14.1, 14.6

Southwood, T. R. E., May, R. M., Hassell, M. P. & Conway, G. R. 1974. Ecological strategies and population parameters. *Amer. Natur.*, 108, 791–804.
14.1, 14.6

Spencer-Booth, Y. 1970. The relationships between mammalian young and conspecifics other than mothers and peers: a review. In Lehrman, D. S., Hinde, R. A. & Shaw, E. (eds.), *Adv. Study Behav.* Vol. 3, pp. 119–194. New York, Academic Press.
9.8

Stallcup, J. A. & Woolfenden, G. E. (In press). Family status and contributions to breeding by Florida Scrub Jays. *Anim. Behav.*
9.2, 9.4

Stamps, J. A. 1977. The relationship between resource competition, risk and aggression in a tropical territorial lizard. *Ecology*, 58, 349–358.
11.2.5

Stearns, S. C. 1976. Life-history tactics: a review of the ideas. *Q. Rev. Biol.*, 51, 3–47.
14.1, 14.3

Stearns, S. C. 1977. The evolution of life history traits. *A. Rev. Ecol. Syst.*, 8, 145–171.
14.1

Stenger, J. 1958. Food habits and available food of ovenbirds in relation to territory size. *Auk*, 75, 335–346.
11.2.1

Stevenson, J. G. 1967. Reinforcing effects of chaffinch song. *Anim. Behav.*, 15, 427–432.
10.5.4

Stiles, F. G. 1973. Food supply and the annual cycle of the anna hummingbird. *Univ. Calif. Publ. Zool.*, 97.
11.2.2

Stimson, J. 1970. Territorial behaviour of the owl limpet, *Lottia gigantea*. *Ecology*, 51, 113–118.
11.2.1, 11.2.5

Stimson, J. 1973. The role of territory in the ecology of the intertidal limpet *Lottia gigantea* (Gray). *Ecology*, 54, 1020–1030.
11.2.1, 11.2.5, 11.3

Stoltz, L. P. & Saayman, G. S. 1970. Ecology and behaviour of baboons in the Northern Transvaal. *Ann. Transv. Mus.*, 26, 99–143.
3.2.3

Stonehouse, B. 1953. The emperor penguin, *Aptenodytes forsteri* Gray. I. Breeding behavior and development. *Falkld Isl. Depend. Surv. Scient. Rep.*, No. 6, 1–33.
3.4

Struhsaker, T. T. 1969. Correlates of ecology and social organisation among African cercopithecines. *Folia Primatol.*, 11, 80–118.
1.3.2

Strum, S. C. 1975. Primate predation: interim report on the development of a tradition in a troop of olive baboons. *Science*, 187, 755–757.
3.5.5, 3.5.7

Stubbs, M. 1977. Density dependence in the life-cycles of animals and its importance in *K*- and *r*-strategies. *J. Anim. Ecol.*, 46, 677–688.
14.1, 14.6

Sudd, J. H. 1960. The transport of prey by an ant, *Pheidole crassinoda*. Em. *Behaviour*, 16, 295–308.
4.3.1

Sudd, J. H. 1967. *An Introduction to the Behaviour of Ants*. Arnold, London.
4.3.1, 4.3.3

Sugiyama, Y. 1967. Social organization of Hanuman langurs. In Altmann, S. A. (ed.), *Social Communication among Primates*, pp. 221–236. Chicago, Chicago University Press.
3.5.4

Svärdson, G. 1949. Competition and habitat selection in birds. *Oikos*, 1, 156–174.
8.3.1

Teleki, G. 1973. *The Predatory Behaviour of Wild Chimpanzees*. Lewisburg, Bucknell University Press.
3.5.5, 3.5.7

Tenaza, R. R. & Tilson, R. L. 1977. Evolution of long-distance alarm calls in Kloss's gibbon. *Nature*, 268, 233–235.
5.5, 5.5.3

Thoday, J. M. 1972. Disruptive selection. *Proc. Roy. Soc. Lond. B.*, 182, 109–143.
12.8

Thoday, J. M. & Boam, T. B. 1959. Effects of disruptive selection. II. Polymorphism and divergence without isolation. *Heredity*, 13, 205–218.
12.8

Thompson, W. A., Vertinsky, I. & Krebs, J. R. 1974. The survival value of flocking in birds: a simulation model. *J. Anim. Ecol.*, 43, 785–820.
2.8

Thomsen, E. & Thomsen, M. 1937. Über das Thermopräferendum der Larven einiger Fliegenarten. *Z. vergl. Physiol.*, 24, 343–380.
12.3

Thornhill, R. 1976. Sexual selection and nuptial feeding behaviour in *Bittacus apicalis* (Insecta: Macoptera). *Amer. Natur.*, 110, 529–548.
7.9.2

Thorpe, W. H. 1939. Further studies on pre-imaginal olfactory conditioning in insects. *Proc. Roy. Soc. Lond. B.*, 127, 424–433.
12.5, 12.8

Thorpe, W. H. 1940. Ecology and the future of systematics. In Huxley, J. S. (ed.), *The New Systematics*. Oxford, Oxford University Press.
12.8

Thorpe, W. H. 1945. The evolutionary significance of habitat selection. *J. Anim. Ecol.*, 14, 67–70.
12.8

Thorpe, W. H. & Hall-Craggs, J. 1976. Sound production and perception in birds as related to the general principles of pattern perception. In Bateson, P. P. G. & Hinde, R. A. (eds.), *Growing Points in Ethology*, pp. 171–189. Cambridge, Cambridge University Press.
10.5.4

Tinbergen, L. 1960. The natural control of insects in pine woods. I. Factors influencing the intensity of predation in song birds. *Arch. Néerl. Zool.*, 13, 265–343.
5.2, 5.3.2

Tinbergen, N. 1951. *The Study of Instinct*. Oxford, Clarendon Press.
3.2.3, 3.2.6, 10.3.1

Tinbergen, N. 1952. Derived activities; their causation, biological

significance, origin and emancipation during evolution. *Q. Rev. Biol.*, 27, 1–32.
10.1.5, 10.2.3

Tinbergen, N. 1953. *Social Behaviour in Animals*. London, Methuen.
10.1.4

Tinbergen, N. 1964. The evolution of signaling devices. In Etkin, W. (ed.), *Social Behavior and Organization Among Vertebrates*, pp. 206–230. Chicago and London, University of Chicago Press.
10.1.5, 10.2.3

Tinbergen, N., Impekoven, M. & Franck, D. 1967. An experiment on spacing out as defence against predators. *Behaviour*, 28, 307–321.
1.1, 11.2.3

Toates, F. 1975. *Control Theory in Biology and Experimental Psychology*. London, Hutchinson.
13.7.2

Tomlinson, J. 1966. The advantage of hermaphroditism and parthenogenesis. *J. theoret. Biol.*, 11, 54–58.
6.2

Tomlinson, P. B. & Soderholm, P. K. 1975. The flowering and fruiting of *Corypha elata* in South Florida. *Principes*, 19, 89–99.
14.3

Treisman, M. 1975. Predation and the evolution of gregariousness. I. Models for concealment and evasion. *Anim. Behav.*, 23, 779–800.
3.2.1

Trivers, R. L. 1971. The evolution of reciprocal altruism. *Q. Rev. Biol.*, 46, 35–57.
1.6, 5.5.1, 9.9.2

Trivers, R. L. 1972. Parental investment and sexual selection. In Campbell, B. (ed.), *Sexual Selection and the Descent of Man*. Chicago, Aldine.
1.5.1, 3.5.4, 5.6, 6.4, 7.1.2, 7.2.2, 7.2,5, 7.9.2, 8.1, 10.2.3

Trivers, R. L. 1974. Parent-offspring conflict. *Amer. Zool.*, 14, 249–264.
1.5.1, 9.6, 10.2.3

Trivers, R. L. 1976. Sexual selection and resource-accruing abilities in *Anolis garmani*. *Evolution*, 30, 253–269.
7.2.4, 11.2.2

Trivers, R. L. & Hare, H. 1976. Haplodiploidy and the evolution of the social insects. *Science*, 191, 249–263.
1.5.1, 10.2.3, 13.5

Trune, D. R. & Slobodchikoff, C. N. 1976. Social effects of roosting on the metabolism of the pallid bat (*Antrozous pallidus*). *J. Mammal.*, 57, 656–663.
3.4

Turner, F. B., Jennrich, R. I. & Weintraub, J. D. 1969. Home ranges and body size of lizards. *Ecology*, 50, 1076–1081.
11.2.1

Turner, J. R. G. 1971. Studies of Müllerian mimicry and its evolution in burnet moths and heliconid butterflies. In Creed, R. (ed.), *Ecological Genetics and Evolution*, pp. 224–260. Oxford, Blackwell.
5.2, 5.8

Turner, J. R. G. 1975. A tale of two butterflies. *Nat. Hist.*, 84, 28–37.
5.2

Turner, J. R. G. 1977. (In press). Butterfly mimicry: the genetical evolution of an adaptation. *Evol. Biol.*, 10, 163–206.
5.1

van den Assem, J. 1967. Territory in the three-spined stickleback *Gasterosteus aculeatus* L. *Behaviour* Suppl., 16, 1–164.
11.2.2

van Lawick-Goodall, J. 1968. The behaviour of free-living chimpanzees in the Gombe Stream Reserve. *Anim. Behav. Monogr.*, 1, 161–311.
3.5.7, 9.8

van Lawick-Goodall, J. 1969. Mother-offspring relationships in free-ranging chimpanzees. In Morris, D. (ed.), *Primate Ethology:*

Essays on the Socio-sexual Behavior of Apes and Monkeys, pp. 364–436. Chicago, Aldine.
9.8

van Rhijn, J. G. 1973. Behavioural dimorphism in male ruffs *Philomachus pugnax* (L.). *Behaviour*, 47, 153–229.
11.2.2

Van Valen, L. 1965. Morphological variation and width of ecological niche. *Amer. Natur.*, 99, 377–390.
12.4.2, 12.7, 12.8

Van Valen, L. 1971. Group selection and the evolution of dispersal. *Evolution*, 25, 591–598.
14.5

Van Valen, L. & Grant, P. R. 1970. Variation and niche width reexamined. *Amer. Natur.*, 104, 589–590.
12.7

Vehrencamp, S. L. 1976. The evolution of communal nesting in Groove-billed Anis. Unpublished Dissertation, Cornell University, 181 pp.
9.7

Vehrencamp, S. L. 1977. Relative fecundity and parental effort in communally nesting Anis, *Crotophaga sulcirostris*. *Science*, 197, 403–405.
9.7

Verner, J. 1964. Evolution of polygamy in the long-billed marsh wren. *Evolution*, 18, 252–261.
11.2.2

Verner, J. 1977. On the adaptive significance of territoriality. *Amer. Natur.*, 111, 769–775.
11.2.1

Verner, J. & Engleson, G. H. 1970. Territories, multiple nest building and polygamy in the long-billed marsh wren. *Auk*, 87, 557–567.
11.2.2

Verner, J. & Willson, M. F. 1966. The influence of habitats on mating systems of North American passerine birds. *Ecology*, 47, 143–147.
11.2.2

Vine, I. 1973. Detection of prey flocks by predators. *J. theor. Biol.*, 40, 207–210.
3.2.1

von Haartman, L. 1949. Der Trauerfliegenschnapper. I. Ortstreue und Hassenbildung. *Acta Zool. Fenn.*, 56, 1–104.
12.6.3

von Haartman, L. 1969. Nest-site and evolution of polygamy in European passerine birds. *Ornis Fenn*, 46, 1–12.
11.2.2

von Haartman, L. 1971. Population dynamics. In Farner, D. S. & King, J. R. (eds.), *Avian Biology*, Vol. I. Academic Press, New York and London.
12.4.1

Waage, J. 1977. Behavioural aspects of foraging in the parasitoid *Nemeritis canescens* (Grav.). Unpublished Ph.D. thesis, London University.
2.3.5, 2.4.2

Wallace, B. 1973. Misinformation, fitness and selection. *Amer. Natur.*, 107, 1–7.
10.4

Walther, F. R. 1969. Flight behaviour and avoidance of predators in Thomson's gazelle (*Gazella thomsoni* Guenther 1884). *Behaviour*, 34, 184–221.
5.4

Ward, P. & Zahavi, A. 1973. The importance of certain assemblages of birds as 'information-centres' for food-finding. *Ibis*, 115, 517–534.
3.3.1

Watson, A. 1967. Territory and population regulation in the red grouse. *Nature*, 215, 1274–1275.
11.3.1

Watts, C. R. & Stokes, A. W. 1971. The social order of turkeys. *Sci. Am.*, 224 (6), 112–118.
3.5.3, 3.5.4, 3.5.6

Weaver, N. 1957. The foraging behavior of honeybees on hairy vetch. *Insect Soc.*, 12, 231–240, 321–326.
4.2.3

Weber, N. A. 1966. Fungus-growing ants. *Science*, 153, 587–604.
4.3.3

Wecker, S. C. 1963. The role of early experience in habitat selection by the prairie deermouse *Peromyscus maniculatus bairdi*. *Ecol. Monogr.*, 33, 307–325.
12.2.2, 12.5, 12.6.3

Weeden, J. S. & Falls, J. B. 1959. Differential response of male ovenbirds to recorded songs of neighboring and more distant individuals. *Auk*, 76, 343–351.
11.4.1

Weihs, D. 1973. Hydromechanics of fish schooling. *Nature*, 241, 290–291.
3.4

Wellington, W. G. 1960. Qualitative changes in natural populations during changes in abundance. *Can. J. Zool.*, 38, 289–314.
14.4

Wells, K. D. 1977a. Territoriality and male mating success in the green frog (*Rana clamitans*). *Ecology*, 58, 750–762.
11.2.2

Wells, K. D. 1977b. The social behaviour of anuran amphibians. *Anim. Behav.*, 25, 666–693.
7.5.3

Werner, E. E. 1976. Niche shift in sunfishes: experimental evidence and significance. *Science*, 191, 404–406.
2.7

Werner, E. E. 1977. Species packing and niche complementarity in three sunfishes. *Amer. Natur.*, 111, 553–578.
2.7, 12.4.2

Werner, E. E. & Hall, D. J. 1974. Optimal foraging and the size selection of prey by the Bluegill Sunfish (*Lepomis macrochirus*). *Ecology*, 55, 1216–1232.
2.2.2, 2.2.3

Whitham, T. G. 1977. Coevolution of foraging in *Bombus*—nectar dispensing *Chilopsis*: a last dreg theory. *Science*, 197, 593–596.
2.3.4

Whitney, C. L. & Krebs, J. R. 1975a. Mate selection in Pacific Tree Frogs. *Nature*, 255, 325–326.
7.9.2

Whitney, C. L. & Krebs, J. R. 1975b. Spacing and calling in Pacific Tree Frogs, *Hyla regilla*. *Can. J. Zool.*, 53, 1519–1527.
11.1, 11.4.1

Whittaker, R. H., Levin, S. A. & Root, R. B. 1973. Niche, habitat, and ecotope. *Amer. Natur.*, 107, 321–338.
12.1

Whittaker, R. H., Levin, S. A. & Root, R. B. 1975. On the reason for distinguishing 'Niche, habitat and ecotope'. *Amer. Natur.*, 109, 479–482.
12.1

Wickler, W. 1957. Vergleichende Verhaltenstudien an Grundfischen. I. Berträge zur Biologie, besonders zur Ethologie von *Blennius fluviatilis* Asso im Vergleich zu einigen anderen Bodenfischen. *Z. Tierpsychol.*, 14, 393–428.
10.3.2

Wickler, W. 1968. *Mimicry in Plants and Animals*. New York, McGraw Hill.
5.1, 10.4

Wiens, J. A. 1970. Effects of early experience on substrate pattern selection in *Rana aurora* tadpoles. *Copeia*, 1970, 543–548.
12.5

Wigglesworth, V. B. 1965. *The Principles of Insect Physiology*. London, Methuen.
12.3

Wilbur, H. M. 1977. Propagule size, number, and dispersion pattern in *Ambystoma* and *Asclepias*. *Amer. Natur.*, 111, 43–68.
14.6

Wilbur, H. M. Tinkle, D. W. & Collins, J. P. 1974. Environ-

mental certainty, trophic level, and resource availability in life history evolution. *Amer. Natur.*, 108, 805–817.
14.6

Wiley, R. H. 1973. Territoriality and non-random mating in the Sage grouse *Centrocercus urophasianus*. *Anim. Behav. Monogr.*, 6, 87–169.
11.2.2

Wiley, R. H. 1974. Evolution of social organisation and life history patterns among grouse. *Q. Rev. Biol.*, 49, 201–227.
11.2.2

Williams, G. C. 1966. *Adaptation and Natural Selection.* Princeton, N.J., Princeton University Press.
1.4, 5.5.1, 6.1, 7.9.1, 10.2.3, 10.5.4, 13.2.3, 13.3.1, 13.8

Williams, G. C. 1975. *Sex and Evolution.* Princeton University Press.
6.1, 6.1.2, 14.3

Williams, G. C., Koehn, R. K. & Mitton, J. B. 1973. Genetic differentiation without isolation in the American Eel, *Anguilla rostrata*. *Evolution*, 27, 192–204.
12.8

Willson, M. F. 1969. Avian niche size and morphological variation. *Amer. Natur.*, 103, 531–542.
12.7

Wilson, E. O. 1968. The ergonomics of caste in the social insects. *Amer. Natur.*, 102, 41–66.
4.5

Wilson, E. O. 1971. *The Insect Societies.* Harvard, Belknap Press.
3.5.5, 4.3.1, 4.5

Wilson, E. O. 1975. *Sociobiology.* Havard, Belknap Press.
1.7, 3.1, 3.5.1, 5.4, 5.6, 5.7, 7.1.2, 7.10, 9.1, 9.9.2, 10.1.1, 10.2.2, 10.4, 11.1, 13.5

Winn, H. E. 1958. The comparative ecology and reproductive behaviour of fourteen species of darter (*Percidae*). *Ecol. Monogr.*, 28, 155–191.
1.3.1

Wolf, L. L. & Hainsworth, F. R. 1971. Time and energy budgets of territorial hummingbirds. *Ecology*, 52, 980–988.
11.2.1, 13.2.3

Wolf, L. L., Hainsworth, F. R. & Gill, F. B. 1975. Foraging efficiencies and time budgets in nectar-feeding birds. *Ecology*, 56, 117–128.
11.2.2

Wolf, L. L. & Stiles, F. G. 1970. Evolution of pair cooperation in a tropical hummingbird. *Evolution*, 24, 759–773.
11.2.2

Woolfenden, G. E. 1973. Nesting and survival in a population of Florida Scrub Jays. *Living Bird*, 12, 25–49.
9.2, 9.6

Woolfenden, G. E. 1975. Florida Scrub Jay helpers at the nest. *Auk*, 92, 1–15.
9.2, 9.3

Woolfenden, G. E. 1976a. Cooperative breeding in American birds. *Proc. XVI. Int. Orn. Congr.* Canberra, Australia, pp. 674–684.
9.1, 9.3

Woolfenden, G. E. 1976b. A case of bigamy in the Florida Scrub Jay. *Auk*, 93, 443–450.
9.6

Woolfenden, G. & Fitzpatrick, J. W. 1978a. Dominance in the Florida scrub jay. *Condor*, 79, 1–12.
9.2

Woolfenden, G. E. & Fitzpatrick J. W. 1978b. The inheritance of territory in group-breeding birds. *Bioscience*, 28, 104–108.
9.2, 9.9.2, 14.6

Wrangham, R. W. 1977. Feeding behaviour of chimpanzees in Gombe National Park, Tanzania. In Clutton-Brock, T. H. (ed.), *Primate Ecology*, pp. 504–538. London, Academic Press.
3.5.5

Wyman, J. 1967. The jackals of the Serengeti. *Animals*, 10, 79–83.
3.2.4, 3.3.2

Wynne-Edwards, V. C. 1962. *Animal Dispersion in Relation to Social Behaviour*. Edinburgh and London, Oliver and Boyd.
1.4, 5.5.1, 9.9.2, 11.3.1

Wynne-Edwards, V. C. 1977. In Ebling, F. J. & Stoddart, D. M. (eds.), *Population Control by Social Behaviour*, Inst. Biol., London.
1.4

Yamasaki, K., Boyse, E. A., Miké, V., Thaler, H. T., Mathieson, B. J., Abbott, J., Boyse, J., Zayas, Z. A. & Thomas, L. 1976. Control of mating preferences in mice by genes in the major histocompatibility complex. *J. exp. Med.*, 144, 1324–1335.
7.9.2

Zahavi, A. 1974. Communal nesting by the Arabian babbler: a case of individual selection. *Ibis*, 116, 84–87.
3.5.2, 3.5.3, 3.5.4, 9.3, 9.4, 9.6, 9.9.1, 9.9.2

Zahavi, A. 1975. Mate selection—a selection for a handicap. *J. theor. Biol.*, 53, 205–214.
7.2.3

Zahavi, A. 1976. Cooperative nesting in Eurasian birds. *Proc. XVI Int. Orn. Congr.* (Canberra, Australia), pp. 685–693.
9.3, 9.6, 9.9.1, 9.9.2

Zahavi, A. 1977a. The cost of honesty. (Further remarks on the handicap principle.) *J. theor. Biol.*, 67, 603–605.
7.2.3

Zahavi, A. 1977b. Reliability in communication systems and the evolution of altruism. In Stonehouse, B. & Perrins, C. M. (eds.), *Evolutionary Ecology*, pp. 253–259. London, Macmillan.
10.3.2

Zayan, R. C. 1975. Modification des effets liés a la priorité de residence chez *Xiphophorus* (Pisces: Poecilidae): le rôle des manipulations experimentales. *Z. Tierpsychol.*, 39, 463–491.
10.3.2

Zimen, E. 1976. On the regulation of pack size in wolves. *Z. Tierpsychol.*, 40, 300–341.
3.5.2, 3.5.4

Zimmerman, J. L. 1971. The territory and its density dependent effect in *Spiza americana*. *Auk*, 88, 591–612.
11.2.2

Organism Index

Subject Index